T0178116

Communications in Computer and Information Science   1734

More information about this series at https://link.springer.com/bookseries/7899

Anban Pillay · Edgar Jembere ·
Aurona Gerber (Eds.)

Artificial Intelligence Research

Third Southern African Conference, SACAIR 2022
Stellenbosch, South Africa, December 5–9, 2022
Proceedings

 Springer

Editors
Anban Pillay (iD)
University of KwaZulu-Natal
Durban, South Africa

Edgar Jembere (iD)
University of KwaZulu-Natal
Durban, South Africa

Aurona Gerber (iD)
University of Pretoria
Pretoria, South Africa

ISSN 1865-0929 ISSN 1865-0937 (electronic)
Communications in Computer and Information Science
ISBN 978-3-031-22320-4 ISBN 978-3-031-22321-1 (eBook)
https://doi.org/10.1007/978-3-031-22321-1

This Springer imprint is published by the registered company Springer Nature Switzerland AG
The registered company address is: Gewerbestrasse 11, 6330 Cham, Switzerland

Preface

This, volume 1734 of the Springer Communications in Computer and Information Science (CCIS) series, contains the revised accepted papers of the third South African Conference of Artificial Intelligence (SACAIR 2022). The second SACAIR 2022 proceedings volume is published as an online proceedings that will be available on the SACAIR 2022 website.

The inter- and trans-disciplinary nature of the SACAIR series of conferences in artificial intelligence is unique in providing a venue for researchers from a diverse set of disciplines that include computer science, engineering, information systems, law, philosophy, and other humanities. The organization of such a conference has to carefully consider the differing research methods, interests, publication standards, and cultures of these disciplines. The conference was thus organized around the following four tracks: Algorithmic, Symbolic and Data-Driven AI (Computer Science and Engineering - CSE), Socio-technical and Human-centered AI (Information Systems - IS), Responsible and Ethical AI (Philosophy and Law - PHIL), and Inter- and Trans-disciplinary AI Research.

The Program Committee comprised 112 members (representing some 43 research institutions), 28 of whom were from outside Southern Africa. Each paper was reviewed by at least two members of the Program Committee in a rigorous double-blind process. Great care was taken to ensure the integrity of the conference including careful attention to avoid conflicts of interest. The following criteria were used to rate submissions and to guide decisions: relevance to SACAIR, significance, technical quality, scholarship, and presentation, which included quality and clarity of writing.

We received just under 100 abstracts, and after a first round of evaluation, 73 submissions were sent to our SACAIR Program Committee members for review. The papers consisted of 54 in the CSE track, 11 in the IS track, and seven in the PHIL track. In total, 26 full research papers were selected for publication in this Springer CCIS volume (which translates to an acceptance rate of 35.6%), whilst a further 18 papers were accepted for inclusion in the online volume (24.7%). The total acceptance rate for publication in the two SACAIR 2022 proceedings volumes was 60.2% for reviewed submissions. In total, four papers from the Responsible and Ethical AI track, eight papers from Socio-technical and Human-centered AI track, and 32 papers from the Algorithmic, Symbolic and Data-Driven AI track were accepted for publication in the two volumes.

Thank you to all the authors who submitted work of an exceptional standard to the conference and congratulations to the authors whose work was accepted for publication. We place on record our gratitude to the Program Committee members whose thoughtful and constructive comments were well received by the authors.

October 2022

Anban Pillay
Edgar Jembere
Aurona Gerber

Message from the General Chairs

It is with great pleasure that we write this foreword to the proceedings of the third Southern African Conference for Artificial Intelligence Research (SACAIR 2022), held as a hybrid online and in-person event during December 5–9, 2022. The program included an unconference for students on December 5 (a student-driven event allowing students to interact with each other and with sponsors and potential employers), a day of tutorials on December 6, and the main conference during December 7–9, 2022.

SACAIR 2022 is the third international conference focused on artificial intelligence (AI) hosted by the SACAIR Steering Committee, an affiliate of the Centre for AI Research (CAIR). CAIR[1] is a South African distributed research network established in 2011 that aims to build world-class artificial intelligence research capacity in South Africa. CAIR conducts foundational, directed, and applied research into various aspects of AI through its nine research groups based at six South African universities (the University of Pretoria, the University of KwaZulu-Natal, the University of Cape Town, Stellenbosch University, the University of the Western Cape, and North-West University).

The inaugural CAIR conference, the Forum for AI Research (FAIR 2019) was held in Cape Town, South Africa, in December 2019, SACAIR 2020 was held in February 2021 after being postponed due to the COVID-19 pandemic, and SACAIR 2021 was an online event hosted by the University of KwaZulu-Natal, in Durban, in December 2021.

We are pleased that this, our third annual Southern African Conference for Artificial Intelligence Research (SACAIR), continued to enjoy the confidence of the South African artificial intelligence research community. The 2022 conference attracted support from both authors, who submitted high-quality research papers, and researchers, who supported the conference by serving on the international Program Committee.

Any sufficiently advanced technology has the potential to transform society for better or worse. Artificial intelligence technologies, particularly in their current data-driven form, have the potential to transform our world for the better. However, especially in the context of machine learning applications, there are well-founded concerns around fairness, structural bias and amplification of existing social stereotypes, privacy, transparency, accountability and responsibility, and trade-offs among all these concerns, especially within the context of security, robustness, and accuracy of AI systems. These issues talk directly to concerns around social justice that have become ever more important in the modern age. It was decided that the theme for SACAIR 2022 should be "AI for Social Justice". The choice of conference theme was intended to ensure multidisciplinary contributions that focus both on the technical aspects and the social impact and consequences of AI technologies. To give expression to this, the conference was organized as a multi-track conference covering broad areas of artificial intelligence:

---

[1] https://www.cair.org.za/.

- Algorithmic, Data Driven and Symbolic AI (Computer Science and Engineering)
- Socio-technical and Human-centered AI (Information Systems)
- Responsible and Ethical AI (Philosophy and Law)
- Inter- and Transdisciplinary AI Research

The accepted papers show a healthy balance between contributions on logic-based AI and those on data-driven AI, as the focus on knowledge representation and reasoning remains an important ingredient of studying and extending human intelligence. In addition, important contributions from the fields of social-technical and human-centred AI and responsible and ethical AI are reported in this volume.

We expect this multi- and interdisciplinary conference to grow into the premier AI conference in Southern Africa as it brings together nationally and internationally established and emerging researchers from disciplines including computer science, engineering, mathematics, statistics, informatics, philosophy, and law. The conference is also focused on cultivating and establishing a network of talented students working in AI from across Africa.

A conference of this nature is not possible without the hard work and contributions of many stakeholders. We extend our sincere gratitude to our sponsors: the Artificial Intelligence Journal (AIJ), the National Institute of Computational Sciences (NiTHeCS), the Centre for Artificial Intelligence Research (CAIR) and the BMW IT Hub South Africa. These sponsors enabled us to offer generous scholarships to students and emerging academics to participate in the conference. We sincerely thank the program chairs for their work in overseeing technical aspects of the conference and the publication of the two volumes of the proceedings, the international panel of reviewers, our keynote speakers, authors, and participants for their contributions. Finally, we extend our gratitude to the track chairs, the local organizing committee, student organizers, and the conference organizer for their substantive contributions to the success of SACAIR 2022.

October 2022   Alta de Waal
                                                               Bruce Watson

Organization

General Chairs

Alta de Waal   BMW IT Hub South Africa and University of Pretoria, South Africa

Bruce Watson   Stellenbosch University, South Africa

Program Committee Chairs

Aurona Gerber   University of Pretoria, South Africa

Edgar Jembere   University of KwaZulu-Natal, South Africa

Anban Pillay   University of KwaZulu-Natal, South Africa

Organizing Committee

Alta de Waal   BMW IT Hub South Africa and University of Pretoria, South Africa

Bruce Watson   Stellenbosch University, South Africa

Aurona Gerber   University of Pretoria, South Africa

Danie Smit   BMW IT Hub South Africa, South Africa

Edgar Jembere   University of KwaZulu-Natal, South Africa

Emile Engelbrecht   Stellenbosch University, South Africa

Emma Ruttkamp-Bloem   University of Pretoria, South Africa

Anban Pillay   University of KwaZulu-Natal, South Africa

Program Committee

Algorithmic, Data-Driven and Symbolic AI Track

Track Chairs

Alta de Waal   BMW IT Hub South Africa and University of Pretoria, South Africa

Anban Pillay   University of KwaZulu-Natal, South Africa

Arina Britz   Stellenbosch University, South Africa

Edgar Jembere   University of KwaZulu-Natal, South Africa

Ivan Varzinczak   Université d'Artois, France

Terence Van Zyl   University of Johannesburg, South Africa

Program Committee

Abiodun Modupe	University of Pretoria, South Africa
Albert Helberg	North-West University, South Africa
Allan De Freitas	University of Pretoria, South Africa
Andrew Paskaramoorthy	University of Cape Town, South Africa
Anna Sergeevna Bosman	University of Pretoria, South Africa
Bruce Watson	Stellenbosch University, South Africa
Bubacarr Bah	African Institute for Mathematical Sciences (AIMS), South Africa
Charis Harley	University of Johannesburg, South Africa
Colin Chibaya	Sol Plaatje University, South Africa
David Toman	University of Waterloo, Canada
Deon Cotterrell	University of Johannesburg, South Africa
Deshendran Moodley	University of Cape Town, South Africa
Duncan Coulter	University of Johannesburg, South Africa
Dustin Van Der Haar	University of Johannesburg, South Africa
Eduan Kotzé	University of the Free State, South Africa
Etienne Barnard	North-West University, South Africa
Fabio Cozman	University of São Paulo, Brazil
Febe de Wet	Stellenbosch University, South Africa
Fred Nicolls	University of Cape Town, South Africa
Gift Khangamwa	University of the Witwatersrand, South Africa
Giovanni Casini	ISTI-CNR, Italy
Guillermo R. Simari	Universidad del Sur in Bahia Blanca, Argentina
Hairong Wang	University of the Witwatersrand, South Africa
Herman Kamper	Stellenbosch University, South Africa
Hima Vadapalli	University of the Witwatersrand, South Africa
Iliana M. Petrova	Inria, France
Inger Fabris-Rotelli	University of Pretoria, South Africa
Jaco Versfeld	Stellenbosch University, South Africa
Jan Buys	University of Cape Town, South Africa
Jesse Heyninck	Open Universiteit, The Netherlands
Jiahao Huo	University of the Witwatersrand, South Africa
Jules-Raymond Tapamo	Univesity of KwaZulu-Natal, South Africa
Justice Emuoyibofarhe	Ladoke Akintola University of Technology, Nigeria
Karim Tabia	Université d'Artois, France
Laura Giordano	Università del Piemonte Orientale, Italy
Laurent Perrussel	IRIT, Universite de Toulouse, France
Leopoldo Bertossi	SKEMA Business School, Canada
Louise Leenen	University of the Western Cape, South Africa
Makhamisa Senekane	University of Johannesburg, South Africa

Mandlenkosi Gwetu	University of KwaZulu-Natal, South Africa
Marelie Davel	North-West University, South Africa
Mohamed Variawa	University of Johannesburg, South Africa
Olawande Daramola	Cape Peninsula University of Technology, South Africa
Paul Amayo	University of Cape Town, South Africa
Pieter de Villiers	University of Pretoria, South Africa
Ramon Pino Perez	Université d'Artois, France
Richard Booth	Cardiff University, UK
Richard Klein	University of the Witwatersrand, South Africa
Roald Eiselen	North-West University, South Africa
Rudzani Mulaudzi	University of the Witwatersrand, South Africa
Shakuntala Baichoo	University of Mauritius, Mauritius
Sihem Belabbes	LIASD, Université Paris 8, France
Stefan Woltran	TU Wien, Austria
Steven James	University of the Witwatersrand, South Africa
Sunday Oladejo	Stellenbosch University, South Africa
Sunday Olatunji	Imam Abdulrahman Bin Faisal University, Saudi Arabia
Tevin Moodley	University of Johannesburg, South Africa
Thembinkosi Semwayo	University of the Witwatersrand, South Africa
Thomas Meyer	University of Cape Town, South Africa
Willie Brink	Stellenbosch University, South Africa
Zainoolabadien Karim	University of the Witwatersrand, South Africa

Socio-Technical and Human-Centered AI Track

Track Chairs

Aurona Gerber	University of Pretoria, South Africa
Knut Hinkelmann	FHNW University of Applied Sciences and Arts Northwestern Switzerland, Switzerland
Sunet Eybers	University of Pretoria, South Africa

Program Committee

Andrea Martin	FHNW University of Applied Sciences Northwestern Switzerland, Switzerland
Catherine S. Price	University of KwaZulu-Natal, South Africa
Corné Van Staden	University of South Africa, South Africa
Danie Smit	BMW IT Hub South Africa, South Africa
Douglas Parry	Stellenbosch University, South Africa
Henk Pretorius	University of Pretoria, South Africa

Johan Breytenbach	University of the Western Cape, South Africa
Machdel Matthee	University of Pretoria, South Africa
Marie Hattingh	University of Pretoria, South Africa
Patrick Mikalef	Norwegian University of Science and Technology (NTNU), Norway
Phil van Deventer	University of Pretoria, South Africa
Rennie Naidoo	University of Pretoria, South Africa
Riana Steyn	University of Pretoria, South Africa
Ridewaan Hanslo	University of Pretoria, South Africa
Zola Mahlaza	University of Pretoria, South Africa

Responsible and Ethical AI Track

Track Chairs

| Emma Ruttkamp-Bloem | University of Pretoria, South Africa |
| Tanya de Villiers-Botha | Stellenbosch University, South Africa |

Program Committee

Andrea Palk	Stellenbosch University, South Africa
Ann-Katrien Oimann	KU Leuven and Royal Military Academy, Belgium
Arzu Formánek	University of Vienna, Austria
Ashley Coates	University of the Witwatersrand, South Africa
Attlee Munyaradzi Gamundani	Namibia University of Science and Technology, Namibia
Cindy Friedman	Utrecht University, The Netherlands
Dilara Boga	Central European University, Austria
Fabio Tollon	Stellenbosch University, South Africa
Helen Robertson	University of the Witwatersrand, South Africa
Karabo Maiyane	University of Pretoria and Nelson Mandela University, South Africa
Mbangula Lameck Amugongo	Namibia University of Science and Technology, Namibia
Niël Conradie	RWTH Aachen University, Germany
Rosemann Achim	De Montfort University, UK
Ryan Nefdt	University of Cape Town, South Africa
Sven Nyholm	Eindhoven University of Technology, The Netherlands
Zach Gudmunsen	University of Leeds, UK

Sponsors

National Institute for
Theoretical and Computational Sciences

#bmwithubsouthafrica

CENTRE FOR ARTIFICIAL
INTELLIGENCE RESEARCH

Contents

Responsible and Ethical AI

Algorithmic, Data Driven and Symbolic AI

Adversarial Training for Channel State Information Estimation in LTE Multi-antenna Systems

Andrew J. Oosthuizen[1,2]([✉]) [iD], Albert S. J. Helberg[1] [iD],
and Marelie H. Davel[1,2,3] [iD]

[1] Faculty of Engineering, North-West University, Potchefstroom, South Africa
aj.oosthuizen.ao@gmail.com, albert.helberg@nwu.ac.za,
marelie.davel@nwu.ac.za
[2] Centre for Artificial Intelligence Research (CAIR), Pretoria, South Africa
[3] National Institute for Theoretical and Computational Sciences (NITheCS),
Stellenbosch, South Africa

Abstract. Deep neural networks can be utilised for channel state information (CSI) estimation in wireless communications. We aim to decrease the bit error rate of such networks without increasing their complexity, since the wireless environment requires solutions with high performance while constraining implementation cost. For this reason, we investigate the use of adversarial training, which has been successfully applied to image super-resolution tasks that share similarities with CSI estimation tasks. CSI estimators are usually trained in a Single-In Single-Out (SISO) configuration to estimate the channel between two specific antennas and then applied to multi-antenna configurations. We show that the performance of neural networks in the SISO training environment is not necessarily indicative of their performance in multi-antenna systems. The analysis shows that adversarial training does not provide advantages in the SISO environment, however, adversarially trained models can outperform non-adversarially trained models when applying antenna diversity to Long-Term Evolution systems. The use of a feature extractor network is also investigated in this study and is found to have the potential to enhance the performance of Multiple-In Multiple-Out antenna configurations at higher SNRs. This study emphasises the importance of testing neural networks in the context of use while also showing possible advantages of adversarial training in multi-antenna systems without necessarily increasing network complexity.

Keywords: Channel state information · Deep learning · Adversarial training · Multiple-in multiple-out systems · Long-term evolution

1 Introduction

Channel state information (CSI) estimation is a wireless communication technique used to estimate channel conditions and equalise channel impairments

A. Pillay et al. (Eds.): SACAIR 2022, CCIS 1734, pp. 3–17, 2022.
https://doi.org/10.1007/978-3-031-22321-1_1

experienced by received data. Within the Long-Term Evolution (LTE) standard, CSI estimation uses pilot signals known to both transmitter and receiver to deliver a single estimation describing channel impairments [1]. In recent years, statistical CSI estimation and deep learning methods have obtained satisfactory results on this task [2]. However, less than optimal linear methods, such as least squares (LS) [3] and linear minimum mean squared error (MMSE) [4], are frequently applied due to the computational expense of more complex CSI estimation methods, making them infeasible to implement on user hardware.

For this reason, we investigate the performance of adversarially trained models since, while their training times are much higher than their non-adversarially trained counterparts, their computational expense remains the same when implemented. Recently Zhao et al. [5] implemented a generative adversarial network (GAN) architecture for CSI estimation and reported improved results due to the addition of adversarial training. We continue the work of Zhao et al. by investigating the effects of adversarial training on multi-antenna (MA) implementations. MA implementations are investigated since using antenna diversity techniques has become the standard in modern-day wireless communication and adds additional resilience to the impairments caused by fading channels [1,6].

The main contributions of this paper are:

- Investigating the difference in performance between adversarial and non-adversarial training for CSI estimation tasks.
- Providing results that show how antenna diversity techniques obtain unexpected performance gains using different adversarial training methods.
- As a side note, we also discuss the importance of model selection in the context of use rather than solely on training environment performance.

The rest of the paper is organised as follows: Sect. 2 provides background on the main techniques referred to, before closely related work is discussed in Sect. 3. The experimental setup is described in Sect. 4. This setup is used for all experiments conducted and reported on in Sect. 5. Results and conclusions are summarised in Sect. 6.

2   Background

We briefly discuss the CSI estimation task, the super-resolution GAN (SRGAN) architecture and antenna diversity techniques in the context in which these are utilised.

2.1   Channel State Information

For the current analysis, CSI estimation is implemented using the pilot-based method described in the LTE standard [1]. Pilot signals are placed in specific sub-carriers of orthogonal frequency division multiplexing (OFDM) modulated signals and are data points known to both transmitter and receiver. OFDM signals are two-dimensional transmissions that transmit data in the frequency and time domain and are thus represented as grids with frequency as the first dimension and time as the second. These signals are used according to Eq. 1 to obtain the initial CSI H from the transmit antenna Tx and receive antenna Rx, where the sub-carrier k represents the position of the pilot signals:

$$H[k] = \frac{Rx[k]}{Tx[k]} \tag{1}$$

Using the initial CSI, we obtain an LS estimation by extrapolating from the known CSI pilot signals to obtain the LS CSI estimation $\hat{H}_{LS}$ [3,6,7]. This method is the simplest form of channel estimation and can be used to equalise received data, resulting in the equalised received data $\hat{T}x$ according to Eq. 2.

$$\hat{T}x = \frac{Rx}{\hat{H}} \tag{2}$$

The process of equalisation is illustrated in Fig. 1 where received data is being equalised using complete knowledge of the CSI, i.e. perfect CSI. Typically, the equalised data does not return to its original transmitted state since some of the receiver antenna noise cannot be estimated but only reduced using denoising methods.

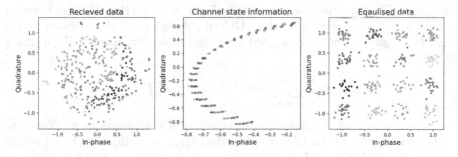

Fig. 1. An illustration of the process of equalising received data with 20 dB antenna noise (left), by using perfect CSI (middle) to obtain equalised data (right) that can be demodulated without error.

Applying LS to the pilot signals to estimate the CSI has a low computational cost but does not provide state-of-the-art performance. Alternatively, methods such as MMSE provide improved performance at the cost of computational efficiency as second-order statistics of the channel need to be calculated and implemented using costly matrix multiplication methods.

Many wireless devices, such as internet of things devices, require real-time computation to ensure no delays are caused in time-sensitive applications. Soltani et al. [8] compare a neural network (NN) CSI estimator to a classic CSI estimation method, finding that the NN outperforms the classic method. However, the NN requires 50 times the floating-point operations of the classic method to achieve this result. This encourages us to find ways of improving NN performance without increasing complexity.

2.2   Super Resolution GAN

Adversarial training is a training method in which a second neural network is introduced in the training process to compete with the primary or generator network. The second neural network, commonly called the discriminator, attempts to identify generated samples from ground truth samples. These networks are trained simultaneously; thus, one network's ability affects the training of the other. Adversarial training is implemented to train generator networks capable of generating samples closer to the ground truth than non-adversarially trained networks [9].

SRGAN [10] is a GAN designed for image super-resolution (SR) tasks, specifically image upsampling tasks. The adversarial training used by Ledig et al. for SRGAN is reported to push the generated SR images closer to the original images' manifold. This SR implementation results in a higher signal-to-noise ratio (SNR) and sharper high-frequency feature generation. SRGAN achieves these results by moving away from using the mean square error (MSE) to compare individual pixels to each other. Instead, SRGAN compares feature maps of the images obtained by a visual geometry group (VGG) network [11] in the adversarial training stage. Applying this VGG loss $l_{content}$ to images generated, the euclidean distance is measured between the feature maps of real and generated images. The loss function, described in Eq. 3, averages the MSE loss between the feature maps, generated by the VGG-19 network ϕ, of the high-resolution image I^{HR} and the image generated from low resolution using the generator $G(I^{LR})$:

$$l_{content} = \frac{1}{ij} \sum_{y=1}^{i} \sum_{x=1}^{j} (\phi(I^{HR})_{x,y} - \phi(G(I^{LR}))_{x,y})^2 \qquad (3)$$

Here i and j represent the respective dimensions of the feature maps. The content loss forms part of the final loss function l_{SRGAN} in Eq. 4 by being added to the adversarial loss used to evaluate the generator using the discriminator $l_{adversarial}$:

$$l_{SRGAN} = l_{content} + 1e^{-3} * l_{adversarial} \qquad (4)$$

SRGAN's generator consists of a convolution layer using a parametric rectified linear unit (PReLU) [12] activation function followed by several residual blocks and another convolution layer, all using batch normalisation [13]. Two 2× upsampler layers are added to upscale the physical dimensions of the feature maps before feeding the maps to an output convolution layer. All residual blocks use 64 channels before these are increased to 256 channels in the upscaling blocks.

The discriminator's structure closely follows a VGG network with eight convolution layers. Each of these layers reduces the feature maps' size while increasing the number of channels until the output is passed to a 1 024 node dense layer followed by a single node layer connected to a sigmoid activation function. This sigmoid output is then used with binary cross-entropy (BCE) loss to determine whether samples are real or generated.

2.3   Diversity Techniques

The principle applied in antenna diversity is to suppress the effects of fading channels by creating spatial diversity between antennas. The idea behind antenna diversity is that the probability that multiple independent channels are experiencing aggressive fading conditions at the same time is unlikely. We call this utilisation of different channels 'diversity gain', which is used to improve the bit error rate (BER) performance of MA systems.

In this work, we focus on both receiver and transmitter antenna diversity methods. For receiver diversity, we implement equal gain combining (EGC) [6], where a single transmit antenna transmits to multiple receiver antennas. Each antenna calculates the CSI between itself and the transmit antenna and equalises the received data. These streams of equalised data are then combined by averaging each corresponding sub-frame over the other receiver antennas.

To implement transmitter diversity, we utilise multiple transmit antennas with single or multiple receive antennas. We use orthogonal space-time block codes (OSTBC), specifically an Alamouti encoder [6] to enable communication in this MA environment. Alamouti encoders utilise the multiple transmitter antennas available by creating space-time block codewords with complex orthogonality. The Alamouti encoder deconstructs a single data stream into multiple data streams that, when transmitted, create a complex orthogonal matrix of the original symbols over the channel. Alamouti encoders require accurate CSI of all possible channel combinations to provide accurately equalised data. Link-level implementations use these MA methods to convert time-varying channels to additive white Gaussian noise (AWGN)-like channels.

3   Related Work

The CSI estimation problem has been approached using several deep learning architectures such as recurrent neural networks, long short-term memory networks and convolutional neural networks (CNNs) [14]. Specifically, CNN architectures have regularly appeared in CSI estimation papers due to the image-like nature of OFDM grids and the fading characteristics of many wireless channels. These CNN implementations outperform LS and are competitive with methods such as MMSE without calculating second-order channel conditions [15, 16]. The computational cost of methods is an important consideration in wireless communications as methods with lower computational cost and adequate performance are regularly selected over better-performing methods. In the LTE standard this is the case as an adapted LS method is used for CSI estimation over methods such as MMSE [1].

In the image processing field, CNNs have obtained state-of-the-art results in denoising and upsampling tasks when paired with adversarial training methods [10]. In the telecommunications domain, we have also seen increased use of GANs, primarily to model wireless channels. By modelling channels with GANs, autoencoders and other end-to-end neural networks can pass gradients through the propagation channel model, simplifying the training and design process of end-to-end implementations [17].

Zhao et al. [5] applies GANs to the CSI estimation problem, hypothesising that the GAN architecture generates CSI predictions closer to the target samples' manifold than traditional methods. Zhao et al. use an adaptation of SRGAN to generate high-resolution CSI from pilot signals. However, several changes are made to the original SRGAN architecture to provide improved results for the CSI estimation task. As illustrated in Fig. 2, the upsampling layers are removed, and a pre-upsampling methodology is implemented. The pre-upsampling is applied by using pilot signals to construct the LS CSI estimate, which would represent the low-quality input in the original SRGAN. By doing this, the upsampling of the pilot signals is essentially removed from the neural network. The authors show improved results when implementing this pre-upsampling method. It is, however, unknown how much of the improved result can be accredited to the adversarial training or is due to the effect of other architectural changes. Furthermore, this architecture is only implemented in a SISO environment, with the architecture's applicability in MA environments still unknown.

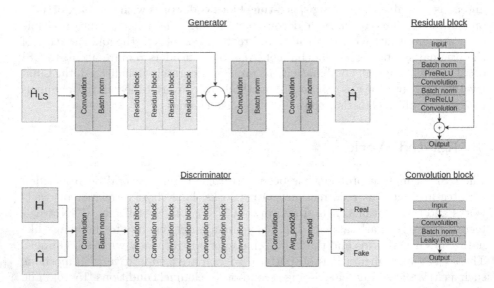

Fig. 2. Illustration of the modified SRGAN network structure proposed by Zhao et al. [5].

4   Experimental Setup

The same experimental setup is used for all experiments. Here we discuss the dataset, different system setups and the training protocol used.

4.1 Dataset

The training dataset is generated over a SISO channel using a simulated delay profile that is typical of traveling in a moving vehicle. Specifically, we use the Extended Vehicular A model delay profile [18] and 50 Hz maximum Doppler shift, similar to Zhao et al. AWGN is added after the data has been transmitted over the channel to model receiver noise. We use the LTE toolbox in Matlab® to simulate the transmission used. The transmitted OFDM dataframe, containing 16 quadrature amplitude modulated quadrature and in-phase (QI) symbols, has a bandwidth of 10 MHz and uses short cyclic prefixes. For the initial training set, no link-level techniques are used to ensure that the channel impairments are not obscured.

We provide the LS estimation $\hat{H}_{LS}$ of the CSI over a set sample size for the inputs to the network. The target of the modified SRGAN generator is the CSI before AWGN is added; an example of CSI being translated to a data sample can be found in Fig. 3. We train and evaluate the networks using a 20 dB AWGN range starting at 1 dB and ending at 20 dB SNR. The inputs and targets are represented using three features extracted from the demodulated OFDM signals: the absolute value of the QI symbols, the normalised quadrature component and the normalised in-phase component. We note that we have a three-channel representation in two dimensions, making our samples similar to RGB images presented to CNN architectures. We train the architectures using a training dataset of 40 000 samples and a validation set of 10 000 samples. An unseen test set for the SISO environment containing 10 000 samples is also generated and used for analysis in the input sample size selection and adversarial network training Sections. In this research, we apply several antenna diversity techniques to measure the performance of different CSI estimators. Due to each technique having a different number of channel paths that interfere with each other at the receiving antennas, the same dataset cannot be used for each method. However, each technique's evaluation uses identical simulation setups and consists of 400 data frames transmitted using the same channel and data generation seeds.

4.2 System Description

The neural networks used are derived from the SRGAN architecture discussed in Sect. 3. Thus, the generator is the same as the residual network (ResNet) architecture used by Zhao et al., with four residual blocks, but reverts back to element wise MSE for the content loss. We select this architecture as it is competitive with other architectures in the field, as well as top performing statistical methods [5]. When we apply adversarial training, the discriminator consists of a VGG-like architecture with an output that can indicate if a real or fake sample has been presented. Both normal element-wise adversarial training and feature-based adversarial training are used to train networks, by either passing CSI or CSI features to the discriminator. The implemented feature-based adversarial training uses a pre-trained VGG-19 network with the decision layers removed as a feature extractor. We thus have a single architecture but three different

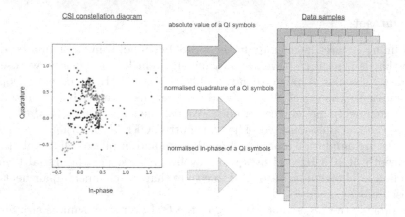

Fig. 3. An illustration of an CSI constellation diagram with 10 dB antenna noise being translated to an input data sample with three features. Each QI symbol translates to a 1×3 point in the data sample and the data sample dimensions correspond to the sub-frame size selected, e.g. 12×7.

training approaches: ResNet with no adversarial training, ResNet with a discriminator and ResNet with an additional feature extractor connected to its discriminator. We refer to these networks as ResNet, adversarial ResNet and SRGAN, respectively.

4.3 Training Protocol

All networks are trained using the Adam optimiser [19], with generators using MSE loss to ensure CSI estimation points are as close to the targets as possible. The discriminators use BCE loss, which, when used in conjunction with the sigmoid layer, gives the discriminator the ability to output the real or fake prediction. Finally, all networks are trained to convergence and final models for each run are selected using early stopping (selecting the model from the training run with best validation accuracy).

Due to the number of trainable hyperparameters and the computational expense of training adversarial networks, we only search over generator learning rate {1e−3, 1e−4, 1e−5}, discriminator learning rate {1e−3, 1e−4, 1e−5, 1e−6}, the size of the discriminator's contribution to the gradient update {1e−2, 1e−3, 1e−4} and the amount of added noise to samples before being sent to the discriminator {noise variance: 0, 1} to use in the hyperparameter grid search. The best hyperparameter combination is selected by evaluating the loss of the trained network on the validation set. Based on the results of this hyperparameter sweep, we generate two additional networks of the best hyperparameter combination using different initialisation seeds. All results reported in Sect. 5 are an average of the results over these three seeds. This hyperparameter search protocol is performed for every network reported on in Sect. 5.

5 Analysis

As the first analysis step, we select an input sample size for the networks by comparing the BER for the different sample sizes. After choosing an input sample size, adversarial networks are trained and compared in a SISO environment. The same networks are then applied to different antenna diversity environments and analysed to determine if adversarial training has affected the BER performance of these systems.

5.1 Sample Size Selection Using ResNet

Input sample size is selected by creating a baseline using the ResNet architecture without adversarial training. To ensure that an adequate sample input size is used, we train three networks with different input sizes: 12×7, 24×14 and 120×14. We train these networks using the training protocol described in Sect. 4.3 to obtain the best-performing networks. Finally, the best performing networks' performance is compared to the LS implementation given as input to the NN and plotted against each other using percentage increase of correctly classified bits (PICCB) as metric. PICCB is calculated by dividing the rate of correctly classified bits from one training method with another.

From the results shown in Fig. 4 we observe that the use of larger samples produces better performance than the use of smaller samples over the entire BER range, but that the increase in performance from 24×14 to 120×14 becomes small. We note that the ResNet generator provides the highest performance gain over LS at lower SNRs, where CSI is notoriously difficult to predict due to noise on the pilot signals masking true channel conditions.

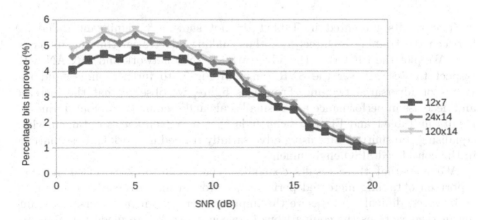

Fig. 4. PICCB of the CSI generated by three ResNet networks utilising different input sample sizes with respect to LS CSI estimation. This analysis is conducted over 20 dB of AWGN using the unseen test set in a SISO environment.

From these results, we select the 24 × 14 sample size as input for the following sections, as the training and inference time is lower than the larger 120 × 14 sample size while not sacrificing significant estimation performance. This decision can further be motivated as the two best-performing input sample sizes display nearly identical behaviours over the SNR range.

5.2   Adversarial Network Training

Comparing the results of adversarially trained networks to non-adversarially trained networks in a SISO environment, we analyse the best performing networks from each training method. We compare the results of the adversarial ResNet and SRGAN networks to the ResNet network. We observe the average loss over the entire test set for each network, followed by the average BER and, finally, the average PICCB. These metrics are displayed in Table 1 for each training approach and are averaged over three network seeds, as described in Sect. 4.3.

Table 1. Performance metrics of three generators utilising different training methods in a SISO environment. The PICCB metric is calculated with respect to ResNet (the non-adversarially trained network) over the entire SNR range.

	ResNet	Adversarial ResNet	SRGAN
Loss	**1.71e−2**	1.76e−2	1.73e−2
BER	1.42e−1	**1.41e−1**	1.41e−1
PICCB	0	**−1.20e−2**	−9.00e−3

The results presented in Table 1 do not show any significant difference between the training architectures, thus making adversarial training unnecessary. We plot the PICCB of the adversarial ResNet network and SRGAN with respect to ResNet over the SNR range in Fig. 5, to further understand the effects of adversarial training. From this figure, we observe that the increase and decrease in performance are of small scale and seemingly random. From the results in Table 1 and Fig. 5 we conclude that no advantage is obtained when equalising transmitted data using adversarially trained network CSI estimations in the considered SISO environment.

We note that these results contradict the results of Zhao et al. as they report on obtaining increased performance from using adversarial training. It is, however, difficult to measure the improvement proven by adversarial training in their work as the comparisons between networks are made using an upsampling method and the current implementation of SRGAN as opposed to non-adversarial and adversarial training methods.

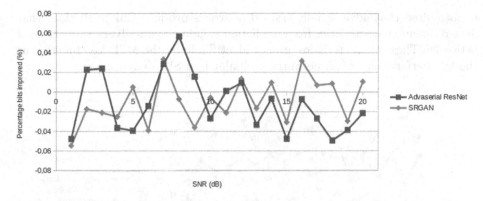

Fig. 5. PICCB of the CSI generated by adversarial training methods with respect to the trained ResNet CSI estimators. This analysis is conducted over 20 dB of AWGN using the unseen test set in a SISO environment.

5.3 Receiver Diversity

We now move on to MA setups to evaluate the performance of the adversarially trained network in common wireless communication systems. The receiver diversity environment is analysed by implementing two systems, the first with two receiver antennas and the second with four receiver antennas. Both of these systems utilise the EGC algorithm to equalise the received datastreams. Using the same NNs trained in the previous sub-section, we evaluate the PICCB of the adversarial networks with respect to the ResNet network architecture for receiver diversity implementations. Examining the average PICCB over the SNR range in Table 2 we see that both adversarial training methods outperform the traditional trained ResNet, with the adversarial ResNet performing the best of the two networks. We also note that the BER performance increases as the number of receivers increases.

Table 2. PICCB over the entire test set SNR range of adversarially trained network with respect to non-adversarially trained ResNet, for a receiver diversity system.

Number of receiver antenna	Adversarial ResNet	SRGAN
2	**1.02e−1**	2.56e−2
4	**1.38e−1**	5.65e−2

Plotting the PICCB over the SNR range in Fig. 6 we confirm this observation as the adversarial ResNet consistently outperforms the traditional ResNet over all trained SNRs in the test set. However, the same can not be claimed for the SRGAN implementation as it performs slightly worse than traditional ResNet at lower SNRs regardless of its average PICCB score. Based on these results it is hypothesised that adversarially trained networks produce CSI predictions that have different characteristics from predictions made by non-adversarially trained networks. These characteristics are not identifiable by the MSE loss function as the loss performance of all networks is similar in a SISO environment.

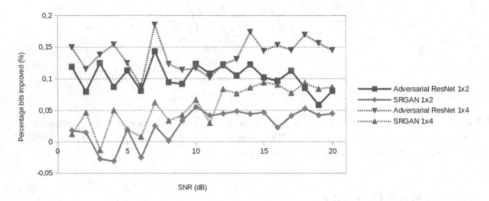

Fig. 6. PICCB of the CSI generated by adversarial training methods with respect to the trained ResNet CSI estimators. This analysis is conducted over 20 dB of AWGN in a multiple receiver environment utilising the EGC method.

5.4   Transmit Diversity

Continuing the MA implementations analysis, we compare the adversarially trained models to the non-adversarially trained models in a transmitter diversity environment. Using two transmit antennas, we construct two environments, one with a single receiving antenna and the second with two receiving antennas. For transmitter diversity, we implement Alamouti OSTBC and once again compare the adversarial networks to the traditionally trained ResNet. Examining the PICCB of this comparison in Table 3 we once again obtain results that differ from the results in the SISO environment. Firstly we observe that the adversarial ResNet no longer outperforms the traditional ResNet. Secondly, we observe that the SRGAN outperforms adversarial ResNet, but not traditional ResNet, again hinting at the possibility that these identical architectures predict differently because of their training method.

Table 3. PICCB of the entire test set SNR range of adversarially trained network with respect to non-adversarially trained ResNet, for a transmitter diversity system.

Number of receiver antenna	Adversarial ResNet	SRGAN
1	**−1.93e−2**	−1.85e−3
2	−3.66e−2	**−1.06e−2**

Further examination of the PICCB over the SNR range of the test set, shown in Fig. 7, displays how SRGAN outperforms adversarial ResNet at higher SNR levels. SRGAN, however, does not outperform traditional ResNet for most of the SNR range. While adversarial ResNet does compete with SRGAN at lower SNRs, this advantage is lost when the PICCB declines as the SNR increases.

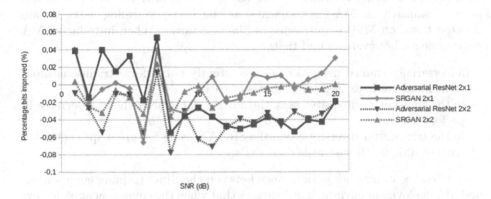

Fig. 7. PICCB of the CSI generated by adversarial training methods with respect to the trained ResNet CSI estimators. This analysis is conducted over 20 dB of AWGN in multiple receivers and transmitters environments utilising the Alamouti method.

Once again, we observe a difference in performance from networks that seemingly performed the same in the training phase of this analysis. These differences are of such a small scale that they are comparable to the differences in the SISO channel. However, in this instance, traditional ResNet consistently performs the worst of the three networks and SRGAN still performs similarly to traditional ResNet at higher SNRs.

From this analysis of different MA environments, we see that adversarial training provides unexpected increases and decreases in performance. Adversarial training seems to introduce characteristics into predictions that affect different MA environmetns in different ways. These results further lead us to believe that the performance indicators used in the training phase are inadequate in selecting networks for MA CSI estimation tasks.

6   Conclusion

This paper implements three methods of training using the same generator network to estimate high-quality CSI from low-quality LS estimations in an LTE environment. We ensure that an adequate input sample size is selected for our baseline non-adversarially trained network by considering three input sample sizes. When analysing the results of our ResNet implementation as baseline we observe:

- improvement over the low-quality LS input over the entire SNR range.
- the ResNet architecture achieves the most BER gain at lower SNR levels.
- the 24×14 input sample performance closely matches the performance of the larger 120×14 sample size.

All training methods perform similarly in the SISO training environment. This behaviour would typically be seen as an indication that the networks will perform similarly in MA environments, as the best-performing networks are selected based on MSE performance in SISO networks. When introducing MA environments, however, we find that:

- adversarially trained networks can consistently outperform traditional non-adversarial methods in receiver diversity cases.
- in the transmitter diversity cases, the adversarial ResNet shows a decrease in performance.
- in the transmitter diversity cases, the SRGAN using feature maps contended with traditional ResNet at higher SNRs.

The difference in network performance between the SISO training environment and MA deployment environment indicates that when the environment of deployment and the effects of adversarial training is well understood, increased BER performance can be obtained. We obtain this increase in performance at the same number of floating point operations as non-adversarially trained networks.

Acknowledgements. The authors gratefully acknowledge the financial support of this study by the Telkom CoE at the NWU and Hensoldt South Africa. The authors acknowledge the Centre for High Performance Computing (CHPC), South Africa, for providing computational resources to this research project.

References

1. Evolved universal terrestrial radio access E-UTRA; base station BS conformance testing. 3GPP TS 36.141, July 2018
2. Sure, P., Bhuma, C.M.: A survey on OFDM channel estimation techniques based on denoising strategies. Eng. Sci. Technol. Int. J. **20**(2), 629–636 (2017)
3. Karami, E.: Tracking performance of least squares MIMO channel estimation algorithm. IEEE Trans. Commun. **55**(11), 2201–2209 (2007)
4. Edfors, O., Sandell, M., Van de Beek, J.-J., Wilson, S.K., Borjesson, P.O.: OFDM channel estimation by singular value decomposition. IEEE Trans. Commun. **46**(7), 931–939 (1998)
5. Zhao, S., Fang, Y., Qiu, L.: Deep learning-based channel estimation with SRGAN in OFDM systems. In: 2021 IEEE Wireless Communications and Networking Conference (WCNC), pp. 1–6 (2021)
6. Cho, Y.S., Kim, J., Yang, W.Y., Kang, C.G.: MIMO-OFDM Wireless Communications with MATLAB. IEEE Press (2011)
7. Carrera, D.F., Vargas-Rosales, C., Yungaicela-Naula, N.M., Azpilicueta, L.: Comparative study of artificial neural network based channel equalization methods for mmWave communications. IEEE Access **9**, 41 678–41 687 (2021)

8. Soltani, N., et al.: Neural network-based OFDM receiver for resource constrained IoT devices. arXiv preprint arXiv:2205.06159 (2022)
9. Goodfellow, I., et al.: Generative adversarial nets. Adv. Neural Inf. Process. Syst. **27** (2014)
10. Ledig, C., et al.: Photo-realistic single image super-resolution using a generative adversarial network. In: Proceedings of the ICVPR, pp. 4681–4690 (2017)
11. Simonyan, K., Zisserman, A.: Very deep convolutional networks for large-scale image recognition. arXiv preprint arXiv:1409.1556 (2014)
12. He, K., Zhang, X., Ren, S., Sun, J.: Delving deep into rectifiers: surpassing human-level performance on ImageNet classification. In: Proceedings of the ICVPR, pp. 1026–1034 (2015)
13. Ioffe, S., Szegedy, C.: Batch normalization: accelerating deep network training by reducing internal covariate shift. In: International Conference on Machine Learning, pp. 448–456. PMLR (2015)
14. Luo, C., Ji, J., Wang, Q., Chen, X., Li, P.: Channel state information prediction for 5G wireless communications: a deep learning approach. IEEE Trans. Netw. Sci. Eng. **7**(1), 227–236 (2018)
15. Yang, Y., Gao, F., Ma, X., Zhang, S.: Deep learning-based channel estimation for doubly selective fading channels. IEEE Access **7**, 36 579–36 589 (2019)
16. Balevi, E., Andrews, J.G.: Deep learning-based channel estimation for high-dimensional signals. arXiv preprint arXiv:1904.09346 (2019)
17. Yang, Y., Li, Y., Zhang, W., Qin, F., Zhu, P., Wang, C.-X.: Generative-adversarial-network-based wireless channel modeling: challenges and opportunities. IEEE Commun. Mag. **57**(3), 22–27 (2019)
18. Evolved universal terrestrial radio access E-UTRA; user equipment (UE) radio transmission and reception. 3GPP TS 36.101, July 2018
19. Kingma, D.P., Ba, J.: Adam: a method for stochastic optimization. arXiv preprint arXiv:1412.6980 (2014)

Content-Based Medical Image Retrieval Using a Class Similarity-Aware Cross-Entropy Loss

Anicet Hounkanrin[✉], Paul Amayo, and Fred Nicolls

Department of Electrical Engineering, University of Cape Town,
Cape Town, South Africa
hnkmah001@myuct.ac.za, {paul.amayo,fred.nicolls}@uct.ac.za

Abstract. This paper addresses the problem of content-based image retrieval (CBIR) in a database of medical images using a two-step strategy. The first step consists in building a classification model using a state-of-the-art convolutional neural network for a preliminary screening of the query images. The classification model is trained using a weighted cross-entropy cost function that accounts for the similarity between classes. The second step of our CBIR method consists in searching for similar images in the database given the predicted class from the previous step. A histogram of oriented gradients (HOG) feature descriptor is used to reduce all images to lower dimensional feature vectors, and the similarity between a query image and the images in the database is evaluated by computing the Euclidean distance between the HOG feature vectors. The proposed method achieved an error score of 123.02 on the IRMA dataset, which represents an improvement of 7.12% over the state-of-the-art result.

Keywords: Content-based image retrieval · Image classification · Convolutional neural networks · Cross-entropy loss

1 Introduction

With the development of more sophisticated imaging technologies, there has been a surge in the integration of various imaging modalities into the daily routine of clinicians. In addition to the old imaging modalities (X-ray imaging), newer modalities such as magnetic resonance imaging (MRI) and positron emission tomography (PET) are now being used to help medical practitioners (radiologists, pathologists, surgeons, etc.) make more advised decisions. In effect, images are omnipresent in the clinical routine from diagnosis to treatment planning and monitoring, resulting in enormous amount of medical image data generated daily around the world. To manage this huge amount of data, medical images are usually stored using a picture archiving and communication system (PACS) where images are annotated with tags corresponding to patient information, imaging modalities, pathology, etc. To retrieve useful information from these archives,

A. Pillay et al. (Eds.): SACAIR 2022, CCIS 1734, pp. 18–30, 2022.
https://doi.org/10.1007/978-3-031-22321-1_2

specific knowledge about the archiving system and the case at hand is required, which restricts usage to experts only. An alternative to this search mechanism is content-based image retrieval (CBIR). Given a query image, CBIR aims at finding similar cases in the archive using information from the image content only. This helps the user make a more accurate diagnosis by taking into account diagnoses from the most relevant cases.

Thanks to the publication of the image retrieval in medical applications (IRMA) dataset in the ImageCLEFmed 2009 challenge, many researchers have proposed methods to help tackle the CBIR problem for medical images. With recent developments in machine learning, a few methods have tried to solve the CBIR task for medical images using learning-based methods. In this paper, we propose a method in line with those recent works. We propose to divide the CBIR problem into two sub-problems: classification and image search. Given a query image, a classification is first done to determine the category of the image using a convolutional neural network (CNN) trained with a class similarity-aware cross-entropy cost function. The predicted class is then used to search the relevant partition of the database, and the most similar cases are returned to the user. This approach turns out to be not only effective for the task, but also very efficient as we show in the next sections.

In Sect. 2, we present the most relevant approaches proposed in the literature. Our proposed method is discussed in Sect. 3. Section 4 summarizes the experiments done and the results obtained, before the conclusion in Sect. 5.

2 Related Work

The growing availability of medical image datasets has triggered interest in content-based image retrieval for medical images. Many researchers have proposed methods for CBIR in medical image databases. In this section, we present the most relevant works to the method proposed in this paper.

Babaie et al. [1] use Radon barcodes for image feature extraction to retrieve similar images in the IRMA dataset. They show that a single Radon projection is sufficient to generate reliable image features for searching and matching. The method performs well compared to other similar approaches, although it does not perform as well as learning-based methods. Similarly, Erfankhah et al. [5] use Gabor filters and Radon transform to extract features from images. The extracted features are used to train a support vector machine (SVM) classification model. Given a query image, its category is determined with the SVM classifier, and similar images are retrieved based on the predicted class for the query image.

Liu et al. [11] combine features extracted from a CNN model (termed "neural codes") with Radon barcodes to get feature description of the images. The CNN model is trained to classify images in the IRMA dataset. They then use the second to last fully connected layer of the CNN model to extract the neural codes, which are subsequently combined with Radon barcodes. For each query image, the retrieval is done by computing the Hamming distance between feature vectors of the query image and the images in the database. Images corresponding to the lowest distances are retrieved. Moreover, they propose a region of interest (ROI) matching method to identify salient regions in the retrieved images. However, the user needs to identify the ROI in the query image before salient regions can be predicted in the retrieved images.

Khatami et al. [9] also combine CNN and Radon projection for feature extraction. They use a shrinking search space strategy whereby the CNN model is trained on a sub-set of the dataset and fine-tuned on another. Similar images are retrieved by comparing the feature vector of the query image and the images in the database using the Manhattan metric. A saliency-based data folding approach is proposed by Camlica et al. [2], where salient regions are detected in the image during training. A local binary pattern (LBP) descriptor is used to extract features from the most salient regions in the images. Karthik and Kamath [8] proposed a multi-view classification method based on a CNN model. They achieved better results than previous methods, and establish the state-of-the-art result on the IRMA dataset.

The method presented in the current work is related to the approach proposed by Karthik and Kamath [8]. However, the performance of the classification model used here is further improved using a class similarity-aware cross-entropy cost function and the similarity of images is evaluated using the Euclidean distance between HOG feature vectors.

3 Proposed Method

In this section, we present an overview of the proposed method, as depicted on Fig. 1. After a brief description of the dataset, we present the two steps of the proposed CBIR method, followed by the definition of the evaluation metrics.

3.1 IRMA Dataset

The IRMA dataset[1] is a collection of radiographs selected from imaging routine at the Department of Diagnostic Radiology, Aachen University of Technology (RWTH), Aachen, Germany. The dataset consists of 14,410 images, belonging to 293 different classes, split into 12,677 training images and 1,733 test images. Each image in the dataset is labeled according to the coding system used by the image retrieval in medical applications (IRMA) project. The labels, termed IRMA codes [10], are strings of 13 alphanumeric characters. The set of the first four characters of the code, called technical axis (T), refers to the image modality. The next three characters of the code constitute the directional axis (D), and refer to the orientation of the object in the image. The directional axis is followed by the anatomical axis (A) consisting of three characters that refer to the body region examined. The last three characters constitute the biological axis (B), which correspond to the biological system examined. Thus any image

---

[1] http://publications.rwth-aachen.de/record/667225.

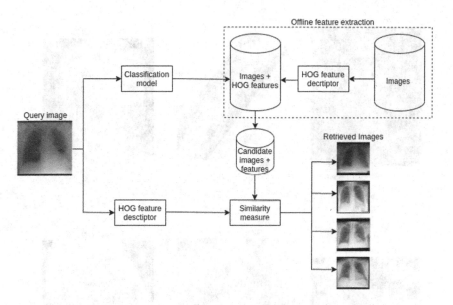

Fig. 1. Overview of the proposed CBIR system.

in the dataset has a label of the form: TTTT–DDD–AAA–BBB. Figure 2 shows a sample of the images in the dataset with their corresponding labels. We also illustrate the class distribution of the dataset on Fig. 3. This shows a high class imbalance in the dataset, making a classification task on this dataset very challenging.

3.2 CNN-Based Classification

The first step towards our CBIR system is the classification of the query images into relevant categories. This serves as a pre-screening to make the retrieval process more efficient. Given the diversity observed in the dataset (image size, viewpoints, body regions), appropriate preprocessing needs to be done before classification.

Image Preprocessing. As the images in the dataset are of different sizes, we use the zero-padding technique to bring all images to the same size of 512×512, preserving the original image aspect ratio. We then downscale each image to 256×256 before feeding them to the CNN model for computational efficiency. Moreover, the pixel values of each image are normalized to fit in the range $[0, 1]$. This helps stabilize the loss during model training.

Model Architecture. CNNs have proven to be very effective at classifying natural images. Here we leverage the classification power of CNNs to classify X-ray images. Given the relatively small size of the IRMA dataset, training a

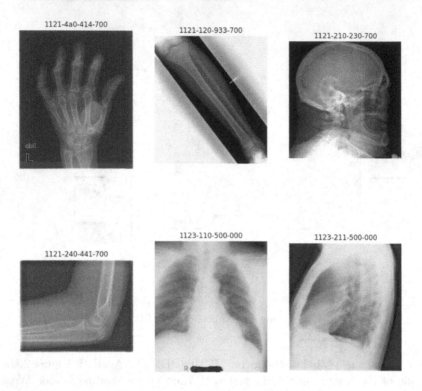

Fig. 2. Sample images from the IRMA dataset.

deep CNN model from scratch could lead to overfitting. To avoid this issue, we make use of transfer learning. Thus CNN models pretrained on ImageNet [4] for a natural image classification task are fined-tuned on the IRMA dataset for medical image classification. A block diagram of the DenseNet [7] model which is used as one of our base models for transfer learning is presented in Fig. 4. Each dense block represents a series of convolutional layer, a batch normalization layer, and a rectified linear unit (ReLU) activation. The last layer of this model is made of a fully connected layer with 1,000 neurons representing the classes for the ImageNet classification challenge. As in a classical transfer learning paradigm, the last fully connected (FC) is removed and the output of the pooling layer is used as image features. The image features are then fed to the classification block made of two FC layers. The first FC layer of this classification block is made of 1,024 neurons, and the second FC layer is made of 193 neurons representing the number of classes in our dataset.

Cross-Entropy Cost Function. The cross-entropy loss is a popular cost function that is used to optimize CNN models for image classification. Given a training image, with label y, the cross-entropy loss for a prediction p is defined by

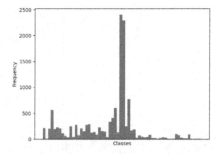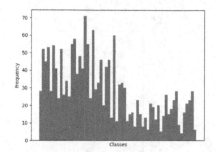

Fig. 3. Class distribution in the training set (left) and the test set (right) of the IRMA dataset.

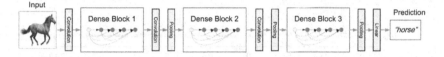

Fig. 4. Block diagram of the DenseNet CNN model used for image classification [7].

$$L(y, p) = -\sum_{k=1}^{N} y_k \log(p_k), \tag{1}$$

where N is the number of classes, $y = (y_k)_{1 \leq k \leq N}$ is the one-hot-encoded target vector, and $p = (p_k)_{1 \leq k \leq N}$ is the probability vector at the output of the softmax layer of the classification model. The cross-entropy loss represents how similar the input probability distribution is to the output distribution. Here we propose to boost the performance of our classification model by using a class similarity-aware cross-entropy cost function instead of the usual cross-entropy loss.

Class Similarity-Aware Cost Function. In the usual cross-entropy loss, labels are one-hot-encoded i.e. the true class is given a weight of 1 whereas all other classes are equally weighted with 0. Predicting a wrong class would result in the same penalty in the cost function regardless of the similarity of the predicted class with target class. This is not ideal when there exists some similarity between classes, since a wrong prediction that is similar to the ground truth class is more tolerable than a prediction that is very dissimilar.

Here we replace the one-hot-encoded label vector by an exponential decay weights vector, where classes similar to the target are given non-zero weights according to their degree of similarity with the target class. Thus for each image with a given ground truth class, the corresponding weight vector is obtained by applying an exponential decay function to the distances between the ground truth class and all classes in the dataset. Since this done during training, it has no effect on inference time. The resulting class similarity-aware cost function is a weighted version of the cross-entropy loss. This cost function is defined by

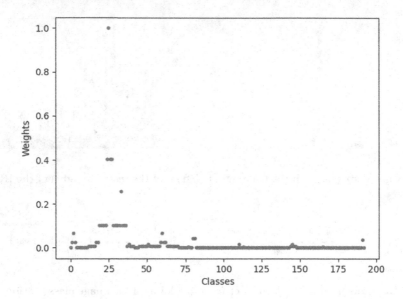

Fig. 5. Exponential decay class weights.

$$L_{WCE}(y, p) = -\sum_{k=1}^{N} \exp\left(-\frac{\Delta(k_{gt}, k)}{\sigma}\right) \log(p_k), \qquad (2)$$

where N is the number of classes, k_{gt} is the ground truth class, σ is a hyperparameter (we used $\sigma = 0.01$ in our experiments), and p_k is an entry in the output probability distribution vector p. The distance $\Delta(.)$ in Eq. 2 measures the similarity between two classes (the IRMA code corresponding to class k and the IRMA code corresponding to class k_{gt}). Thus the distance between two classes corresponding to the IRMA codes l and $\hat{l}$ is defined by

$$\Delta(l, \hat{l}) = \sum_{k=1}^{m} \sum_{i=1}^{n} \frac{1}{i} \delta_k(l_i, \hat{l}_i), \qquad (3)$$

where m is the total number of axes in an IRMA code ($m = 4$), n is the number of characters in an axis ($n = 4$ for the technical axis and $n = 3$ for the other three axes), and $\delta_k(.)$ is an indicator function returning 0 if the predicted character at position i of the kth axis matches the ground truth character at position i, and 1 otherwise. Figure 5 shows an illustration of the exponential decay weights for the target class.

This weighted cross-entropy cost function is used to fine-tune the CNN model to classify the images in the IRMA dataset. After classification, each query image is passed to the image retrieval step to search for similar images in the dataset.

Image Retrieval. In order to retrieve similar images to a query image, we must compare images in the dataset to the query image. Since these images

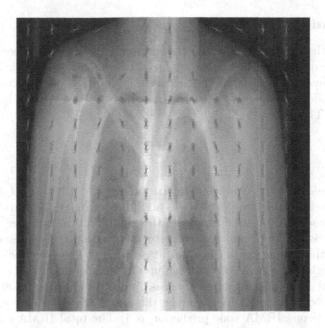

Fig. 6. Histogram of oriented gradients of an X-ray image displayed on the original image.

are high-dimensional data, comparing the raw images could be computationally prohibitive. We therefore need to convert the raw images into some feature representation for efficient comparison. Here we use a histogram of oriented gradients [3] feature descriptor for feature extraction.

Histogram of oriented gradients (HOG) is an image feature descriptor that extracts features from images by computing local pixel gradients. Given a image, the HOG features are extracted by dividing the images into cells (groups of pixels) of the same size and computing gradient of each cell. The histogram of gradient orientation of each cell is stored as a cell feature vector. The feature vector of the entire image is obtained by concatenating all cells feature vectors. In our experiments, the images are rescaled to 128×128 in order to have feature vectors of the same length for all images. Each image is divided into cells of 16×16 pixels and the gradient orientation of each cell is stored into a histogram of eight bins. This results in a feature vector of length 512 for each image. Figure 6 illustrates the block feature extraction from an X-ray image.

To compare images in the dataset to the query image, we only need to compare their HOG feature vectors. Given a query image, the Euclidean distance of its HOG feature vector and the feature vectors of the images in the database are computed. The images with the smallest distances to the query image are returned to the user.

3.3   Evaluation Metrics

We evaluate the performance of the trained model using two different metrics: the IRMA Error as described in the ImageCLEF image retrieval challenge [14], and the retrieval accuracy that measures the proportion of correctly classified images. Considering one axis of the IRMA code, the IRMA Error is defined as presented by Thommasi et al. [14]:

$$Error = \sum_{i=1}^{n} \frac{1}{b_i} \frac{1}{i} \delta(l_i, \hat{l}_i), \tag{4}$$

where n is the number of characters in the axis i.e. $n \in \{3; 4\}$; b_i is the number of possible labels at position i; and $\delta(.)$ is an indicator function returning 0 if the predicted character at position i matches the ground truth character at position i, and 1 otherwise. The predicted character at position i is represented by $\hat{l}_i$, whereas l_i is the ground truth character at position i. The error for each axis is normalized such that a completely wrong prediction has an error of 0.25 for that axis. The total error for a given test image is obtained by adding the errors across all four axes, so that for each test image the maximum error corresponding to a completely wrong IRMA code prediction is 1. The total IRMA error is then obtained by aggregating the errors across all the 1,733 test images. Intuitively, the total IRMA error represents the number of images that are completely misclassified out of the 1,733 test images.

The retrieval accuracy is derived from the IRMA error, and defined by

$$Accuracy = \frac{N - Error}{N} \times 100, \tag{5}$$

where N is the total number of test images.

4   Experiments and Results

In this section, we present the experiments conducted and the results obtained.

4.1   Experiments

All our classification models are implemented using the TensorFlow machine learning library. We ran four sets of experiments, each with one state-of-the-art deep learning model: DenseNet-121 [7], Inception-v3 [13], ReseNet-101 [6], or VGG-16 [12]. For each set of experiments, we establish a baseline by training the model with the normal cross-entropy loss. We then train another model with the proposed weighted cross-entropy cost function, and the results are compared to the baseline model. Our experiments were done on a machine with an NVIDIA Kepler40M (12GB DDR5 RAM) GPU. All classification models were trained using a stochastic gradient descent (SGD) optimizer with initial learning rate $l_r = 0.01$, momentum $\mu = 0.9$, and a batch size of 32 images. The models are trained on 80% of the 14410 training images, while the remaining 20% were used for model validation. Most models converged within 50 epochs of training. The best performing models on the validation set is used to evaluate the performance on the test set of 1,733 images.

4.2   Results

We report the results of classification and image retrieval separately.

Image Classification. For each set of experiments, we compare the performance of the model trained with the weighted cross-entropy loss (WCE) to the corresponding baseline model. Our best performing model is the DenseNet-121-WCE model which achieves an accuracy of 92.90% and an IRMA error of 123.02. This is followed by the ReseNet-101 model with an accuracy of 92.63% and an error of 127.79. The results of our four sets of experiments are summarized in Table 1. We also compare or results to the best reported results in the literature in Table 2. The DenseNet-121-WCE outperforms the best published results by Karthik and Kamath [8], bringing the error score on the IRMA dataset from 132.45 to 123.02. This represents an improvement of 7.12% over the best published result on this dataset.

Table 1. The models trained with the class similarity-aware cross-entropy loss consistently perform better than the baseline models.

Method	Accuracy (%)	Error
Inception-v3	91.62	145.23
Inception-v3-WCE	92.46	130.62
DensenNet-121	92.40	131.71
Proposed method (DenseNet-121-WCE)	**92.90**	**123.02**
VGG-16	89.97	173.79
VGG-16-WCE	91.18	152.92
ReseNet-101	92.24	134.48
ReseNet-101-WCE	92.63	127.79

Image Retrieval. We randomly selected six images from the test set, and retrieved the five most similar images in the train set for each query image. Figure 7 gives an illustration of the retrieval results for our proposed method. As it can be seen in Fig. 7, the retrieved images are mostly very similar to the query images, showing the effectiveness of the proposed method.

Table 2. Comparison of the retrieval accuracy and error with state-of-the-art methods.

Method	Accuracy (%)	Error
Proposed method (DenseNet-121-WCE)	**92.90**	**123.02**
Karthik and Kamath [8]	92.35	132.45
Camlica et al. [2]	91.54	146.55
TAUbiomed [14]	90.22	169.50
Idiap [14]	89.68	178.93
Khatami et al. [9]	90.30	168.05
Liu et al. [11]	87.07	224.13
FEITIJS [14]	86.01	242.46
Erfankhah et al. [5]	85.69	248.03

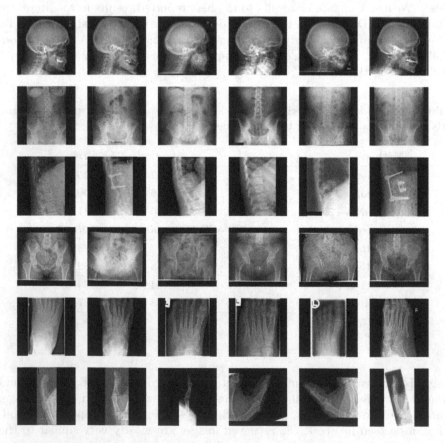

Fig. 7. Retrieval results for six test images in the IRMA dataset. The first column represents the query images taken from the test set. For each query image, the five most similar images in database are retrieved.

5   Conclusion

In this paper, we presented a method to retrieve similar images in a database of medical images. We proposed a retrieval method based on a classification model trained with a weighted cross-entropy cost function that accounts for class similarities. The evaluation of the method on the IRMA dataset reveals an improvement of 7.12% over previously published results. Thanks to its efficiency, the proposed CBIR method could be an integrant part of a computer-assisted diagnosis (CAD) system, and help radiologists make more accurate diagnosis and treatment planning.

Acknowledgements. We acknowledge financial support from the German Academic Exchange Service (DAAD) and the University of Cape Town.

Computations were performed using facilities provided by the University of Cape Town's ICTS High Performance Computing team: https://ucthpc.uct.ac.za/.

References

1. Babaie, M., Tizhoosh, H.R., Zhu, S., Shiri, M.E.: Retrieving similar X-ray images from big image data using radon barcodes with single projections. arXiv preprint arXiv:1701.00449 (2017)
2. Camlica, Z., Tizhoosh, H.R., Khalvati, F.: Medical image classification via SVM using LBP features from saliency-based folded data. In: 2015 IEEE 14th International Conference on Machine Learning and Applications (ICMLA), pp. 128–132. IEEE (2015)
3. Dalal, N., Triggs, B.: Histograms of oriented gradients for human detection. In: 2005 IEEE computer society conference on computer vision and pattern recognition (CVPR 2005), vol. 1, pp. 886–893. IEEE (2005)
4. Deng, J., Dong, W., Socher, R., Li, L.J., Li, K., Fei-Fei, L.: ImageNet: a large-Scale Hierarchical Image Database. In: 2009 IEEE Conference on Computer Vision and Pattern Recognition, pp. 248–255. IEEE (2009)
5. Erfankhah, H., Yazdi, M., Tizhoosh, H.R.: Combining real-valued and binary gabor-radon features for classification and search in medical imaging archives. In: 2017 IEEE Symposium Series on Computational Intelligence (SSCI), pp. 1–5. IEEE (2017)
6. He, K., Zhang, X., Ren, S., Sun, J.: Deep residual learning for image recognition. In: Proceedings of the IEEE Conference on Computer Vision and Pattern Recognition, pp. 770–778 (2016)
7. Huang, G., Liu, Z., Van Der Maaten, L., Weinberger, K.Q.: Densely connected convolutional networks. In: Proceedings of the IEEE Conference on Computer Vision and Pattern Recognition, pp. 4700–4708 (2017)
8. Karthik, K., Kamath, S.S.: A deep neural network model for content-based medical image retrieval with multi-view classification. Vis. Comput. **37**(7), 1837–1850 (2021)
9. Khatami, A., Babaie, M., Tizhoosh, H.R., Khosravi, A., Nguyen, T., Nahavandi, S.: A sequential search-space shrinking using CNN transfer learning and a radon projection pool for medical image retrieval. Expert Syst. Appl. **100**, 224–233 (2018)

10. Lehmann, T.M., Schubert, H., Keysers, D., Kohnen, M., Wein, B.B.: The IRMA code for unique classification of medical images. In: Medical Imaging 2003: PACS and Integrated Medical Information Systems: Design and Evaluation, vol. 5033, pp. 440–451. International Society for Optics and Photonics (2003)
11. Liu, X., Tizhoosh, H.R., Kofman, J.: Generating binary tags for fast medical image retrieval based on convolutional nets and radon transform. In: 2016 International Joint Conference on Neural Networks (IJCNN), pp. 2872–2878. IEEE (2016)
12. Simonyan, K., Zisserman, A.: Very deep convolutional networks for large-scale image recognition. arXiv preprint arXiv:1409.1556 (2014)
13. Szegedy, C., Vanhoucke, V., Ioffe, S., Shlens, J., Wojna, Z.: Rethinking the inception architecture for computer vision. In: Proceedings of the IEEE Conference on Computer Vision and Pattern Recognition, pp. 2818–2826 (2016)
14. Tommasi, T., Caputo, B., Welter, P., Güld, M.O., Deserno, T.M.: Overview of the CLEF 2009 medical image annotation track. In: Peters, C., et al. (eds.) CLEF 2009. LNCS, vol. 6242, pp. 85–93. Springer, Heidelberg (2010). https://doi.org/10.1007/978-3-642-15751-6_9

Jacobian Norm Regularisation and Conditioning in Neural ODEs

Shane Josias[1,2]([envelope])[iD] and Willie Brink[1][iD]

[1] Applied Mathematics, Stellenbosch University, Stellenbosch, South Africa
{josias,wbrink}@sun.ac.za
[2] School for Data Science and Computational Thinking, Stellenbosch University,
Stellenbosch, South Africa

Abstract. A recent line of work regularises the dynamics of neural ordinary differential equations (neural ODEs), in order to reduce the number of function evaluations needed by a numerical ODE solver during training. For instance, in the context of continuous normalising flows, the Frobenius norm of Jacobian matrices are regularised under the hypothesis that complex dynamics relate to an ill-conditioned ODE and require more function evaluations from the solver. Regularising the Jacobian norm also relates to sensitivity analysis in the broader neural network literature, where it is believed that regularised models should be more robust to random and adversarial perturbations in their input. We investigate the conditioning of neural ODEs under different Jacobian regularisation strategies, in a binary classification setting. Regularising the Jacobian norm indeed reduces the number of function evaluations required, but at a cost to generalisation. Moreover, naively regularising the Jacobian norm can make the ODE system more ill-conditioned, contrary to what is believed in the literature. As an alternative, we regularise the condition number of the Jacobian and observe a lower number of function evaluations without a significant decrease in generalisation performance. We also find that Jacobian regularisation does not guarantee adversarial robustness, but it can lead to larger margin classifiers.

Keywords: Neural ODEs · Regularisation · Sensitivity

1 Introduction

Neural ordinary differential equations (neural ODEs) [4] are a class of implicit deep learning layers, or continuous-depth models, in which the solution procedure (i.e. the forward pass) is separated from the definition of the layers (the neural network). This separation allows for flexibility in controlling the error of the solution procedure to trade-off against accuracy, and the use of adaptive solvers and correctors for ODE systems that are difficult to solve. Through the use

This work is based on research supported by the National Research Foundation of South Africa (grant number 138341).

of the adjoint method for gradient calculations, neural ODEs become memory efficient with a cost that grows constant in the number of layers, but sacrifices on computational complexity since the forward and reverse trajectories for a given point must be re-evaluated [4]. Mitigating the computational complexity inherent in neural ODEs can make them more tractable for larger datasets and a wider variety of modelling problems [9].

Neural ODEs define a continuous transformation on the data where the forward pass through the ODE is represented by an integral. The computational complexity of a neural ODE is represented by the number of function evaluations (NFE) required by the ODE solver to determine the solution trajectory [7,9,15]. For a mini-batch of data considered during training, NFE describes the number of times points along a solution trajectory are passed through the neural network. During training, the dynamics of the data transformation process become increasingly complex such that the solver must take finer steps to determine a solution [7]. Figure 1 shows how the NFE increases during the training of an unregularised neural ODE on a dataset (details of this experiment will follow in Sect. 3). It is believed that the complexity of the dynamics is related to the conditioning of the ODE system [7,9]. Simpler dynamics on the other hand, and a lower NFE, encourage faster convergence for neural ODEs and make them more practically feasible [9,15]. However, it is not clear whether generalisation and robustness of the solution are maintained under simplified dynamics, or whether there is a trade-off.

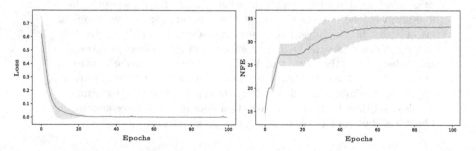

Fig. 1. As the loss is minimised during the training of an unregularised neural ODE on a 2-dimensional dataset (left), the number of function evaluations (NFE) required by the ODE solver increases (right). Here the solver requires on average about 34 function evaluations per batch. For continuous normalising flows on CIFAR-10, the NFE can go up to about 1400 [9].

Following Finlay et al. [9], we are interested in encouraging simple dynamics through regularising the Jacobian of a neural ODE, in classification problems instead of the normalising flows context. The Jacobian is evaluated at the input data, serving as initial conditions for the neural ODE. The focus on Jacobian regularisation provides a link to sensitivity analysis found in the conventional neural network literature [11,13,14,23,26] where both Frobenius and spectral norm regularisation have been employed. Conventional neural networks define an explicit data transformation, and the Jacobian represents the change in output with respect to changes in the input data. The Jacobian of a neural ODE, on the other hand, gives an indication of how the vector field describing the data flow changes with respect to changes in initial conditions. In this paper we borrow sensitivity analysis tools from the neural network literature, such as perturbation analysis and distance to decision boundaries, in order to evaluate Jacobian regularisation strategies in the training of neural ODEs. Our contributions can be summarised as follows.

1. We empirically investigate the effect of simplified dynamics on generalisation, adversarial vulnerability, and conditioning through Jacobian regularisation, with a view towards making neural ODE training more practical without sacrificing performance.
2. We show that regularising the Jacobian norm can make the ODE system more ill-conditioned, contrary to what is suggested in the literature [9], and while it reduces the NFE, it does so at the cost of test accuracy.
3. We show that regularising the condition number of the Jacobian reduces the NFE without sacrificing test accuracy.
4. Finally, we show that Jacobian norm regularisation can increase distance to the decision boundary for correctly classified data points, but does not guarantee robustness against adversarial perturbations.

Section 2 provides background on neural ODEs and the regularisation methods we investigate. Section 3 describes experiment methodology, and Sect. 4 discusses results on a running example. Section 5 extends the experiments to additional datasets for verification. Section 6 relates existing literature to this study, and Sect. 7 provides conclusions and ideas for future work.

2   Background and Definitions

Neural ODEs specify the dynamics of a continuous data transformation process by parameterising the ODE vector field with a neural network. Suppose that f_θ is a neural network with parameter set θ, and $h(t)$ the transformed data at time t, for $t \in [t_0, t_1]$. The neural ODE is then defined as

$$\frac{dh(t)}{dt} = f_\theta(h(t), t), \tag{1}$$

with an initial condition $h(t_0)$ represented by the input data. We are interested in a solution $h(t_1)$ at some time t_1, with (t_0, t_1) normally chosen as $(0, 1)$ in the literature [4,7,12]. The solution itself is unique provided that the neural network is Lipschitz continuous, due to Picard's existence theorem [5]. Linear and convolutional layers, and nonlinearities such as ReLU and tanh satisfy this property. The forward pass of the data is determined by integration:

$$h(t_1) = h(t_0) + \int_{t_0}^{t_1} f_\theta(h(t), t) \, dt. \tag{2}$$

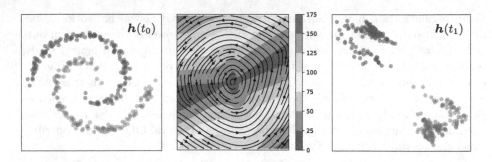

Fig. 2. The vector field defined by a neural ODE and the effect on a 2-dimensional dataset. The network in this neural ODE was trained to linearly separate the two classes shown on the left, leading to dynamics described by an unstable spiral fixed point in the vector field (middle). The colours in the vector field plot represent vector magnitude.

The initial condition $h(t_0)$ and solution at t_1, $h(t_1)$, define the input and output of the neural ODE, both of which having the same dimension as the input data. The network f_θ itself defines the gradient (vector field) of the solution trajectory at any given point in space. A loss function $\mathcal{L}$ to be optimised accepts the solution determined by a standard numerical ODE solver:

$$\mathcal{L}\left(h(t_1)\right) = \mathcal{L}\left(h(t_0) + \int_{t_0}^{t_1} f_\theta(h(t), t)\, dt\right) = \mathcal{L}\left(\text{ODESolve}(h(t_0), f_\theta, t_0, t_1)\right).$$
(3)

While it is possible to backpropagate through the operations of the ODE solver, such a strategy incurs a high memory cost and introduces additional numerical error [4] as the computational graph can become quite large. Instead, it is preferred to compute $\frac{\partial \mathcal{L}}{\partial \theta}$ by the adjoint sensitivity method, which considers $\frac{\partial \mathcal{L}}{\partial h(t)}$, $\frac{\partial f_\theta}{\partial \theta}$ and $\frac{\partial f_\theta}{\partial h(t)}$. These quantities can be computed efficiently and the integrals for constructing the forward and reverse trajectories can be found through standard ODE solvers. This approach scales more efficiently, has a low memory cost and explicitly controls numerical error [4].

Figure 2 provides an example of how a neural ODE trained jointly with a linear classifier at the end transforms input data at time t_0 to become linearly separable at time t_1.

NFE represents the number of times points along a solution trajectory are passed through the network f_θ that defines a neural ODE. An ODE solver that makes finer discretisations will require a higher NFE, and indicates more complex dynamics. In our experiments, NFEs will be averaged across the batches of an epoch, with batch size kept constant across regularisation methods. Investigating the impact of batch size on NFE is left for future work.

Jacobian regularisation is motivated by the fact that during training, NFE and the Frobenius norm of Jacobian matrices both increase [9]. The Jacobian $J \in \mathbb{R}^{d \times d}$ we consider is that of the neural network defining the vector field, with respect to the input data:

$$J = \nabla_{h(t_0)} f_\theta(h(t), t). \tag{4}$$

We will experiment with regularising the Frobenius norm $\|J\|_F$, the spectral norm $\|J\|_2$, and the condition number $\kappa(J)$. These are defined as

$$\|J\|_F = \sqrt{\sum_{i=1}^{d} \sum_{j=1}^{d} |J_{i,j}|^2}, \quad \|J\|_2 = \sigma_{\max}(J), \quad \kappa(J) = \frac{\sigma_{\max}(J)}{\sigma_{\min}(J)}, \tag{5}$$

where $\sigma_{\max}$ and $\sigma_{\min}$ refer to the largest and smallest singular values of a matrix. Regularising $\kappa(J)$ serves as a preliminary investigation into the claim that NFE increases due to ill-conditioning of the ODE [7,9].

For the purposes of regularisation, we will evaluate the Jacobian at the initial conditions (the input data). Thus, we make an assumption that regularising dynamics at the initial conditions can lead to regularised dynamics across the entire trajectory. Regularising $\|J\|_2$ controls the rate at which the input space is stretched along the first principal axis. Regularising $\|J\|_F$ effectively scales the Jacobian matrix, and should induce the same effect as controlling $\|J\|_2$, as is also apparent from the fact that $\|J\|_2 \leq \|J\|_F$. Exploring means of explicitly regularising dynamics across the entire solution trajectory is left for future work.

3   Methodology

We consider a supervised classification setting, where given a neural ODE defined by f_θ we aim to find parameters θ such that a linear classifier g_ϕ built on the solution of the ODE can associate an output $h(t_1)$ with a one-hot encoded label y. Given a labelled training set $\mathcal{D} = \{(x_i, y_i)\}_{i=1}^{N}$ with $x_i \in \mathbb{R}^d$, and letting $h_i(t_0) = x_i$, we construct the classifier g_ϕ to accept solutions $h_i(t_1)$ of a neural ODE system and to output confidence scores over labels. For a sample $(x, y) \in \mathcal{D}$, we denote the cross-entropy loss of g_ϕ as $\mathcal{L}_{\text{base}}(g_\phi(x), y)$ and construct additional loss functions with Jacobian regularisation terms as follows:

$$\mathcal{L}_*(x, y) = \mathcal{L}_{\text{base}}(g_\phi(x), y) + \lambda_n \|J(x)\|_*, \tag{6}$$

$$\mathcal{L}_{\text{cond}}(x, y) = \mathcal{L}_{\text{base}}(g_\phi(x), y) + \lambda_c \|\kappa(J(x)) - 1\|^2. \tag{7}$$

We set $\lambda_n = 1$ and $\lambda_c = \frac{1}{2}$ in later experiments, and add 10^{-6} to the denominator of $\kappa(J)$ to avoid underflow in the early stages of training. The total loss to be minimised is then computed as the average over samples in a mini-batch.

The neural ODE is implemented using the torchdiffeq framework [3]. The neural ODE and linear classifier are trained jointly, end-to-end. The addition of a linear classifier encourages the neural ODE vector field to linearly separate the classes. Training is done for 300 epochs, and repeated for 10 random seeds to determine stability across runs.

In high dimensions, direct computation of the Jacobian becomes computationally expensive. For now, as a test-bed for our hypotheses, the 2-dimensional intertwining moons dataset shown in Fig. 3 is used as a running example. Other datasets are considered in Sect. 5, and more efficient methods to scale to higher dimensions will be mentioned in Sect. 7.

Fig. 3. The intertwining moons dataset (red and blue indicate class labels). (Color figure online)

We are interested in the relationships between NFE as defined in Sect. 2, generalisation performance, and the sensitivity of a trained model's output under perturbations. The remainder of this section discusses the metrics used in our experiments.

Generalisation and sensitivity are investigated by means of performance on a hold-out test set, as well as input perturbations. The input perturbations include varying levels of Gaussian noise, and adversarial perturbations created by the fast gradient sign method [10]:

$$x_p = x + \epsilon \operatorname{sign}\left(\nabla_x \mathcal{L}(x, y)\right), \tag{8}$$

where ϵ is varied to increase the severity of the adversarial attack. This kind of perturbation moves a test sample x in a direction that increases the value of $\mathcal{L}(x, y)$, such that it is more likely for a trained model to predict an incorrect class label. Standard test set performance corresponds to $\epsilon = 0$.

Jacobian norms and condition numbers are averaged over all data points for each epoch. Keeping track of the norms and condition numbers can aid in determining the effect of each regularisation strategy. For instance, we will show in Sect. 4 that regularising the Frobenius and spectral norm of the Jacobian can make the ODE system more ill-conditioned.

Distance to decision boundary is a common metric in neural network sensitivity analysis, with the hypothesis that larger distances correspond to a more robust model. To determine the distance to a decision boundary, we generate points on d-dimensional spheres uniformly at random. The spheres are centered at a data point, with increasing radii. We perform a linear search over the spheres to determine the largest radius for which points are still labelled consistently. This can be made more efficient using a binary search algorithm.

4   Results

4.1   Generalisation and Sensitivity

Regularising $\|J\|_F$ and $\|J\|_2$ more than halves the NFE during training, as shown in Fig. 4. However, Fig. 5 shows a reduced test accuracy at $\epsilon = 0$ (no perturbations), revealing that complex dynamics might be a prerequisite for linearly separable representations of the input data. The steeper slope for Jacobian norm regularisation in Fig. 5 suggests that norm regularisation leads to classifiers that are more vulnerable to adversarial perturbations than the other regularisation strategies, for small values of ϵ. Even though Jacobian norm regularisation does not lead to adversarially robust classifiers, they do provide models that are more stable across runs, as is evident from the lower standard deviations in Figs. 4 and 5. On the contrary, Fig. 6 shows a slower decline in performance under random (Gaussian) perturbations. This could relate to our observation that Jacobian norm regularisation leads to larger classification margins (as detailed in Sect. 4.3). The effect of random noise perturbation is mitigated by a larger margin, since those perturbations are less likely to cross decision boundaries. Adversarial perturbations specifically move the input data in a direction that will lead to a miss-classification.

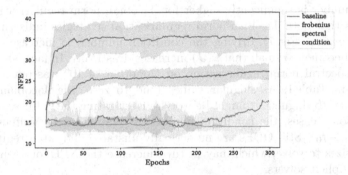

Fig. 4. Regularising the Jacobian norm (both Frobenius and spectral) and condition number controls the NFE required by the ODE solver during training. The baseline trains with no regularisation. Solid curves and shaded regions indicate mean and standard deviation over 10 runs.

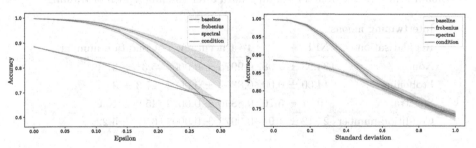

Fig. 5. Accuracy as a function of adversarial perturbations. $\epsilon = 0$ corresponds to standard test set performance. The baseline seems to be the most robust.

Fig. 6. Accuracy as a function of Gaussian perturbations. Standard deviation of 0 corresponds to standard test set performance.

Regularising the condition number to be closer to 1 also reduces the NFE, though not as significantly as Jacobian norm regularisation. On the other hand, condition number regularisation achieves a reduced NFE at virtually no cost to test accuracy. Training converges more stably when compared to the baseline (no regularisation). Interestingly, the baseline remains the least sensitive to adversarial perturbations. It is possible that the degree of complexity for unregularised dynamics in two dimensions is higher than the impact of adversarial perturbations, especially considering that the data coverage here is fairly dense.

4.2   Jacobian Norms and Condition Numbers

Table 1 shows that while Jacobian norm regularisation reduces the NFE, the condition number of the Jacobian increases compared to both the baseline and Jacobian conditioning regularisation. That is, Jacobian norm regularisation can make the ODE more ill-conditioned. Moreover, the standard deviation on the condition number is significantly higher for Jacobian norm regularisation. The issue here is that both Jacobian norm regularisation strategies indiscriminately push σ_{min} to zero (Fig. 7), driving the condition number to increase. In some cases, σ_{min} becomes so small that $\kappa(\boldsymbol{J})$ increases drastically. In fact, an extreme outlier for spectral norm regularisation on the order of 10^6 was omitted from the results in Table 1. As singular values tend to zero, the Jacobian matrix becomes closer to singular. It might be worth investigating the stiffness of neural ODEs. In some cases, the condition number is related to the stiffness index $S = \kappa(\boldsymbol{J})(t_1 - t_0)$. Stiff ODEs are numerically unstable and often require very small step sizes to solve, which may in turn increase the NFE, or even call for the use of implicit solvers.

Table 1. Measures of NFE, test accuracy and Jacobian condition number, for the different regularisation strategies investigated. The Jacobian condition number in the last column is an average over the training data after the final epoch of training.

Intertwining moons			
Regularisation	NFE	Test accuracy	Condition number
None	34.98 ± 2.98	0.9975 ± 0.0008	5.31 ± 3.51
Frobenius	14.00 ± 0.00	0.8862 ± 0.0003	27.3 ± 34.1
Spectral	19.81 ± 5.76	0.8846 ± 0.0022	45.9 ± 70.6
Condition number	27.12 ± 1.94	0.9973 ± 0.0007	6.10 ± 5.22

Figure 7 shows that Jacobian condition number regularisation controls the spectral norm to some extent. The two graphs also indicate that Jacobian condition number regularisation effectively pushes $\kappa(\boldsymbol{J})$ closer to 1.

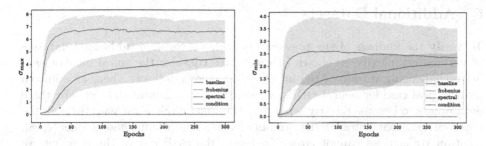

Fig. 7. Regularising the Jacobian condition number controls $\sigma_{\max}$ (left) and $\sigma_{\min}$ (right).

4.3 Distance to Decision Boundary

Figure 8 shows average distance to decision boundaries, for different regularisation strategies. Jacobian condition number regularisation results in little change from the baseline (no regularisation). Jacobian norm regularisation, on the other hand, does increase the average distance to the decision boundary for correctly classified points, and is consistent with prior work [14]. There seems to be a trade-off between having a classifier with a large margin and sufficiently complex dynamics for effective classification.

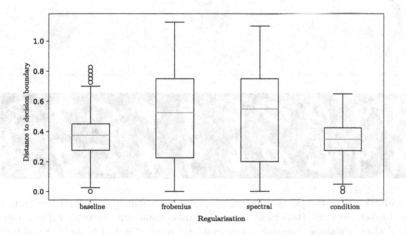

Fig. 8. Box-and-whisker plot of distance to decision boundary for different regularisation strategies, over the training data points. Jacobian norm regularisation increases distance to decision boundary on average, consistent with the literature.

5   Additional Datasets

To verify the findings of the previous section, we consider additional datasets of varying complexity, as shown in Fig. 9. We consider the spiral dataset as a more challenging version of the intertwining moons dataset. Then, as a known pathological case for neural ODEs, we consider the annulus dataset. For these data points to become linearly separable, the inner circle must pass through the outer circle. Since neural ODEs can only learn smooth homeomorphisms [7,18], solution trajectories cannot cross. To lessen the challenge of the annulus, we also consider the broken annulus dataset where part of the outer circle has been removed, allowing a valid trajectory for points in the inner circle. We train neural ODE classifiers for each of these three datasets, over 100 epochs. Finally, we consider 5-dimensional latent representations of the 10-class MNIST dataset, from a trained autoencoder. Neural ODE classifiers for the MNIST dataset are trained over 25 epochs.

We again compare NFEs and test accuracy of an unregularised neural ODE and ones trained with Jacobian norm and condition number regularisation. Training of all models is repeated for 10 random seeds.

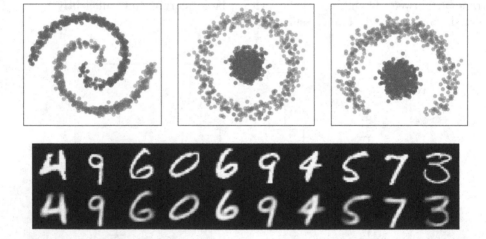

Fig. 9. Additional datasets used in our experiments: spiral (top left), annulus (top middle), broken annulus (top right), and 5-dimensional autoencoder representations of MNIST (here we show samples of original and reconstructed images in the bottom row).

NFE, accuracy and condition numbers are reported in Table 2. The finding that Jacobian condition number regularisation reduces NFE without a significant cost to generalisation performance, is consistent for all datasets. Surprisingly, Jacobian norm regularisation leads to test performance worse than random on the broken annulus dataset. It is possible that Jacobian norm regularisation

Table 2. NFEs, test accuracies and condition number for regularisation strategies investigated.

Spiral dataset			
Regularisation	NFE	Test accuracy	Condition number
None	40.27 ± 7.28	0.9978 ± 0.0030	20.6 ± 11.6
Frobenius	14.00 ± 0.00	0.7425 ± 0.0108	18.6 ± 13.7
Spectral	14.02 ± 0.06	0.7517 ± 0.0159	30.0 ± 20.0
Condition number	26.94 ± 11.3	0.8620 ± 0.1006	7.88 ± 5.88
Annulus			
Regularisation	NFE	Test accuracy	Condition number
None	38.08 ± 4.50	0.9528 ± 0.0251	23.2 ± 16.6
Frobenius	14.02 ± 0.06	0.6506 ± 0.0058	39.1 ± 38.7
Spectral	18.11 ± 10.8	0.6541 ± 0.0058	40.5 ± 58.8
Condition number	29.36 ± 3.18	0.9335 ± 0.0090	5.88 ± 5.53
Broken annulus			
Regularisation	NFE	Test accuracy	Condition number
None	34.40 ± 3.03	0.9965 ± 0.0032	11.7 ± 4.46
Frobenius	16.10 ± 4.66	0.4562 + 0.1246	29.6 ± 43.7
Spectral	14.96 ± 1.90	0.3984 ± 0.1032	70.2 ± 61.9
Condition number	30.01 ± 3.96	0.9917 ± 0.0068	15.0 ± 29.7
5-dimensional MNIST			
Regularisation	NFE	Test accuracy	Condition number
None	29.59 ± 2.20	0.9522 ± 0.0012	2.47 ± 0.40
Condition number	19.19 ± 1.38	0.9364 ± 0.0011	1.54 ± 0.18

asserts too strong an inductive bias on trajectories learnt by the neural ODE, such that some types of dynamics are suppressed regardless of how simple a dataset may be. For the MNIST dataset, which is higher in dimension and has more classes, results remain consistent: Jacobian condition number regularisation can reduce NFE without a significant cost to test accuracy.

In Fig. 10, we report adversarial and Gaussian robustness for the additional 2-dimensional datasets. Again, the baseline seems to be most robust to adversarial perturbations, except for the annulus and broken annulus datasets, where there is not much difference between the baseline and condition number regularisation. The slower decline in accuracy under Gaussian perturbations (spiral and annulus dataset) is again evident for Jacobian norm regularisation, especially for the spiral dataset.

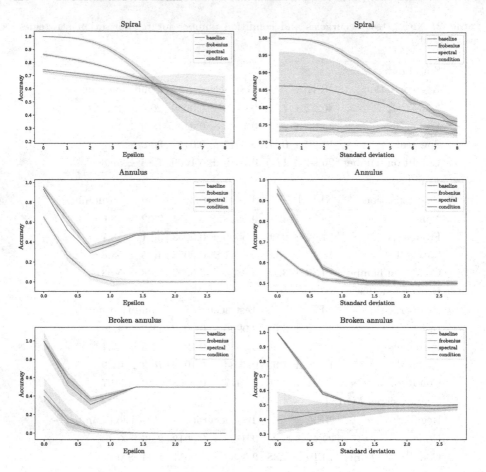

Fig. 10. Accuracy as a function of adversarial perturbations (left), and accuracy as a function of Gaussian perturbations (right), for the additional 2-dimensional datasets.

6   Related Work

As mentioned earlier, neural ODEs can be regarded as continuous depth neural networks that are more memory efficient and separate the definition of the model from the forward pass. Neural ODEs may also be suitable for modelling continous-time data from systems such as physical systems [20,22,25,27], continuous-time time series [6,16,21] and continuous normalising flows [4,8,9, 17,19]. Neural ODEs also appear in score-based generative modelling where a stochastic differential equation is used to model the data generation process, and have been shown to generate GAN-level quality samples without the need for adversarial training (which can suffer from issues such as mode collapse) [24]. Song et al. [24] convert the stochastic differential equation into an ordinary differential equation whose solution allows for the generation of samples.

It is important to acknowledge that there are modelling disadvantages to neural ODEs. In fact, neural ODEs are not universal function approximators: they can only learn smooth homeomorphisms [7,18] due to the fact that solution trajectories cannot cross one another. Dupont et al. [7] circumvent this by augmenting the dimensions of the ODE, akin to the kernel trick in SVMs. Problems requiring complex dynamics and a high number of function evaluations then become simpler to solve with lower NFEs, much like the aim of our work. Yan et al. [12] suggest that the restriction of smooth homeomorphism can lead to more robust models, and perform perturbation vulnerability experiments by comparing neural ODEs with CNNs. Most related to our paper are the works of Finlay et al. [9] who regularise the Frobenius norm of the Jacobian in the context of normalising flows, and Kelly et al. [15] who encourage simpler dynamics by introducing a differentiable proxy for the time cost of a numerical solver. Bai et al. [2] observe increasing instability during the training of deep equilibrium models [1], a class of deep implicit layers where the solution procedure is defined by a fixed point iteration. This instability leads to an increased NFE and a higher likelihood of divergence from a fixed point, all of which are reduced by regularising the Frobenius norm of the Jacobian at the fixed point.

There also exists works in the broader machine learning literature that regularise the Frobenius and spectral norm of a neural network's weights or input-output Jacobian, towards improving generalisation and robustness [11,13,14,23, 26]. Similar to our paper, these works operate under the hypothesis that norm regularisation can reduce sensitivity to perturbations in the input. Yoshida et al. [26] provide an upper bound to the spectral norm of a neural network's weight matrices that allows for more efficient regularisation. Johansson et al. [14] extend this by providing an exact method to compute the spectral norm of a neural network's input-output Jacobian, through a power iteration procedure. Our findings on an increased distance to decision boundary for Frobenius and spectral norm regularisation are consistent with those of Johansson et al. [14]. We further expect that their approach can be applied to neural ODEs, so that our experiments can be scaled to higher dimensional datasets. Hoffman et al. [13] provide an efficient computation for the Frobenius norm of a Jacobian through random projections. Their method, however, is not applicable to spectral norms or condition numbers, as those cannot be written in terms of a matrix trace operation.

7   Conclusion

We investigated the required number of function evaluations (NFE) and conditioning of a neural ODE system under different Jacobian regularisation strategies. Our experiments show that while Jacobian norm regularisation is effective at reducing the NFE of neural ODE solvers, they do so at the cost of test accuracy and may make the Jacobian matrix ill-conditioned. Jacobian condition number regularisation reduces NFE without a significant loss in accuracy, and controls the norm and condition number of the Jacobian. We acknowledge that this study involved small datasets, and should be regarded as an initial exploration into neural ODE conditioning. Moreover, Jacobian condition number regularisation adds a significant computational overhead to the training process that might not be balanced by a reduced NFE. Future work will therefore investigate a more efficient computation of the condition number, similar to the power iteration already attempted for convolutional neural networks [14,26]. This will allow experiments to be scaled to higher dimensional datasets, such as images, and may also allow for the regularisation of dynamics across the entire solution trajectory [15]. Higher dimensional datasets are important to consider given that the goal of reducing NFE through Jacobian regularisation is to make neural ODEs more practically relevant. Another possible direction would be to characterise the stiffness of the ODE at the solutions, as stiff ODEs often require higher NFEs or an implicit solver.

References

1. Bai, S., Kolter, J.Z., Koltun, V.: Deep equilibrium models. In: Advances in Neural Information Processing Systems, vol. 32, pp. 690–701 (2019)
2. Bai, S., Koltun, V., Kolter, J.Z.: Stabilizing equilibrium models by Jacobian regularization. In: International Conference on Machine Learning, vol. 139, pp. 554–565 (2021)
3. Chen, R.T.: Torchdiffeq: PyTorch implementation of differentiable ODE solvers (2018). https://github.com/rtqichen/torchdiffeq
4. Chen, R.T., Rubanova, Y., Bettencourt, J., Duvenaud, D.K.: Neural ordinary differential equations. In: Advances in Neural Information Processing Systems, vol. 31, pp. 6572–6583 (2018)
5. Coddington, E.A., Levinson, N.: Theory of Ordinary Differential Equations. Tata McGraw-Hill Education (1955)
6. De Brouwer, E., Simm, J., Arany, A., Moreau, Y.: GRU-ODE-Bayes: continuous modeling of sporadically-observed time series. In: Advances in Neural Information Processing Systems, vol. 32, pp. 7377–7388 (2019)
7. Dupont, E., Doucet, A., Teh, Y.W.: Augmented neural ODEs. In: Advances in Neural Information Processing Systems, vol. 32, pp. 3134–3144 (2019)
8. Falorsi, L., Forré, P.: Neural ordinary differential equations on manifolds. arXiv:2006.06663 (2020)
9. Finlay, C., Jacobsen, J.H., Nurbekyan, L., Oberman, A.: How to train your neural ODE: the world of Jacobian and kinetic regularization. In: International Conference on Machine Learning, vol. 119, pp. 3154–3164 (2020)
10. Goodfellow, I., Shlens, J., Szegedy, C.: Explaining and harnessing adversarial examples. In: International Conference on Learning Representations (2015)
11. Gu, S., Rigazio, L.: Towards deep neural network architectures robust to adversarial examples. In: International Conference on Learning Representations, ICLR Workshop Track Proceedings (2015)
12. Hanshu, Y., Jiawei, D., Vincent, T., Jiashi, F.: On robustness of neural ordinary differential equations. In: International Conference on Learning Representations (2020)
13. Hoffman, J., Roberts, D.A., Yaida, S.: Robust learning with Jacobian regularization. arXiv:1908.02729 (2019)

14. Johansson, A., Strannegård, C., Engsner, N., Mostad, P.: Exact spectral norm regularization for neural networks. arXiv:2206.13581 (2022)
15. Kelly, J., Bettencourt, J., Johnson, M.J., Duvenaud, D.K.: Learning differential equations that are easy to solve. In: Advances in Neural Information Processing Systems, vol. 33, pp. 4370–4380 (2020)
16. Kidger, P., Morrill, J., Foster, J., Lyons, T.: Neural controlled differential equations for irregular time series. In: Advances in Neural Information Processing Systems, vol. 33, pp. 6696–6707 (2020)
17. Lou, A., et al.: Neural manifold ordinary differential equations. In: Advances in Neural Information Processing Systems, vol. 33, pp. 17548–17558 (2020)
18. Massaroli, S., Poli, M., Park, J., Yamashita, A., Asama, H.: Dissecting neural ODEs. In: Advances in Neural Information Processing Systems, vol. 33, pp. 3952–3963 (2020)
19. Mathieu, E., Nickel, M.: Riemannian continuous normalizing flows. In: Advances in Neural Information Processing Systems, vol. 33, pp. 2503–2515 (2020)
20. Miles, C., Sam, G., Stephan, H., Peter, B., David, S., Shirley, H.: Lagrangian neural networks. In: International Conference on Learning Representations, Workshop on Integration of Deep Neural Models and Differential Equations (2020)
21. Rubanova, Y., Chen, R.T., Duvenaud, D.K.: Latent ordinary differential equations for irregularly-sampled time series. In: Advances in Neural Information Processing Systems, vol. 32, pp. 5321–5331 (2019)
22. Sanchez-Gonzalez, A., Bapst, V., Cranmer, K., Battaglia, P.: Hamiltonian graph networks with ODE integrators. arXiv:1909.12790 (2019)
23. Sokolić, J., Giryes, R., Sapiro, G., Rodrigues, M.R.: Robust large margin deep neural networks. IEEE Trans. Signal Process. **65**(16), 4265–4280 (2017)
24. Song, Y., Sohl-Dickstein, J., Kingma, D.P., Kumar, A., Ermon, S., Poole, B.: Score-based generative modeling through stochastic differential equations. In: International Conference on Learning Representations (2020)
25. Wang, W., Axelrod, S., Gómez-Bombarelli, R.: Differentiable molecular simulations for control and learning. In: International Conference on Learning Representations, Workshop on Integration of Deep Neural Models and Differential Equations (2020)
26. Yoshida, Y., Miyato, T.: Spectral norm regularization for improving the generalizability of deep learning. arXiv:1705.10941 (2017)
27. Zhong, Y.D., Dey, B., Chakraborty, A.: Symplectic ODE-net: learning Hamiltonian dynamics with control. In: International Conference on Learning Representations (2019)

Improving Cause-of-Death Classification from Verbal Autopsy Reports

Thokozile Manaka[1]([⊠]) [ID], Terence van Zyl[2] [ID], and Deepak Kar[3] [ID]

[1] School of Computer Science and Applied Mathematics,
University of the Witwatersrand, Johannesburg, South Africa
thokozilemanaka@wits.ac.za
[2] Institute for Intelligent Systems, University of Johannesburg,
Johannesburg, South Africa
tvanzyl@uj.ac.za
[3] School of Physics, University of the Witwatersrand,
Johannesburg, South Africa
deepak.kar@wits.ac.za

Abstract. In many lower-and-middle income countries including South Africa, data access in health facilities is restricted due to patient privacy and confidentiality policies. Further, since clinical data is unique to individual institutions and laboratories, there are insufficient data annotation standards and conventions. As a result of the scarcity of textual data, natural language processing (NLP) techniques have fared poorly in the health sector. A cause of death (COD) is often determined by a verbal autopsy (VA) report in places without reliable death registration systems. A non-clinician field worker does a verbal autopsy (VA) report using a set of standardized questions as a guide to uncover symptoms of a COD. This analysis focuses on the textual part of the VA report as a case study to address the challenge of adapting NLP techniques in the health domain. We present a system that relies on two transfer learning paradigms of monolingual learning and multi-source domain adaptation to improve VA narratives for the target task of the COD classification. We use the Bidirectional Encoder Representations from Transformers (BERT) and Embeddings from Language Models (ELMo) models pretrained on the general English and health domains to extract features from the VA narratives. Our findings suggest that this transfer learning system improves the COD classification tasks and that the narrative text contains valuable information for figuring out a COD. Our results further show that combining binary VA features and narrative text features learned via this framework boosts the classification task of COD.

Keywords: Natural language processing · Transfer learning · Monolingual learning · Multi-domain adaptation · Cause of death

1 Introduction

Natural language processing (NLP) methods have played an important role in extracting useful information from unstructured narrative text for classification tasks. Most of these applications have been in the general English domain with convolutional neural networks (CNNs) [1–3], recurrent neural networks (RNNs) [4] and attention based methods [5,6]. The medical domain also saw attention based methods used in [7,8], RNNs [9] and CNNs [10].

The development of NLP models in the health domain has however shown to be advancing quite slowly in comparison to the general English domain. This is due to a limited access to shared annotated datasets across health institutions and laboratories because of patient privacy and confidentiality policies. There are also insufficient common conventions and standards of annotating clinical data for training and bench-marking NLP applications in this domain [11]. Sometimes data is not available at all as is the case with many lower-and-middle-income countries where the bulk of fatalities occur outside of medical facilities and no physical autopsies are done [12]. The World Health Organization (WHO) has endorsed the use of a verbal autopsies (VAs) in these countries to find the cause of death (COD). These are records of interviews about the events surrounding an uncertified cause of death. As with the language of health reports, this public health report necessitates its own domain-specific development and training as NLP models developed for the general English language text do not generalize well on its narratives.

Transfer learning [13–17] allows for knowledge derived from tasks rich in data to be applied to tasks, languages, or domains where data is limited. It consist of two steps, pretraining on one task or domain (source) and domain adaptation where the learned representations are used in a different task, domain or language (target). Kim [1] shows that transfer learning has advanced the development of NLP techniques through natural language modeling as a source task and Pikuliak et al. [18] reports that the limited data access that results in declined performances of NLP techniques can be addressed with cross-lingual and monolingual learning techniques.

While domain adaptation assumes that the source domain has abundant training data and aims to use the knowledge learned here to aid the tasks in the target domain which has limited resources, many text classification tasks are domain-dependent in the sense that a text classifier trained on one set of data is likely to under perform on another set as is with the case of labeled and unlabeled or unseen data. Also, while the most used domain adaptation technique is the one-source-one-target approach, Chen and Cardie [19] shows that multi-domain text classification, where labeled data is present for many domains, but in low amounts for effective training of a text classifier is a more plausible reality. By nature of how it is collected and transcribed, we contend that a VA report represents one actual multi-domain context, and the NLP technologies used on it will thus require a sufficient amount of data from a number of domains.

Motivated by the multi-domain text classification task where data from a number of domains is fused together for training on a feature or classifier level, we propose a technique that leverages the two transfer learning paradigms of monolingual learning and multi-domain adaptation via the use of embeddings from ELMo pretrained in the English domain and those from BERT pretrained in the biomedical domain. The idea is that the biomedical domain is a subset of the

English domain [20] and therefore by using character-level information from the English domain when computing VA embeddings, grammatical and syntax errors in VA reports will be well taken care of. Additionally, word-level information from the medical domain will capture biomedical relations like symptom interactions across different diseases [9], information that is crucial for COD classification.

This paper's clinical and technical contributions are:

1. A transfer learning approach that improves the cause-of-death (COD) classification task of narrative text features of a verbal autopsy (VA) report by making the most out of understanding the domain adaption process.
2. Refined verbal autopsy text representations more suited for the COD classification task.
3. We show that multi-domain adaptation for text classification achieves better recall scores than currently used text representation techniques.

2   Background

2.1   Verbal Autopsies

More than a half of the yearly fatalities in lower-and-middle-income nations take place outside of hospitals and due to inadequate death registration systems, this leaves no cause of death (COD) information available [21]. Verbal autopsy (VA) technology was developed to address the requirement for COD information for researchers and legislators. Trained surveyors interview a close relative involved in taking care of the deceased about events surrounding their death and physicians and more recently automated algorithms later code the surveys for a COD [22].

For a variety of CODs, VAs have shown to be 98% specific [23] and while some studies show the textual part of the report to be of very little use in COD classification [24], other works have show that the information in the narrative text helps in making accurate diagnoses [25,26].

2.2   Transfer Learning

For many real life machine learning tasks, it can be difficult to collect a lot of data when taking on a new task. This is especially true in domains such as health where there are data privacy policies for patients in place. Datasets in these domains are also unique to individual institutions and as such it has grown to be a challenge to obtain satisfactory model performances using a small amount of training data.

Baxter [27] proposed the transfer learning technique to tackle this problem. This technique allows for knowledge from languages, tasks, or domains with lots of data to be transferred to areas with less data. Deep learning techniques have played a significant role in developing SOTA transfer learning techniques across numerous areas of applications because of their proved accuracies. Deep learning techniques are data driven and these advances have not shown equal development in the health domain. Those that are emerging in the clinical domain such as ClinicalBERT [28], the publicly available Clinical BERT Embeddings [7], and biomedical and scientific literature like BioBERT [8], BioELMo [29], SciBERT [30] depend on data that is transcribed by medical professionals and is still unique to institutions to some extent.

We define transfer learning using a task and a domain where the latter D, is composed of a feature space χ and a marginal probability distribution $P(X)$ over the feature space with $X = \{x_i, ..., x_n\} \in \chi$.

A task T on the other hand is composed of a label space Y, a prior distribution $P(Y)$ and a conditional probability distribution $P(Y|X)$ for a training pair of $x_i \in X$ and $y_i \in Y$.

Given a source domain D_S, a source task T_S, a target domain D_T and a target task T_T, transfer learning can happen when $D_S \neq D_T$ or $T_S \neq T_T$ and it learns the target conditional probability distribution $P_T(Y_T|X_T)$ in D_T with the information gained from D_S and T_S.

For text classification, χ is the space of sentence representations, x_i the ith term vector corresponding to a sentence and X is the sample of sentences used for training. This study focuses on cross-domain learning, a setting where the source domain is different from the and target domains differ $D_S \neq D_T$.

Domain Adaptation. The marginal probability distributions of the source, P_S and of the target P_T domains can differ, $P_S(XS) \neq P_T(XT)$, a setting which is called domain adaptation.

Although domain adaptation is normally investigated in a single source domain, there is a unique case of domain adaptation where data from numerous sources is available for training, a paradigm known as multi-source domain adaptation. A number of works have been done in this setting including training a single model from combined data from multiple sources [31]. Other techniques involve training separate models for each source domain and combining them by ensemble techniques with self training [32], using a linear combination of the base models [33] and multi-tasking and linear combination [34]. Neural Network-based models are the more recent ones and include attention based models [35,36].

3   Methods

The experimental set up of the framework applied on a verbal autopsy (VA) dataset is presented in this section. The experiment is divided into three parts; part one focuses on the selection of the best vocabulary set for the ELMo [6] language model. The second part compares the strategies of handling class imbalances in text classification and the third part focuses on cause of death (COD) classification of binary VA features, narrative text features and a hybrid of binary and narrative features settings of a verbal autopsy report.

3.1 Algorithms

BERT [37], BERT Experts-PubMed [38], ELMo [6], BioELMo [9] and a feedforward neural network classifier.

ELMo and BERT embeddings were computed on Google Colaboratory and the pretrained models used were from Tensorflow Hub[1].

3.2 Dataset

The verbal autopsy (VA) dataset is from the MRC/Wits Rural Public Health and Health Transitions Research Unit (Agincourt), in South Africa [39], ethics clearance number:M110138. It is a unit that supports investigations into causes and impacts of diseases to social transitions and populations. The data was collected from 1992 to 2015 and consists of 8698 VAs. The VA records were reviewed for features suggestive of uncontrollable hyperglycemia by a clinician with paediatric training and expertise in type-1 diabetes management in high-income and low-and-middle-income countries. 3708 cases had symptoms of uncontrollable hyperglycemia and 77 were identified as positive and 7755 negative cases of death by uncontrollable hyperglycemia. Apart from the closed ended questions the data also has "free text" describing circumstances leading up to death.

Table 1. Verbal autopsy binary features

female	tuber	diabetes	men-con	cough	ch-cough	diarr	exc-urine	exc-drink	diagnosis
0	1	0	0	1	1	0	0	1	0
1	0	0	0	1	1	1	0	0	0
1	1	0	0	1	1	0	0	1	1
1	0	0	0	0	0	0	0	0	0
0	0	0	1	0	0	0	0	1	1
1	1	0	1	1	1	1	0	0	0

Table 2. A sample of verbal autopsy narrative

Narrative

The deceased started by having painful abdomen. The following day she was taken to the clinic. The nurse didn't say what was wrong. Treatment was given but nothing change. After few days. She was feeling cold. She had difficult in breathing and she stop talking and walking. Where she was taken to the health center. Where oxygen was given and referred to the hospital by an ambulance as an out patient. The doctors said it was poison. Water drip, oxygen and treatment was given but no change. She died the same day at the hospital while the doctors where helping her

Diagnosis: 0

---

[1] https://www.tensorflow.org/hub.

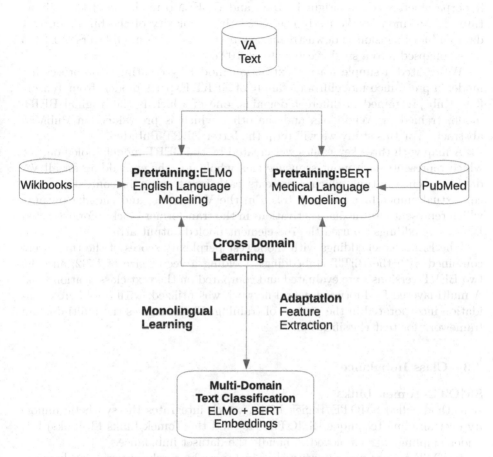

Fig. 1. Text classification framework

Binary features were identified as symptoms for which there was a yes or no response, and these replies were transformed into 1's and 0's. These include the symptoms indicated in Table 1 along with excessive thirst, urination, mental disorientation, and others. Text features were assigned to the narratives outlining the deceased's signs and circumstances leading up to death; an example of one of these elements is shown in Table 2.

We trained three ELMo models from scratch to evaluate the effectiveness of features taken from the English, medical, and public health domains. One vocabulary was built using the English Wikipedia and monolingual news crawl data from WMT 2008–2012, another the VA corpus, and the third one from PubMed abstracts. The datasets were preprocessed by text lowercasing and punctuation removal. The training data was then randomly divided into numerous training files, each including tokenized text with one sentence per line. We cloned the tensorflow implementation of ELMo from github repository[2] and kept the same hyperparameters of the original ELMo and BioELMo models and trained the three ELMo models. We lastly computed the perplexity of the biLMs on test data and for inclusion in downstream task text classification, all layers of ELMo were collapsed into a single vector of size 1024.

We created a simple feature extraction model by creating a preprocessing model, a pretrained model from a family of BERT Experts models from Tensorflow Hub pretrained in different domains, one of which is the original BERT model trained on Wikibooks and the other which is pretrained on PubMed abstracts. For this study we will term the latter BERT-PubMed.

A map with three key values was created by the BERT model:pooled output which represents each input sequence as a whole, i.e. the embedding for all VA data, sequence output which represents each input token in context, i.e. the contextual embedding for every token in the VA corpus, and encoder outputs which represent intermediate activations in the transformer blocks. For extracted BERT embeddings we used the 768-element pooled output array.

The ELMo emebeddings with the lowest perplexity scores of the three were combined with the BERT embeddings, yielding a vector size of 1792, and the two BERT versions were evaluated and compared on the text classification task. A multi-layered feed-forward neural network was utilized, with 5-fold cross validation incorporated in the process of training. Figure 1 gives the multi-domain framework for text classification.

3.3   Class Imbalance

SMOTE-Tomek Links
A method called SMOTE-Tomek Links, which integrates the synthetic minority oversampling technique (SMOTE) [40] and the Tomek Links (T-Links) [41] undersampling, was employed to handle the dataset imbalance.

SMOTE creates new minority-class instances by combining previously existing minority-class examples along the border connecting all of their k-nearest neighbors while the Tomek Links undersampling technique identifies those sets of data points that are close yet fall into different classes [41]. If a and b are instances of classes A and B, respectively, then a and b are referred to as Tomek Links points if for the distance between them $d(a,b)$, $d(a,b) < d(a,c)$ or $d(a,b) < d(b,c)$ for another point c.

Cost-Sensitive Classification
Cost-sensitive learning addresses the class imbalance by changing the model's cost function so that misclassifications of training samples from the minority class are given more weight and hence are more costly.

---

[2] https://github.com/allenai/bilm-tf.

For a single prediction x_i, the weighted cross entropy (CE) loss for class j, is given by

$$CE = -\frac{1}{N}\sum_i \alpha_i \sum_{j\in\{0,1\}} y_{ij} log p_{ij} \qquad (1)$$

where x_i belongs to a set of training instances X, associated with a label y_i and $y_i \in \{0,1\}$, and the predicted probabilities of the classes is p_i, where $p_i \in [0,1]$. $\alpha_i \in [0,1]$ is set by the inverse class frequency and it's value is equal to 1 when the loss function is not weighted.

We examined the two procedures for dealing with class imbalances and compared them to when no sampling was done. Following [42] we increased the weight of incorrectly labeling a VA case by changing the cost function of our model's fully connected layer during the training by multiplying each example's loss by a factor. The computed class weights ratio used is 0.50494305 : 51.07608696.

For data resampling we coupled the undersampling technique of Tomek Link with the oversampling method of SMOTE (SMOTE-Tomek Links). Each fold was sampled, and the classifier was trained on the training folds before being verified on the remaining folds. Classifier evaluation was done using recall, precision, F1-score, and area under the receiver operating characteristic curve (AUC-ROC). We tested the model with binary features, narrative text features, and a hybrid of binary and narrative text features.

4   Results and Discussion

Table 3. ELMo language models evaluation: Perplexity

Technique	Vocabulary	Tokens	Train perplexity	Test perplexity
ELMo	English Wikipedia	5.5B	43.23	37.32
ELMo	Verbal Autopsy	982 495	71.55	50.01
BioELMo	PubMed Abstracts	2.46B	47.44	33.01

The ELMo model trained on a Wikibooks and Book Corpus vocabulary achieved better perplexity scores on both the training and evaluation sets compared to the ELMo models pretrained on PubMed abstracts and a verbal autopsy (VA) vocabulary with the former performing better than the latter, as depicted in Table 3 above. We believe that this is a result of the different sizes of the vocabulary sets derived from the three datasets and the fact that PubMed abstracts and the VA corpus contain numerous mentions of words in the English wikipedia and books where more linguistic knowledge including spellings, syntax and grammar are learned.

 54   T. Manaka et al.

Table 4. Class imbalance strategies on ELMo, BERT and combined ELMo and BERT embeddings on VA text features setting

Model	Sampling	Recall	Precision	F1-Score	AUC-ROC	Accuracy
ELMo	No Resampling	0	0	0	0.5	0.9898
	SMOTE-Tomek Links	0.5000	0.0500	0.1005	0.7064	0.9086
	Weighted Cross Entropy	0.5500	0.0159	0.0309	0.5999	0.7468
BERT	No Resampling	0	0	0	0.5	0.9898
	SMOTE-Tomek Links	0.6500	0.1044	0.1799	0.7667	0.8811
	Weighted Cross Entropy	0.7000	0.1194	0.2040	0.7435	0.8351
BERT-PubMed	No Resampling	0	0	0	0.5	0.9898
	SMOTE-Tomek Links	0.6911	0.1135	0.1950	0.7966	0.8697
	Weighted Cross Entropy	0.8133	0.1219	0.2120	0.7794	0.8800
ELMo +BERT	No Resampling	0	0	0	0.5	0.9898
	SMOTE-Tomek Links	0.7300	0.2000	0.3139	0.5724	0.9862
	Weighted Cross Entropy	0.8455	0.2308	0.3626	0.7130	0.7187
ELMo +BERT-PubMed	No Resampling	0	0	0	0.5	0.9898
	SMOTE-Tomek Links	0.7654	0.2300	0.3537	0.6678	0.9782
	Weighted Cross Entropy	0.8755	0.3146	0.4629	0.8413	0.8032

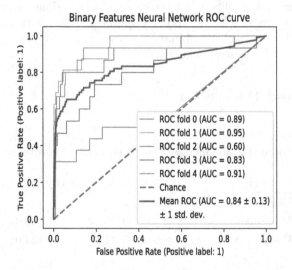

Fig. 2. ROC curves and AUC-ROC of the neural network classifier on binary features of a VA report.

In general, the weighted binary cross entropy loss function produced better results across all metrics compared to the SMOTETomek sampling on the text classification task, as shown in Table 4, and both are better than when no sampling is done at all. The performance scores achieved by weighted cross entropy loss function are however consistent with Madabushi et al. [42] who demonstrated that while BERT is capable of handling imbalanced datasets with no additional data augmentation, when the training and evaluation sets differ, as they do with VA reports, it struggles to generalize effectively.

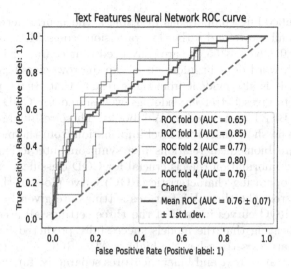

Fig. 3. ROC curves and AUC-ROC of neural network classifier on text features of a VA report.

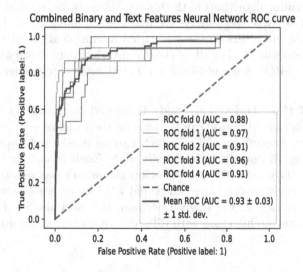

Fig. 4. ROC curves and AUC-ROC of neural network classifier on the combined binary and text features of a VA report.

Madabushi et al. [42] further show through an analysis on BERT with data augmentation and one without that data modifying techniques such as resampling and data augmentation techniques like synonym replacements do not lead to substantial improvements when using pretrained models like BERT. This is also consistent with the work of Wei and Zou [43], who tested the same methods for improving the text classification task.

Across all embedding comparisons in Table 4, the neural network classifier achieved high recall scores and rather low precision scores. However with recall scores of around 0.8466–0.8755, we are convinced that character information in combination with word domain information can improve classification of cause of death (COD). It is also evident from these results that BERT pretrained on PubMed abstracts gives better embeddings well suited for COD classification than those from BERT pretrained on Wikibooks. These results are in line with Qiao et al. [29] who show that word-level information from the medical domain is able to capture biomedical relations like symptom interactions and across different diseases, information that is crucial for COD classification.

The receiver operating characteristic (ROC) curve plots of the neural network classifier across the three VA features settings are given by Fig. 2, Fig. 3 and Fig. 4. The ROC curves in each of the three settings all ascended toward the top left, indicating that the models successfully predicted both the cases. The combined features setting has the highest AUC-ROC score (93%) in comparison to the narrative text and binary features separately, further proving the importance of text features in COD classification.

Our results are consistent with those of Manaka et al. [25], who compared four machine learning algorithms in the binary VA features, textual VA features and a hybrid of textual and binary VA features. Manaka et al. [25] attested that when the narrative text features are used in combination with the binary features, the background and depth of the events relating to death are enhanced, achieving an AUC-ROC score of 97% on a neural network classifier.

Limitations of the Study. This study is limited to the target task of text classification. More investigation can be done on the tasks of relation extraction and named entity recognition (NER) which can be considered important tasks in NLP in the health domain. Additionally, only English and clinical domains are used in this study, since this classification framework has significant impact on results, more experimentation could reveal whether or not similar behaviour occurs for other subsets of the English domain. It would also be interesting to investigate the same architecture with different character and word embedding models.

5   Conclusion

We have experimentally shown that a multi-source domain adaptation can improve the cause of death classification task for verbal autopsy reports. Further, we have shown that this is possible in a setting where one language is a subset of another with the character-based language model used in the English domain and the word-based language model on the subset clinical domain. In our upcoming work, we will examine how well this architecture performs in other English subdomains, such as finance, and we will look into a variety of character-level and word-level embedding techniques.

Acknowledgments. We are grateful to MRC/Wits-Agincourt for providing the verbal autopsy reports and guidance into understanding the dataset. Thokozile Manaka is supported by the United Nation's Organization of Women in Science for The Developing World (OWSD).

References

1. Kim, Y.: Convolutional neural networks for sentence classification. In: Proceedings of the Conference on Empirical Methods in Natural Language Processing, pp. 1746–1751 (2014)
2. Zhang, X., Zhao, J., LeCun, Y.: Character-level convolutional networks for text classification. In: Advances in Neural Information Processing Systems, pp. 649–657 (2015)
3. Conneau, A., Kiela, D., Schwenk, H., Barrault, L., Bordes, A.: Supervised learning of universal sentence representations from natural language inference data. CoRR, abs/1705.02364 (2017). http://arxiv.org/abs/1705.02364
4. See, A., Liu, P.J., Manning, C.D.: Get to the point: summarization with pointer-generator networks. CoRR, abs/1704.04368 (2017). http://arxiv.org/abs/1704.04368
5. Lin, Z., et al.: A structured self-attentive sentence embedding. CoRR, abs/1703.03130 (2017). http://arxiv.org/abs/1703.03130
6. Peters, M., et al.: Deep contextualized word representations. In: NAACL (2018)
7. Alsentzer, E., et al.: Publicly available clinical BERT embeddings. CoRR, abs/1904.03323 (2019). http://arxiv.org/abs/1904.03323
8. Lee, J., et al.: BioBERT: a pre-trained biomedical language representation model for biomedical text mining. CoRR, abs/1901.08746 (2019). http://arxiv.org/abs/1901.08746
9. Jin, Q., Dhingra, B., Cohen, W.W., Lu, X.: Probing biomedical embeddings from language models. CoRR, abs/1904.02181 (2019). http://arxiv.org/abs/1904.02181
10. Zheng, L., Wang, Y., Hao, S., Shin, A.Y., Jin, B., Ngo, A.D.: Web-based real-time case finding for the population health management of patients with diabetes mellitus: a prospective validation of the natural language processing-based algorithm with state-wide electronic medical records. JMIR Med. Inform. 4(4) (2016)
11. Ohno-Machado, L., Nadkarni, P., Chapman, W.: Natural language processing: an introduction. J. Am. Med. Inform. Assoc. 18, 554–551 (2011)
12. Unite Nations. Department of economic and social affairs, population division, united nations. World Population Prospects: The 2012 revision.ST/ESA/SER.A/334 (2013)
13. Kooverjee, N., James, S., Van Zyl, T.L.: Investigating transfer learning in graph neural networks. Electronics 11(8), 1202 (2022)
14. Kooverjee, N., James, S., Van Zyl, T.: Inter-and intra-domain knowledge transfer for related tasks in deep character recognition. In: 2020 International SAUPEC/RobMech/PRASA Conference, pp. 1–6. IEEE (2020)
15. Karim, Z., van Zyl, T.L.: Deep learning and transfer learning applied to Sentinel-1 DInSAR and Sentinel-2 optical satellite imagery for change detection. In: 2020 International SAUPEC/RobMech/PRASA Conference, pp. 1–7. IEEE (2020)
16. Van Zyl, T.L., Woolway, M., Engelbrecht, B.: Unique animal identification using deep transfer learning for data fusion in Siamese networks. In: 2020 IEEE 23rd International Conference on Information Fusion (FUSION), pp. 1–6. IEEE (2020)

17. Variawa, M.Z., Van Zyl, T.L., Woolway, M.: A rules-based and transfer learning approach for deriving the Hubble type of a galaxy from the galaxy zoo data. In: 2020 IEEE 23rd International Conference on Information Fusion (FUSION), pp. 1–7. IEEE (2020)
18. Pikuliak, M., Šimko, M., Bielikova, M.: Cross-lingual learning for text processing: a survey. Expert Syst. Appl. **165** (2021)
19. Chen, X., Cardie, C.: Multinomial adversarial networks for multi-domain text classification. CoRR, abs/1802.05694 (2018). http://arxiv.org/abs/1802.05694
20. Yan, Z., Jeblee, S., Hirst, G.: Can character embeddings improve cause-of-death classification for verbal autopsy narratives? In: Proceedings of the BioNLP 2018 Workshop and Shared Task, vol. 34, no. 19, pp. 234–239 (2019)
21. United Nations. Department of economic and social affairs, population division, United Nations (2013)
22. Danso, S., Atwell, E., Johnson, O.: A comparative study of machine learning methods for verbal autopsy text classification. Int. J. Comput. Sci. Issues **10**(2), 47–60 (2014)
23. Todd, J., Balira, R., Grosskurth, H., Mayaud, P., Mosha, F.: HIV-associated adult mortality in a rural Tanzania population. AIDS **11**, 801–807 (1997)
24. King, C., Zamawe, C., Banda, M., Bar-Zee, N., Bird, J.: The quality and diagnostic value of open narratives in verbal autopsy: a mixed-methods analysis of partnered interviews from Malawi. BMC Med. Res. Methodol. **16**(13) (2016)
25. Manaka, T., Van Zyl, T.L., Wade, A.N., Kar, D.: Using machine learning to fuse verbal autopsy narratives and binary features in the analysis of deaths from hyperglycaemia. In: Proceedings of SACAIR2021, vol. 1, pp. 90–106 (2022)
26. Jeblee, S., Gomes, M., Hirst, G.: Multi-task learning for interpretable cause of death classification using key phrase predictions. In: Proceedings of the BioNLP 2018 Workshop, vol. 34, no. 19, pp. 12–27 (2018)
27. Baxter, J.: A model of inductive bias learning. J. Artif. Intell. Res. **12**, 149–198 (2000)
28. Huang, K., Altosaar, J., Ranganath, R.: ClinicalBERT: modeling clinical notes and predicting hospital readmission. CoRR, abs/1904.05342 (2019). http://arxiv.org/abs/1904.05342
29. Qiao, J., Bhuwan, D., William, C., Xinghua, L.: Probing biomedical embeddings from language models. In: Proceedings of the 3rd Workshop on Evaluating Vector Space Representations for NLP, pp. 82–89 (2019)
30. Beltagy, I., Cohan, A., Lo, K.: SciBERT: pretrained contextualized embeddings for scientific text. CoRR, abs/1903.10676 (2019). http://arxiv.org/abs/1903.10676
31. Aue, A., Gamon, M.: Customizing sentiment classifiers to new domains: a case study. In: Proceedings of Recent Advances in Natural Language Processing (RANLP), vol. 1, pp. 2–11 (2005)
32. Li, S., Zong, C.: Multi-domain sentiment classification. In: Proceedings of the 46th Annual Meeting of the Association for Computational Linguistics on Human Language Technologies: Short Papers, vol. 1, pp. 257–260 (2008)
33. Mansour, Y.: Domain adaptation with multiple sources. In: Neural Information Processing Systems Conference (NIPS) (2009)
34. Wu, F., Huang, Y.: Sentiment domain adaptation with multiple sources. In: Proceedings of the 54th Annual Meeting of the Association for Computational Linguistics, vol. 1, pp. 301–310 (2016)
35. Kim, Y., Stratos, K., Kim, D.: Domain attention with an ensemble of experts. In: Proceedings of the 55th Annual Meeting of the Association for Computational Linguistics, pp. 643–653 (2017)

36. Su, Y., Yan, X.: Cross-domain semantic parsing via paraphrasing. In: Proceedings of the 2017 Conference on Empirical Methods in Natural Language Processing (2017)
37. Devlin, J., Chang, M., Lee, K., Toutanova, K.: BERT: pre-training of deep bidirectional transformers for language understanding. CoRR, abs/1810.04805 (2018). http://arxiv.org/abs/1810.04805
38. Abadi, M., et al.: TensorFlow: large-scale machine learning on heterogeneous systems (2015). https://www.tensorflow.org/. Software available from tensorflow.org
39. Kahn, K., Collinson, M., Gómez-Olivé, F., Mokoena, O., Twine, R., Mee, P.: Profile: agincourt health and socio-demographic surveillance system. Int. J. Epidemiol. **41**(4), 988–1000 (2008)
40. Bowyer, K.W., Chawla, N.V., Hall, L.O., Kegelmeyer, W.P.: SMOTE: synthetic minority over-sampling technique. CoRR, abs/1106.1813 (2011). http://arxiv.org/abs/1106.1813
41. Schmidt-Thieme, L., Thai-Nghe, N., Do, T.N.: Learning optimal threshold on resampling data to deal with class imbalance. In: Proceedings of the 8th IEEE International Conference on Computing (2000)
42. Madabushi, H.T., Kochkina, E., Castelle, M.: Cost-sensitive BERT for generalisable sentence classification with imbalanced data. CoRR, abs/2003.11563 (2020). https://arxiv.org/abs/2003.11563
43. Wei, J.W., Zou, K.: EDA: easy data augmentation techniques for boosting performance on text classification tasks. CoRR, abs/1901.11196 (2019). http://arxiv.org/abs/1901.11196

Real Time In-Game Playstyle Classification Using a Hybrid Probabilistic Supervised Learning Approach

Lindsay John Arendse$^{(\boxtimes)}$ (iD), Branden Ingram(iD), and Benjamin Rosman(iD)

School of Computer Science and Applied Mathematics,
University of the Witwatersrand, Johannesburg, South Africa
Lindsay.Arendse@students.wits.ac.za,
{Branden.Ingram,Benjamin.Rosman1}@wits.ac.za
https://www.wits.ac.za/csam/

Abstract. In interactive digital media, such as video games, bringing about an adaptive or personalised experience requires a mechanism for correctly classifying or identifying the player style, before attempting to modify the experience in some way that improves player interest and immersion. This work presents a framework for solving this problem of in-game real time playstyle classification. We propose a hybrid probabilistic supervised learning approach, using Bayesian Inference informed by a K-Nearest Neighbors based likelihood, that is able to classify players in real time at every step within a given game level using only the latest player action or state observation. This improves on current approaches dependent on previous episodic player action trajectories in order to classify the player. Furthermore, we highlight the effect that this representation of the player state-action observation has on the in-game playstyle classification's accuracy, prediction stability, and generalisability. We apply and test our framework using MiniDungeons, a rogue-like dungeon exploration game, and further evaluate our framework using a natural dataset containing human player action data from the platforming game Super Mario Bros. The experimental results obtained from our approach outperforms existing work in both domains. Furthermore, the evaluation results highlights the ability of our framework to generalise to unseen levels, without the need for additional retraining. Additionally, the Super Mario evaluation results illustrates the scalability of our framework to a more complex game environment with human player data.

Keywords: Game AI · Playstyle identification · Playstyles · Player modeling · Supervised learning · Bayesian inference · K-nearest neighbour · Rogue-like · Platforming · MiniDungeons · Super Mario Bros

1 Introduction

In video games there are sometimes several *ways* or *styles* in which players can play a game. Different players find different parts of a game challenging and rewarding. Catering for this diversity in player playstyle, preference, and skill is quite challenging for game creators [3]. Adapting to player preference and skill is important in achieving higher levels of player engagement and is especially vital in games used for education [2]. Furthermore, from a game design perspective, in order for game developers to maximise the number of target players it is essential to create games which cater for different gameplay experiences or player playstyles [24].

A player playstyle can be seen as a representation of the player's strategy and player profile [1]. In order to understand or react accordingly to a given playstyle a model approximation of the player is needed. Approaches within the research area of Player Behaviour Modeling are centered around the creation of this model approximation of players [27]. In other words, a player model can be defined as an abstracted representation of a player's behaviour in a game environment [1]. Player modelling provides a mechanism which game designers and researchers can use to gain insight and understanding into how players are feeling and how players might act [28]. Player modelling in itself is an interesting challenge and problem domain. Player models are additionally useful when combined with game personalisation or game adaptability (also known as *dynamic difficulty adjustment* [4], *adaptive player experience* or *game balancing* [23]). This application of player models to game personalisation or adaptability is of increasing importance in video games and is especially necessary when the purpose of game AI is to improve the experience or enjoyment of the human player [1].

Before attempting to modify the gameplay experience in some way which improves player interest and immersion, a mechanism for correctly classifying or identifying the player preference or player style is required. We present a framework for solving this problem of in-game *real time* playstyle classification. We propose a hybrid probabilistic supervised learning approach, using Bayesian Inference informed by a K-Nearest Neighbors based likelihood, that is able to classify players in real time at every step within a given game level using only the latest player action or state observation. As part of our experiment we compare our hybrid classifier approach to a comparative approach based on unsupervised clustering of the player action trajectories, using an LSTM-autoencoder [14]. Furthermore, we highlight the effect that this representation of the player state-action observation has on the in-game playstyle classification's accuracy, prediction stability, and generalisability.

We apply and test our framework using the MiniDungeons game domain. MiniDungeons is a turn-based top-down tile based dungeon exploration game, created as a benchmark research domain for modeling and understanding human playstyles [8,9]. From an implementation perspective we make use of a python, OpenAI Gym compatible, re-implementation[1] [15]. Our framework is further evaluated using a natural dataset[2] containing human player action data from the platforming game Super Mario Bros [6]. We respectively obtain accuracies and prediction stability results which highlight the success of our framework

---

[1] https://github.com/ganyariya/gym-md.
[2] http://guzdial.com/datasets.html.

when compared to existing work. Prediction stability is a necessary feature, not present in related studies, for real time playstyle classification since frequent fluctuations in the prediction can adversely affect the game experience. The rest of the work is organised as follows, Sect. 2 and 3 describes related work and background, Sect. 4 outlines our research methodology, and our experiment discussion and analysis is presented in Sect. 5.

2 Related Work

There are a number of closely related areas of work centered around the problem of creating adaptive or personalised games. Several studies focus on the player model creation process, in other words, modelling or mimicking human players [8–10,15]. Other works aim to investigate methods for identifying, classifying, or clustering the player, with the aim of trying to answer the question 'does the observed player belong to a known player type or playstyle?' [5,12,14,17]. There are several other researchers focused on how the game experience can be changed or personalised to the identified player [4,13,22,23]. Our work looks to recover the underlying playstyles present in play logs which could be used to aid all of these studies.

Normoyle and Jensen [17] investigates a method which uses Bayesian semi-parametric clustering for creating player clusters. Normoyle and Jensen take post-match data and cluster based on how the player's choices affect the end game result, in contrast to clustering on the outcomes directly. Iwasaki and Hasebe [15] use a clustering approach, called x-means, to cluster play log data with the aim of evolving different agent playstyles created using a genetic algorithm, also called C-NEAT [12]. We apply this idea of player personas in order to generate our set of rule based player proxies. Additionally, we look to further explore the concept of play logs by analysing the impact different representations have on performance.

Past work in this field has been largely concerned with identifying playstyles at the end of the game episode, in order to evolve or create diverse player persona agents, or to analyse player trace data for analytical or game testing purposes.

The user-study work by Valls-Vargaswork et al. [25] present a player modeling framework to capture non-stationarity within player strategy by using sequential machine learning techniques which incorporate predictions from previous temporal player observations. However, the effect of prediction stability is not featured as a component of their approach. Our work looks to highlight the impact the stability score has when predicting playstyles during gameplay. Additionally, Valls-Vargaswork et al. [25] relied on a manual annotation process for providing playstyle labels to the collected player trace data at fixed intervals. This manual annotation is not only time consuming but may also be infeasible from a cost perspective. Our framework's Trajectory Processing step which utilises player personas and clustering to provide a feasible solution to the lack of labelled player datasets.

Hernandez-Leal et al. [7] also considered the concept of non-stationary strategies whereby they utilised a Bayesian framework to train an agent which learns an optimal policy against an opponent whose strategy switches. In our work we look to apply Bayesian inference to track the belief over a set of playstyles which was inspired by their belief tracking of observed policies.

The work by Ingram et al. [14] looked to remedy the issue of requiring labelled data by utilising an unsupervised lstm autoencoder clustering approach. Their autoencoder model works by projecting trajectory information into a lower dimensional latent representation which could then be clustered using vanilla clustering approaches like Gaussian Mixture Models. This approach, however, does not guarantee that the clusters identified correspond semantically to the playstyles we wish to recover. Additionally the clustering step relies on previous episodic player action trajectories. In contrast we propose a hybrid probabilistic supervised learning approach, using Bayesian Inference informed by a K-Nearest Neighbors based likelihood, that is able to classify players in real time at every step within a given game level using only the latest player action or state observation.

Gow et al. [5] also utilised an clustering approach which incorporates multi-class Linear Discriminant Analysis (LDA) to model players in Snakeotron and Rogue Trooper. Similar to Valls-Vargaswork et al. [25] they relied on play log segmentation and summaries. The ability to classify players at every game step may allow for the game's adaptability to become more responsive and granular than these segmentation based approaches. Our model can accommodate this important capability which is required for assistive companion agents, dynamic difficulty adjustment made during game play, or for tailoring tutoring based games.

The classification of players needs to happen in near real time and concurrently with other computations, such as graphics rendering, non-player character behaviour logic, and game physics [1,27]. Therefore techniques used in real time playstyle identification should strive to be as computationally efficient and inexpensive for applications within game AI [1]. For this reason we have chosen to make use of more computationally efficient algorithms, such as Bayesian Inference and K-Nearest Neighbors search, in building our framework.

Computationally complex classifiers pose a risk at runtime by interrupting or halting the game experience. The user study work by Scott and Khosmood [21] highlights this adverse effect which computationally heavy approaches have on a game's playability and player experience. Scott and Khosmood conclude that approaches which reduces the playability of the game and thus the overall player experience are unacceptable in a commercial video-game setting. Furthermore, from a model explainability perspective understanding the playstyle prediction outcome, identifying which game mechanics or features certain playstyles use, and gaining insight into how the different player type clusters relate are useful to game designers when deciding on which game mechanics to change, and which new game features to implement or remove [14,24]. For this reason, we have chosen explainable machine learning approaches such as Bayesian Inference and K-Nearest Neighbors. As part of our experiment we analyse the player action trajectory data and identify the key generalised behaviours observed.

Our methodology section outlines how we collect, pre-process, analyse, train, and evaluate our approach, as well as our definition of the playstyle in-game classification problem.

3   Background

This section describes the terminology and definitions necessary for defining the in-game playstyle classification problem.

3.1   Play Log Definition

Game metrics are the recorded numeric data generated, during gameplay, as a result of the player interacting with the game [24]. Playstyles in games are identified based on the observed in-game event information [15]. This is a reason why the use of a *play log* [12], which captures such event information, is valuable for identify play styles. Thus our classification approaches uses this concept of a play log, which is a vector containing the essential game metrics required to identify or discriminate between playstyles [12,15].

We define the play log as *pl*, an n-dimensional vector containing k^n game metrics m: $pl = [m_1, m_2, \cdots, m_k]^n$ where $n, k \in \mathbb{N}$ and $m \in \mathbb{R}$. Let Υ be an instance of *pl*, with a specific n, k, and m values. Let Ψ be an instance of *pl* containing unseen play log game metric data with the same n, k, and m values as Υ. It is important to note that the game metrics $m_{1...k}$ are recorded and updated at every step within a given level of a game.

3.2   Playstyle Set Definition

Our aim is to build playstyle classifiers which make use of play log metric data to distinguish between different playstyles. During the early stages of game development, especially when mechanics of a game are being tested, changed, and developed it is challenging for game designers to collect game metric data from human players [15]. Furthermore, playtesting with human players can be time consuming and a costly exercise. This is why quite a number of studies create and make use of agents as a proxy for human players [8,10,15]. Hence, we will also make use of proxy agents, also called personas [9,10], to mimic human playstyles within the given game and thus generate the play log metric data. We define a set Γ of *proxy agent implemented playstyles*, called $\Gamma \subseteq P_g$, where P_g is the global set of all possible playstyles for a given game g.

3.3   Game Levels

It is common for games to have one or more levels. Thus for a given game, we divide all available levels into three sets: the first called *train* which is used for the training of the playstyle classifiers, the second is called *seentest* which is used for the seen levels prediction evaluation, and the third called *unseentest* which is used for the unseen levels prediction evaluation.

3.4   Playstyle In-Game Classification Problem Definition

For a given game g, a player agent α, and a level l from g. We aim to find a model classifier Ω_Υ which is trained on labeled play log data, Υ, obtained using the set, $\Gamma \subseteq P_g$, of proxy agents on the set of training levels, $train$, for the game g which can classify unseen play log entries, Ψ, to a given playstyle in Γ.

$$\Omega_\Upsilon : \Psi \rightarrow \Gamma, \ (\forall \, step \in l, \exists \, \Omega_\Upsilon) \tag{1}$$

In other words, we would like to find a model Ω_Υ which can at every $step$, within a level l, in the game g classify the unseen play log entries in Ψ, of agent α, to a playstyle in Γ. The model classifier Ω_Υ is evaluated using the respective level sets $seentest$ and $unseentest$.

4   Playstyle Classification Method

Our proposed framework provides a method for solving the problem of in-game real time playstyle classification. As illustrated in Fig. 1 our framework involves multiple steps where low-level player action trajectories are processed to generate corresponding playstyle labels. These steps are broken down into: Trajectory Processing and Playstyle Classification.

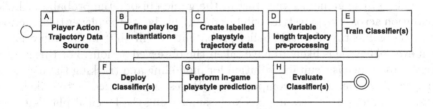

Fig. 1. Proposed framework overview

4.1   Trajectory Processing

Our playstyle identification framework is able to account for two situations related to the availability of existing player play log data. Our framework begins with the *Player Action Trajectory Data Source* step, as illustrated in Fig. 1(A). Regarding the source data, two situations could occur, which we respectively refer to as Case I and Case II. Case I arises when there is no existing player play log data, thus for Case I we make use of rule-based proxy agents to mimic human player playstyles and generate the source player action trajectory data. Case II corresponds to when there already exists player action trajectory data. The next step in our framework is to define the play log instantiations, as shown in Fig. 1(B). As discussed in Subsect. 3.1 the play log contains the essential game metrics required to identify or discriminate between playstyles. For this reason the construction of the play log is important and several play log representations should be considered. As part of this work we highlight some of the key considerations and effects that the play log representation has on the in-game playstyle prediction.

Once the play log representations has been defined, the next step in our framework is to respectively in Case I and II obtain labelled playstyle trajectory data, as shown in Fig. 1(C). In Case I, the proxy agents are used to generate the play log data along with the agent playstyle label. For Case II, traditional unsupervised clustering is used to obtain the playstyle label for the respective play log data. In order to handle the variable length of player action trajectories across different episodes, we pivot the episode play log such that each entry within the play log is assigned the play log playstyle label (Fig. 1(D)). This pre-processing step prepares the data for the training and development of our classifier.

4.2   Playstyle Classification

In this section we present our hybrid probabilistic supervised learning approach, which uses Bayesian Inference informed by a K-Nearest Neighbors (KNN) based likelihood, that is able to classify players in real time at every step. At any given point in a game, there exists a belief over which playstyle the player is currently using. In other words, there is a probability distribution across the playstyles which exist in the game. At the start of the game episode the probability belief distribution across the playstyles will be equally likely. As the player takes actions within the game the playstyle belief probabilities will shift in the direction of which ever playstyle is likely to have taken the observed actions. For this reason, Bayesian Inference is a suitable solution for modelling and tracking these changes in player playstyle. A core component of Bayesian Inference is the likelihood probability function. We train a KNN classifier using the labelled play log data, from steps A-E in Fig. 1, which when given an unseen play log entry will return a playstyle classification as well as the probability belief of this classification across the playstyle label classes. This probability belief distribution of the KNN classification is what we use as the likelihood probability during the posterior belief update, as shown in Eq. 2 which illustrates our Bayesian Belief Update process:

$$P(playstyle|observation) = \frac{P(observation|playstyle)P(playstyle)}{P(observation)} \quad (2)$$

where $playstyle \in \Gamma$, $observation \in \Psi$, and where $P(observation|playstyle)$ is the likelihood probability distribution, provided by the KNN classification's probability distribution trained using the labeled play log data in Υ. The prior distribution is initialised to equally likely at the start of the Bayesian belief update process, and at every step the highest posterior playstyle belief probability is returned as the playstyle prediction.

5   Evaluation by Experiments

As part of our experiment we develop two variants of our hybrid supervised Bayesian classifier based on two different play log instantiations which we have defined. We make use of MiniDungeons as our Case I domain, and the human player Super Mario Bros dataset as our Case II domain. Furthermore from an evaluation perspective we compare our approach to the comparative unsupervised approach by Ingram et al. [14]. We begin our Case I evaluation of our framework using the MiniDungeons game domain [8,9]. Subsection 5.1 outlines the experiment setup and the implementation details related to the application of our framework method. The Case II human player experiment setup, implementation, and evaluation is presented in Subsect. 5.3.

5.1   Case I: MiniDungeons Experiment

MiniDungeons is a two-dimensional top-down dungeon exploration game, and is a common benchmark research domain for modeling and understanding human playstyles [8,9]. From a play log instantiation perspective we define two types of play logs, each of which measure the player interaction at different granularities:

- Low level visitation grid (LLVG): is a two-dimensional vector which tracks the number of visits made to each position in a level (i.e. models at the player action level [1]). For a given MiniDungeons level the LLVG play log dimensions will be equal to the level's row and column count.
- Tactic level information (TLI): is a one-dimensional vector which stores the number of times a specific action type is taken, as well as the number of times the player visits each level cell or square type [12]. The TLI play log aims to track the higher level player tactics [1].

To generate these two respective play log datasets we define a set of six player proxy agents as described in Fig. 2. The playstyles are created based on the main objectives available in a given level of MiniDungeons, for example: collect treasure, restore hit points (HP) by collecting potions, battling monsters for experience points, and reaching the dungeon exit. Furthermore, the choice of our playstyle proxy agents is informed by the clustering analysis work done by Iwasaki and Hasebe [15] who evolve multiple human proxy agents without predefining the desired playstyle or reward function. The core playstyles generated by their approach are: a *runner* based playstyle, which prioritises exiting the dungeon as quick as possible, a *completionist* player, which aims to interact with treasure, monster, and potion elements within the dungeon, a *treasure centric* based playstyle, which focuses on collecting treasure, and a *safty-first* playstyle, which focuses on collecting treasure and potions. Respectively for our work, as illustrated in Fig. 2, we have the treasure centric agents called *brave treasure hunter* and *pure treasure hunter*. The *pure treasure hunter* only collects unguarded treasure, whilst the *brave treasure hunter* will collect treasure guarded by monsters and replenish HP if needed. The *monster killer* playstyle agent is concerned with battling as many monsters as possible without dying. If

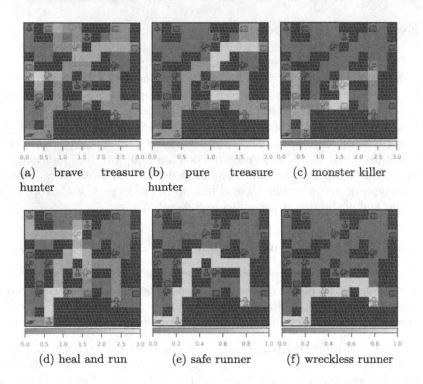

(a) brave treasure (b) pure treasure (c) monster killer
hunter   hunter

(d) heal and run   (e) safe runner   (f) wreckless runner

Fig. 2. Playstyles heatmap for MiniDungeons (Level 9) [9].

the monster killer agent's HP is low, potions will be collected to restore health. The *heal and run* agent aims to collect all the potions in a given level. The *safe runner* and *wreckless runner* playstyles respectively aim to exit the dungeon as fast as possible. The *safe runner* agent takes the shortest available safe path to the dungeon exit (i.e. a path which does not encounter any monsters), whilst the *wreckless runner* agent will take the shortest path to the dungeon exit.

From the training and evaluation level perspective the respective MiniDungeons levels[3] used are:

- *train* (md-test-v0, md-hard-v0, md-random_1-v0, md-random_2-v0, md-gene_1-v0, md-gene_2-v0, md-strand_1-v0)
- *seentest* (md-strand_2-v0, md-holmgard_0-v0, md-holmgard_3-v0, md-holmgard_5-v0, md-holmgard_7-v0, md-holmgard_8-v0)
- *unseentest*: (md-holmgard_1-v0, md-holmgard_2-v0, md-holmgard_4-v0, md-holmgard_6-v0, md-holmgard_9-v0, md-holmgard_10-v0)

The levels assigned to the *train* set were selected in order to reserve the Holmgard levels for the seen and unseen evaluations. Each Holmgard level was randomly assigned to the respective *seentest* and *unseentest* sets. Each level run was repeated ten times in each respective evaluation case.

---

[3] https://github.com/ganyariya/gym-md/blob/main/README/resources/md_stages_screenshots/README.md.

Initially two types of classifiers were developed, namely KNN and Bayes, each of which has a play log variant based on the LLVG and TLI play log instantiations. The classifiers trained on the *train* levels and evaluated on the *seentest* and *unseentest* levels are respectively called: KNN (LLVG), KNN (TLI), Bayes (LLVG), and Bayes (TLI). Each level run was repeated ten times for each playstyle within each respective evaluation case.

For both the KNN (LLVG) and KNN (TLI) classifiers the number of nearest neighbours used is three (i.e. $k = 3$, determined empirically during experiment). The KNN (LLVG) model classifies the observed play log entry using the nearest or most similar log entries in the form of the LLVG play log [20]. Similarly the KNN (TLI) model classifies using the nearest entries in the form of the TLI play log.

For the Bayesian Inference based classifiers, Bayes (LLVG) and Bayes (TLI), the initial prior probability belief distribution is equally likely across the playstyles. The likelihood probability that the in-game player observation belongs to a specific playstyle is informed by the historic LLVG and TLI play logs respectively. The posterior probability is then updated using the likelihood probability and the prior probability distribution. This process of updating the playstyle probability belief distribution is repeated as the new evidence (i.e. the player observation) is observed [20].

In our initial evaluations the KNN classifiers outperformed the Bayes classifiers, and upon careful inspection we attribute the poorer Bayes performance to the likelihood probability function not having enough historic data to discern and provide probabilities which can adequately influence the playstyle belief distribution. The problem with the Bayes classifier was that in unseen situations the default equally likely probability was being returned by the likelihood function. Which had little effect on the belief update because the observation was equally likely. We then improved the likelihood probability function by including a cosine similarity lookup for nearest similar observations. Which results in a more informed probability likelihood being used when compared to the default equally likely probability. The likelihood probability function used to inform the playstyle belief update is core to the success of the Bayes classifier. This insight as well as the KNN and Bayes comparative accuracy and prediction results, presented in Subsect. 5.2, suggested that a well balanced approach would be to combine the two classifier types into a hybrid classifier - which we refer to as the Hybrid Bayesian Supervised Learning classifier. In our hybrid classifier the KNN output prediction probability distribution is used as the Bayesian likelihood function. Two variants of our Hybrid classifier, called Hybrid (LLVG) and Hybrid (TLI), was created based on the LLVG and TLI play log instantiations and are respectively informed by the KNN (LLVG) and KNN (TLI) likelihood probability classifiers.

In order to baseline our hybrid approach, two baselines were used. The first baseline is a simple 'random' classifier which will at each step make a random prediction on the player's playstyle. The second comparative baseline is an unsupervised approach which makes use of an LSTM-autoencoder to cluster the player action trajectories [14]. The Autoencoder is used to project variable length trajectories in a uniform latent space which is then used for the clustering step. Although, this approach is unsupervised we used the ground truth labels generated for both the Mario and the MiniDungeons domain in order to compute the performance of this model.

5.2   Case I: MiniDungeons Experimental Results

As outlined in the MiniDungeons experiment setup, we have run two kinds of evaluations on the respective classifiers. The first evaluation is run on levels which have been seen before by the respective classifiers. The seen case is an appealing evaluation in situations where a game's levels are known and defined. For example, a game developer would like to adapt non-player character attack patterns, during the encounter with the player, within the game's existing levels. In this situation the game developer can train and deploy our classifiers and adapt the game accordingly. Next let us say that the game has been released successfully and our classifiers have been deployed and shipped with the game's initial release. In this situation our second evaluation on unseen levels may assist the game developer. As new levels are added to the game via updates, the classifiers which can generalise to unseen levels do not need to be re-trained or re-deployed. The representation of the play log is key to enabling this generalisability across unseen levels.

The LLVG play log representation is at a lower granularity, than the TLI representation. The LLVG play log is level specific since the play log's dimensions and the positional relevance of the player is coupled to the specific level under consideration. This means that, in this case, the LLVG classifier cannot be applied to unseen levels. This is why the respective LLVG classifier results are absent in the unseen evaluations (as shown in Fig. 4 and Table 1). Since the TLI play log is at a higher 'tactic level' abstraction the same TLI vector can generalise across levels. The results shown in Fig. 3 confirm that in the seen case that each classifier is able to on average accurately predict the correct playstyle, well above the random classifier and our comparative case. The KNN (TLI)

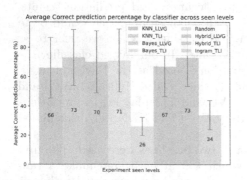

Fig. 3. Seen evaluation

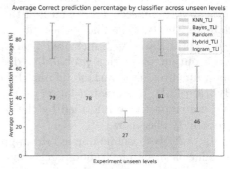

Fig. 4. Unseen evaluation

and Hybrid (TLI) classifiers marginally performs better than the other classifiers. The interesting observation is that in all three classifier types the TLI based representation resulted in better average accuracies when compared to the LLVG variants. This suggests that the TLI play log representation better captured the player interaction in comparison to the LLVG representation. Initially, we expected the LLVG representations to perform better since the play logs are better fit to each level and are at a lower granularity. However, this is not the case since the KNN LLVG performed marginally weaker.

Figure 4 summarises the unseen evaluation of the respective TLI based classifiers. On average our classifiers accurately predict the correct playstyle well above the random classifier and to our comparative case. The results in Fig. 4 confirm that the higher level TLI play log representation is able to generalise to unseen levels. The Hybrid (TLI) classifier marginally outperformed the KNN (TLI) and Bayes (TLI), which we attribute to the difference in stability scores, which is summarised in Table 1. However, before discussing the stability score metric, it is important to note that, in both the seen and unseen evaluations, our classifiers at each step received only the latest play log entry, whilst the Ingram (TLI) classifier received the latest play log entry as well as the previous temporal play log entries. Thus highlighting the success of our framework's variable length pre-processing step in removing the dependency on previous episodic player action trajectories in order to classify the player. As highlighted in Fig. 4 we outperform the comparative case. Although some of the poor performance associated with the approach by Ingram et al. [14] can be attributed to the unsupervised nature of the model, our performance is substantially greater and can be done using only the current play log state rather than requiring the past temporal play log entries.

As mentioned, Table 1 summarises an important result relating to the average playstyle prediction stability for each classifier. The stability of the classifier is a count of how many times, during an episode, the classifier changes its playstyle type prediction (e.g. transitioning from a Monster Killer prediction to a Brave Treasure Hunter prediction). The lower the stability score, the more stable the classifier. Since the random classifier makes a random prediction at every step the

Table 1. Overall average stability score by classifier

Classifier type	Average stability score		
	Seen levels		Unseen levels
	LLVG	TLI	TLI
KNN	8.317	4.5	2.778
BAYES	0.222	0.111	0.444
Random	51.017		54.105
Hybrid	2.583	1.611	1.556
Ingram et al. (TLI)	1.333		1.167

stability score is high. Table 1 highlights the big difference in stability between the KNN and Bayesian approaches. We respectively attribute these to the nearest neighbour distance logic in KNN and the gradual belief update in the Bayesian update. If you need a more stable classifier, then the Bayesian approach may be better. If the rate of prediction type change is not a concern then the KNN approach can be considered. The idea behind the Hybrid classifier is that the classifier should ideally bring about good balance between the rate of change in playstyle prediction type, i.e. the stability, and the accuracy of the believed playstyle prediction. When comparing the unseen evaluation TLI stability scores of the KNN (2.778), Bayes (0.444), and Hybrid (1.556), we see that the Hybrid classifier's stability sits between the KNN and Bayes classifiers, whilst obtaining a higher average accuracy in the unseen case, as seen in Fig. 4.

The choice of classifier, in terms of stability, is dependent on how the playstyle prediction classifier is used to adapt a game accordingly. For example, when adapting a game's difficult at the end of a level or after an enemy encounter, then the prediction stability of the classifier is less of a consideration because the game adaptability occurs less frequently. However, if you are adapting the non-player character (NPC) behaviour in real time in response to player actions then the stability of the classifier is quite important. In this case of playstyle-adaptive NPC agents, players may be confused or the gameplay experience may be adversely affected if the playstyle prediction changes more often, which may result in conflicting or inconsistent NPC agent behaviour. We continue the evaluation of our framework using the Super Mario Bros human player data, in Subsects. 5.3 and 5.4.

5.3 Case II: Super Mario Bros Experiment

The Super Mario Bros evaluation dataset was created as part of a user study, conducted by Guzdial and Riedl, consisting of seventy-four human players which played through twelve levels of Super Mario Bros [6]. Super Mario Bros is a platforming game which involves moving the main player character called Mario through a two-dimensional level traversing it from left to right whilst navigating platforms, jumping gaps, and overcoming enemies, until the end of the level is reached. Mario is able to move left and right, jump, run, and shoot fireballs, if the enabling *fire flower* item is collected [18]. We make use of the Super Mario Bros natural human player action dataset in order to ascertain whether our framework is able to scale to a more complex domain. Furthermore, Super Mario Bros is a suitable domain for studying playstyles since there are a number of different ways to play the game. The global goal of Super Mario Bros is to reach the end of the level, however, there are coins which can be collected and various enemies which can be defeated in different ways based on collecting different items which all affect the final score obtained. There is also an exploratory element where players can try find shortcuts, hidden areas, or bonus rooms accessed by traversing green warp pipes or by breaking platform blocks to enable new pathways.

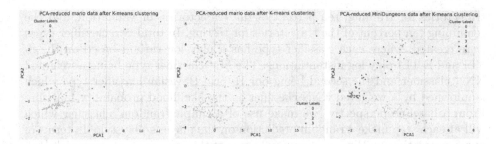

Fig. 5. PCA-Reduced Mario (S-TLI) data after clustering.

Fig. 6. PCA-Reduced Mario (E-TLI) data after clustering.

Fig. 7. PCA-Reduced Mini-Dungeons (TLI) data after clustering.

For this evaluation we define two types of play logs, based on the TLI representation, each of which summarise the player interaction at different granularities:

- Summarised Tactic level information (S-TLI): is a one-dimensional vector which stores the number of times the player jumps, kills, runs, breaks bricks, and dies. The S-TLI is a summarised play log view of the player's in game interactions.
- Extended Tactic level information (E-TLI): is a one-dimensional vector which is an expanded more granular form of the S-TLI tracking the occurrence of forty action and in-game interaction events.

To generate these two respective play log datasets[4] we processed the source Mario data and performed K-means clustering on the respective play log data to obtain the prospective playstyles within the dataset. Figure 5 and 6 respectively show the S-TLI and E-TLI PCA-Reduced Mario data after K-means clustering. This processing step was necessary as the mario dataset is unlabeled. The efficacy of this approach to surfacing prospective playstyles, is demonstrated by applying the same unsupervised K-means clustering to our MiniDungeons dataset. Here we were able to successfully extract the six desired clusters which correspond to the six playstyle proxy agents, as shown in Fig. 7. The respective values of k was determined using the silhouette coefficient and elbow method [16].

It is interesting to see that the K-means clustering on the S-TLI and E-TLI play log representations both resulted in four clusters, despite being at different levels of granularity. Which highlights a possible commonality or relationship in the underlying play log data. The cluster labels obtained will be used as the playstyle labels during training. The idea as shown in the MiniDungeons case, Fig. 7, is that similar play log entries belong to the same playstyle cluster. Thus unseen play log entries should fall within a cluster containing the most similar play log data. From a training and evaluation perspective we took the player action mario dataset and partitioned eighty percent of the player action

---

[4] https://github.com/LJArendse/playstyle_classification_using_hybrid_probabilistic_supervised_learning.

trajectories for training, ten percent of the trajectories for validation, and the remaining ten percent of the trajectories for testing. In total two classifier types were created, where each classifier type has a play log variant based on the S-TLI and E-TLI play log instantiations. The first classifier type being a weighted KNN classifier and the second being our Hybrid Bayesian classifier type which is informed by a weighted KNN classifier as the likelihood probability function. From a baseline perspective we make use of a simple 'random' classifier which will at each step make a random prediction on the player's playstyle. In terms of our comparative baseline we again make use of the Ingram et al. approach [14]. For the unsupervised Ingram et al. classifier we make use of the ground truth cluster labels generated in order to compute the performance of this model.

5.4   Case II: Super Mario Bros Experimental Results

For the Super Mario Bros evaluation the intuition behind our choice of play log representation is that a lower level play log granularity in terms of the player interaction should better surface or distinguish the underlying prospective playstyles. Which in turn should bring about a stronger prediction accuracy performance. The average correct prediction percentage results obtained for the S-TLI and E-TLI evaluations are respectively shown in Figs. 8 and 9. These shown results confirm the ability of our framework to operate and scale within a more complex domain using human player action trajectory data. Secondly, the difference in the average correct prediction percentage results between the S-TLI and E-TLI Hybrid classifiers confirm our intuition related to the player interaction play log granularity. The only difference between the S-TLI and E-TLI evaluations is this play log granularity, which results in a substantial difference in the playstyle classifier's prediction accuracy. When comparing the weighted KNN classifier results to our Hybrid classifier we see a greater difference in average prediction performance, which we attribute to the respective stability scores obtained. As shown in Table 2, our Hybrid classifier's Bayesian belief update results in a better stability score than the Weighted KNN which more rapidly changes the playstyle prediction type due to the nature of the nearest neighbour search. The lower the stability score, the more stable the classifier. In comparison to the MiniDungeons evaluation, the Mario evaluation better highlights the effect that the stability of the classifier has on the end prediction result. The Hybrid

Table 2. Overall average stability score by classifier

Classifier evaluation	Average stability score	
	S-TLI	*E-TLI*
Weighted KNN	0.196	0.080
Hybrid	0.014	0.004
Random	0.745	0.754
Ingram et al.	0.040	0.161

classifier's more gradual Bayesian belief update allows for the classifier output to be more certain about the playstyle under observation. It is important to note that the trade-off between the prediction and the model stability still exists. The MiniDungeons evaluation showed that a better stability does not necessary guarantee a better average prediction result. In the MiniDungeons case both the stability scores of the Ingram et al. (TLI) and Bayes (TLI) were more stable in compared to the Hybrid (TLI), but the average prediction percentage results were not better than the Hybrid (TLI). The strength of our Hybrid classifier and framework is again highlighted as our classifiers at each step received only the latest play log entry, whilst the Ingram classifiers received the latest play log entry as well as the previous temporal play log entries. Thus again confirming the success of the variable length pre-processing step.

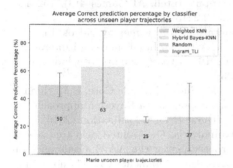

Fig. 8. Mario S-TLI average correct prediction percentage by classifier across unseen player trajectories.

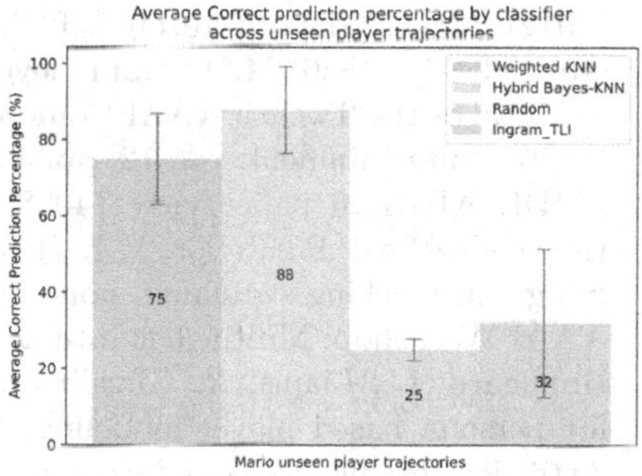

Fig. 9. Mario E-TLI average correct prediction percentage by classifier across unseen player trajectories.

6   Conclusion and Future Work

Our work contributes a hybrid probabilistic supervised learning framework, using Bayesian Inference informed by a K-Nearest Neighbors based likelihood, that is able to classify players in real time at every step within a given game level using only the latest player action or state observation. Furthermore we outperform our comparative baselines whilst using only the latest player observation to make our prediction. Our experiment highlights the success of our framework in a complex human player setting and the effect the play log representation has on the prediction's accuracy, stability, and generalisability. As part of our future work we plan to use our playstyle classification framework to adapt the gameplay experience in a meaningful way. We would also like to apply our work in a competitive online game setting in order to illustrate the potential use and benefit to the e-sports community from a player strategy analysis and game analytics perspective.

References

1. Bakkes, S., Spronck, P., Lankveld, G.V.: Player behavioural modelling for video games. Entertainment Comput. **3**, 71–79 (2012). https://doi.org/10.1016/j.entcom.2011.12.001
2. Bontchev, B.: Holistic player modeling for controling adaptation in video games. In: Proceedings of the 14th International Conference e-Society 2016 (2016)
3. Charles, D., Black, M.: Dynamic player modelling: a framework for player-centred digital games. In: Proceedings of the International Conference on Computer Games: Artificial Intelligence, Design and Education (2004)
4. Gonzalez-Duque, M., Palm, R.B., Risi, S.: Fast game content adaptation through Bayesian-based player modelling. In: 2021 IEEE Conference on Games (CoG), pp. 01–08 (2021). https://doi.org/10.1109/CoG52621.2021.9619018
5. Gow, J., Baumgarten, R., Cairns, P., Colton, S., Miller, P.: Unsupervised modeling of player style with LDA. IEEE Trans. Comput. Intell. AI Games **4**(3), 152–166 (2012). https://doi.org/10.1109/TCIAIG.2012.2213600
6. Guzdial, M., Riedl, M.O.: Game level generation from gameplay videos. In: Proceedings of the Twelfth AAAI Conference on Artificial Intelligence and Interactive Digital Entertainment, vol. 12, pp. 44–50 (2016). https://aaai.org/ocs/index.php/AIIDE/AIIDE16/paper/view/14008/13593
7. Hernandez-Leal, P., Taylor, M.E., Rosman, B., Sucar, L.E., de Cote, E.M.: Identifying and tracking switching, non-stationary opponents: a Bayesian approach. In: AAAI Workshop: Multiagent Interaction without Prior Coordination (2016)
8. Holmgard, C., Liapis, A., Togelius, J., Yannakakis, G.: Monte-Carlo tree search for persona based player modeling. In: Proceedings of the AAAI Conference on Artificial Intelligence and Interactive Digital Entertainment, vol. 11, no. 5, pp. 8–14 (2021). https://ojs.aaai.org/index.php/AIIDE/article/view/12849
9. Holmgard, C., Liapis, A., Togelius, J., Yannakakis, G.N.: Evolving personas for player decision modeling. In: 2014 IEEE Conference on Computational Intelligence and Games, pp. 1–8 (2014). https://doi.org/10.1109/CIG.2014.6932911
10. Holmgård, C., Liapis, A., Togelius, J., Yannakakis, G.N.: Personas versus clones for player decision modeling. In: Pisan, Y., Sgouros, N.M., Marsh, T. (eds.) ICEC 2014. LNCS, vol. 8770, pp. 159–166. Springer, Heidelberg (2014). https://doi.org/10.1007/978-3-662-45212-7_20
11. Hunter, J.D.: Matplotlib: a 2D graphics environment. Comput. Sci. Eng. **9**(3), 90–95 (2007). https://doi.org/10.1109/MCSE.2007.55
12. Iawasaki, Y., Hasebe, K.: Identifying playstyles in games with neat and clustering. In: 2021 IEEE Conference on Games (CoG), pp. 1–4 (2021). https://doi.org/10.1109/CoG52621.2021.9619024
13. Ingram, B.: Generating tailored advice in video games through play-style identification and player modelling. In: Proceedings of the AAAI Conference on Artificial Intelligence and Interactive Digital Entertainment, vol. 17, no. 1, pp. 228–231 (2021). https://ojs.aaai.org/index.php/AIIDE/article/view/18913
14. Ingram, B., Rosman, B., van Alten, C., Klein, R.: Play-style identification through deep unsupervised clustering of trajectories. In: 2022 IEEE Conference on Games (CoG), pp. 393–400 (2022). https://doi.org/10.1109/CoG51982.2022.9893680
15. Iwasaki, Y., Hasebe, K.: A framework for generating playstyles of game AI with clustering of play logs. In: Proceedings of the 14th International Conference on Agents and Artificial Intelligence, ICAART, vol. 3, pp. 605–612. INSTICC, SciTePress (2022). https://doi.org/10.5220/0010869500003116

16. Kodinariya, T., Makwana, P.: Review on determining of cluster in k-means clustering. Int. J. Adv. Res. Comput. Sci. Manag. Stud. **1**, 90–95 (2013)
17. Normoyle, A., Jensen, S.: Bayesian clustering of player styles for multiplayer games. In: Proceedings of the AAAI Conference on Artificial Intelligence and Interactive Digital Entertainment, vol. 11, no. 1, pp. 163–169 (2021). https://ojs.aaai.org/index.php/AIIDE/article/view/12805
18. Pedersen, C., Togelius, J., Yannakakis, G.: Modeling player experience in super mario bros, pp. 132–139 (2009). https://doi.org/10.1109/CIG.2009.5286482
19. Pedregosa, F., et al.: Scikit-learn: machine learning in Python. J. Mach. Learn. Res. **12** (2011)
20. Russell, S.J., Norvig, P.: Artificial Intelligence: A Modern Approach, 3rd edn. Pearson, London (2009)
21. Scott, G., Khosmood, F.: Complementary companion behavior in video games. Master's thesis, School of Computer Science, California Polytechnic State University, San Luis Obispo (2017). https://digitalcommons.calpoly.edu/theses/1744. https://doi.org/10.15368/theses.2017.55. arXiv Short Paper Version https://arxiv.org/abs/1808.09079
22. Togelius, J., Nardi, R.D., Lucas, S.M.: Making racing fun through player modeling and track evolution (2006)
23. Tremblay, J., Verbrugge, C.: Adaptive companions in FPS games. In: FDG (2013)
24. Tychsen, A., Canossa, A.: Defining personas in games using metrics. In: Future Play (2008). https://doi.org/10.1145/1496984.1496997
25. Valls-Vargas, J., Ontañón, S., Zhu, J.: Exploring player trace segmentation for dynamic play style prediction. In: Proceedings of the AAAI Conference on Artificial Intelligence and Interactive Digital Entertainment, vol. 11, no. 1, pp. 93–99 (2021). https://ojs.aaai.org/index.php/AIIDE/article/view/12782
26. Waskom, M.L.: Seaborn: statistical data visualization. J. Open Source Softw. **6** (2021). https://doi.org/10.21105/joss.03021
27. Yannakakis, G., Spronck, P., Loiacono, D., Andre, E.: Player Modeling. Dagstuhl Follow-Ups, vol. 6, pp. 45–59. Dagstuhl Publishing (2013)
28. Zhang, M., Verbrugge, C.: Modelling player understanding of non-player character paths. In: Proceedings of the AAAI Conference on Artificial Intelligence and Interactive Digital Entertainment, vol. 14, no. 1 (2018). https://dl.acm.org/doi/abs/10.5555/3505378.3505415

The Missing Margin: How Sample Corruption Affects Distance to the Boundary in ANNs

Marthinus Wilhelmus Theunissen[1,2]([✉]) [iD], Coenraad Mouton[1,2,3] [iD],
and Marelie H. Davel[1,2,4] [iD]

[1] Faculty of Engineering, North-West University, Potchefstroom, South Africa
tiantheunissen@gmail.com
[2] Centre for Artificial Intelligence Research (CAIR), Cape Town, South Africa
[3] South African National Space Agency (SANSA), Hermanus, South Africa
[4] National Institute for Theoretical and Computational Sciences (NITheCS),
Stellenbosch, South Africa

Abstract. Classification margins are commonly used to estimate the generalization ability of machine learning models. We present an empirical study of these margins in artificial neural networks. A global estimate of margin size is usually used in the literature. In this work, we point out seldom considered nuances regarding classification margins. Notably, we demonstrate that some types of training samples are modelled with consistently small margins while affecting generalization in different ways. By showing a link with the minimum distance to a different-target sample and the remoteness of samples from one another, we provide a plausible explanation for this observation. We support our findings with an analysis of fully-connected networks trained on noise-corrupted MNIST data, as well as convolutional networks trained on noise-corrupted CIFAR10 data.

Keywords: Classification margin · Label corruption · Generalization

1 Introduction

The study of artificial neural networks (ANNs) and their performance on unseen test data embodies many different overlapping themes, such as capacity control [26], loss landscape geometry [6], information theoretical approaches [25], and algorithmic stability [3]. Optimal *classification margins* remains a popular concept, showing both theoretical support and empirical evidence of being related to generalization [7,16]. Simply put, the classification margin of a sample with regard to a specific model is the shortest distance the sample will need to move, in a given feature space, in order to change the predicted output value. Hereafter, we refer to this concept as a *margin*. In the literature it is also called 'minimum distance to the boundary' or 'minimum adversarial perturbation' [14], depending on context.

M. W. Theunissen and C. Mouton—Equal contribution.

Large margins have long been used to indicate good generalization ability in classification models [2,28]. The supporting intuition is simple: With a larger margin, a sample can have more varied feature values (potentially due to noise) while still being correctly classified. It is argued that overparameterized ANNs tend to find large margin solutions [13,27].

Recent studies of the relationship between margin and generalization typically measure the average margins over the training set or some sampling thereof. In this work, we ask whether the margins of individual samples tend to reflect this average behaviour. Specifically, we focus on ANN-based classification models and introduce controlled noise into the training process. Using different types of training sample corruption, we demonstrate a number of intricacies related to input margins and their relation to generalization. Specifically, we contribute the following:

1. By using target and input noise, we point out local margin behaviour that is inconsistent with the global average. We find that, while all margins have a strong tendency to increase, label- and input-corrupted samples maintain significantly smaller margins than uncorrupted samples. We also find that only label-corrupted samples noticeably affect the margins of clean samples.
2. We discuss the implications of these inconsistencies. Our findings suggest that using the average margins as a metric is only fitting if the set of contributing samples has an equal level of diversity for all models being compared.
3. We probe the underlying mechanisms that lead to these inconsistencies. We hypothesize that label-corrupted samples have reduced margins because of their proximity to clean samples, and the input-corrupted samples have smaller margins due to a lack of incentive to increase them.

Our choice of using noise is not arbitrary. Adding artificial noise to a training set and investigating the ability of ANNs to generalize in spite of this corruption is a popular technique in empirical investigations of generalization. A good example of success with such methods is the seminal paper by Zhang et al. [30], where it was shown that overparameterized models can generalize in spite of having enough capacity to fit per-sample random noise such as label corruption or randomized input features. Similar noise has been used extensively to experimentally probe ANNs [15,17,24].

The rest of the paper is structured as follows: Sect. 2 describes related work on classification margins and generalization. In Sect. 3 we define a margin and describe our exact method of measuring it. Following this, Sect. 4 presents details on our experimental setup. The resulting margins, along with notable local inconsistencies, are presented and discussed in Sect. 5. In the final section, we investigate the nature of these local inconsistencies, describing plausible underlying mechanisms.

2 Related Work

Research on margins extends back much earlier than the advent of powerful ANNs. For example, the effect of margins on generalization performance in linear models such as Support Vector Machines (SVMs) are well-studied [26]. An inherent issue with extending this work to modern ANNs is that their decision boundaries are highly non-linear, high dimensional, and contain high curvature.

Finding the closest boundary point to a given sample is often considered intractable, as such, some works opt to rather estimate the margin [7,19] or define bounds on these margins [21]. Elsayed et al. [4] derive a linear approximation of the shortest distance to the decision boundary. This is then formulated as a penalty term which is used during training to ensure that each sample has at least a specific (chosen hyperparameter) distance to the decision boundary, for both the input space and hidden layers. Networks trained using this loss function exhibit greater adversarial sample robustness, and better generalization when trained on data with noisy labels than networks trained using conventional loss functions. Similarly, Jiang et al. [7] further utilize the same approximation to predict the generalization of a large number of Convolutional Neural Networks (CNNs), by training a linear classifier on the margin distributions of the existing models. However, it is shown in [29] show that this approximation likely considerably underestimates the true distance to the decision boundary.

Several authors [5,9,29] use a simple linear interpolation between two samples of different classes to find a point on the decision boundary which separates them. This is however unlikely to result in a distance that is near the true minimum. In a similar approximate fashion, Somepalli et al. [20] take the average of the distance in five random directions around an input sample, where each directional-distance is calculated separately using a simple bisection method.

Karimi et al. [9] introduce 'DeepDIG', an approach based on an auto-encoder that finds samples near the decision boundary that are visually similar to the original training samples. However, this method does not specifically attempt to find the nearest point on the decision boundary. Finally, Youzefsadeh and O'Leary [29] formulate finding the shortest distance to the decision boundary of a given sample as a constrained optimization problem. While this method is highly accurate, it is computationally very expensive, especially for high-dimensional data, e.g. natural images such as MNIST [12] and CIFAR10 [10].

Similar to Youzefsadeh and O'Leary [29], in this work we find actual points in feature space. We have two reasons for this: (a) simplified approximations might lead to misconceptions of the role margins play in a model's ability to generalize, and (b) finding actual points inherently considers the non-linear intricacies in the function mapping. In Sect. 3 we describe our selected method in detail.

3   Formulating the Classification Margin

In the previous section, we have given an overview of different methods of calculating the classification margin. We opt to use the most precise method, which formulates the margin calculation as a non-linear constrained minimization problem.

Let $f : X \to \mathbb{R}^{|N|}$ denote a classification model with a set $N = \{0, 1, ..., n\}$ of output classes. For an input sample $\mathbf{x}$ we search for a point $\hat{\mathbf{x}}$ on the decision boundary between classes i and j, where i is $\arg\max(f(\mathbf{x}))$, with $i, j \in N, j \neq i$, and $\hat{\mathbf{x}}$ is the nearest point to $\mathbf{x}$ on the decision boundary. Formally, for some distance function $dist$, we find $\hat{\mathbf{x}}$ by solving the following constrained minimization problem (CMP):

$$\min_{\hat{\mathbf{x}}} dist(\mathbf{x}, \hat{\mathbf{x}}) \tag{1}$$

such that

$$f(\hat{\mathbf{x}})[i] = f(\hat{\mathbf{x}})[j] \tag{2}$$

for the ith and j^{th} output node, respectively. Finding a point that meets the condition defined in Eq. 2 exactly, is virtually impossible. In practice, a threshold is used, so that a point is considered valid (on the decision boundary) if $|f(\hat{\mathbf{x}})[i] - f(\hat{\mathbf{x}})[j]| \leq 10^{-3}$. In order to find the nearest point on the decision boundary for all j, we search over each class $j \neq i$ separately for each sample and choose the one with the smallest distance. As is convention for margin measurements [20, 29], we use Euclidean distance[1] as metric, meaning the margin is given by:

$$dist(\mathbf{x}, \hat{\mathbf{x}}) = |\mathbf{x} - \hat{\mathbf{x}}|_2 \tag{3}$$

In order to solve each CMP, we make use of the augmented Lagrangian method [1], using the conservative convex separable approximation quadratic (CCSAQ) optimizer [22] for each unconstrained optimization step, implemented with the NLOpt (non-linear optimization) library [8] in Python.

While it is possible to extend the nearest boundary optimization to the hidden layers, it is difficult to compare models with layers of varying dimensionality. Additionally, hidden layer margins can be manipulated and require some form of normalization [7]. As such, we limit our analysis to input-space margins.

4   Experimental Setup

In order to investigate how sample corruption affects margin measurements, we train several networks of increasing capacity to the point of interpolation (close to zero train error) on the widely used classification datasets MNIST [12] and CIFAR10 [10]. 'Toy problems' such as these are used extensively to probe generalization [15, 20, 30]. We corrupt the training data of some models using two types of noise, defined in Sect. 4.1, separately. Capacity and generalization are strongly linked. In the overparameterized regime we expect generalization to improve systematically with an increase in capacity and a corresponding increase in average margins. This setup allows us to determine whether expected behaviour is consistent across all samples.

4.1   Controlled Noise

We use two specific types of noise, inspired by Zhang et al. [30]: Label corruption and Gaussian input corruption. These have been designed to represent

---

[1] In practice, we optimize for the squared Euclidean distance in order to reduce the computational cost of gradient calculations, but report on the unsquared distance in all cases.

two complications that are often found in real world data and could affect the generalization of a model fitted to them. Label corruption represents noise that comes from mislabeled training data (mislabeling is common in large real-world datasets and even in benchmark datasets [18]), inter-class overlap, and general low separability of the underlying class manifolds.

Gaussian input corruption, on the other hand, represents extreme examples of out-of-distribution samples. These are 'off-manifold' samples displaying a high level of randomness. Such samples do not necessarily obscure the true underlying data distribution, but still require a significant amount of capacity to fit the excessive complexity that needs to be approximated when fitting samples with few common patterns.

Given a training sample $(\mathbf{x}, c)$ where $\mathbf{x} \in \mathbb{R}^d$ and $c \in N$ for a set of classes N, the corruption of a sample can be defined as follows:

- *Label corruption*: $(\mathbf{x}, c) \rightarrow (\mathbf{x}, \hat{c})$ where $\hat{c} \neq c, \hat{c} \in N$.
- *Gaussian input corruption*: $(\mathbf{x}, c) \rightarrow (\mathbf{g}, c)$ where $\mathbf{g} \in \mathbb{R}^d$ and each value in $\mathbf{g}$ is sampled from $\mathbb{N}(\mu_{\mathbf{x}}, \sigma_{\mathbf{x}})$.

Alternative labels are selected at random and $\mathbb{N}(\mu_{\mathbf{x}}, \sigma_{\mathbf{x}})$ is a normal distribution, with $\mu_{\mathbf{x}}$ and $\sigma_{\mathbf{x}}$ the mean and standard deviation of all the features in the original sample $\mathbf{x}$. Henceforth, we will drop the 'Gaussian' when referring to 'Gaussian input corruption'.

4.2 MNIST Models

For the MNIST dataset, we train three distinct sets of MLPs, with each set containing models of identical depth and identically varied width:

- **MNIST:** A set of clean MNIST models. These serve as baselines, showing the level of generalization and margin sizes to be expected should the models not have been trained on any corrupted data.
- **MNISTlc** (MNIST-label-corrupted): Models with the same capacities as the previous set, but where 20% of the training set is label corrupted.
- **MNISTgic** (MNIST-Gaussian-input-corrupted): Models with the same capacities as the clean models, however, 20% of the training set is input corrupted.

All models for these tasks have the following hyperparameters in common. They use a 55 000/5 000 train-validation split of the training data. They are all single hidden layer ReLU-activated MLPs with widths ranging from 100 to 10 000 hidden layer nodes, and a single bias node. Stochastic gradient descent (including momentum terms) is used to minimize the cross-entropy loss on mini-batches of size 64 selected at random. The initial learning rate is set to 0.01 and then multiplied by 0.99 every 5 epochs. For each set, we train three random initializations. Note that we train the **MNISTlc** models for 1 000 epochs and models from the other two sets for 100 epochs. This is because the label-corrupted dataset required more epochs to interpolate.

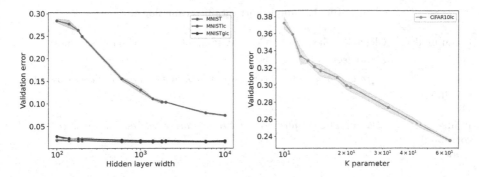

Fig. 1. Validation error for MNIST models (left) and CIFAR10 models (right). Values are averaged over three random seeds and shaded areas indicate standard deviation. Note that all models interpolated except for the smallest two capacities for **MNISTlc** and **CIFAR10lc**. The maximum train error (across all models) was 0.0573.

The resulting generalization ability of all three sets are depicted in Fig. 1 (left). Note that, as expected for all three tasks, with more capacity we see an improvement in validation set performance. Also note that only the label corruption results in any significant reduction in validation performance, as also previously reported in [24].

4.3   CIFAR10 Models

In order to verify our main findings surrounding label corruption we also replicate the **MNISTlc** task using CNNs on the CIFAR10 dataset, with 10% label corruption. We use a similar architecture to the 'standard' CNN used by Nakkiran et al. [15]. Each CNN consists of four ReLU-activated convolutional layers, with $[k, 2k, 4k, 8k]$ output channels respectively, where we choose various values of k between 10 and 64 to create a group of models with varying capacity. Each model also includes max and average pooling layers, and a final fully connected layer of 400 nodes. This set of models is referred to as **CIFAR10lc**.

All models are trained on a 45 000/5 000 train-validation split of the CIFAR10 dataset where the training set contains 10% label corruption. These models are trained for 500 epochs in order to minimize a cross-entropy loss function on mini-batches of 256 samples using the Adam optimizer. The initial learning rate of 0.001 is multiplied by 0.99 every 10 epochs. Three initialization seeds are used. From the relevant validation errors in Fig. 1 (right), we again observe that more capacity is accompanied by better generalization performance.

4.4   Terminology

When describing margin behaviour, we refer to different subselections of margins based on the type of sample (clean or corrupted) as well as the type of model. To prevent confusion, we always refer to these subselections using a name constructed from the type of sample and then the type of model, separated by a

Table 1. Sample corruption terminology.

Clean model	Clean samples	Corrupt samples	Overall samples
	clean:clean	N/a	*overall:clean*
Label-corrupted model	*clean:label-corrupted*	*corrupt:label-corrupted*	*overall:label-corrupted*
Input-corrupted model	*clean:input-corrupted*	*corrupt:input-corrupted*	*overall:input-corrupted*

colon, as shown in Table 1. For example, if we are referring to the margins for the uncorrupted samples with regard to a label-corrupted model we will refer to them as the *clean:label-corrupted* margins.

5 Results

We calculate the margins for 10 000 randomly selected training samples, for all of the models defined in Sect. 4, using the method described in Sect. 3. Only correctly classified samples are considered. This amounts to solving 11 capacities × 3 random seeds ×3 datasets ×9 class pairs ×10k samples $= 8\,910k$ individual CMPs for the MNIST models and $11 \times 3 \times 9 \times 10k = 2\,970k$ for the CIFAR10 models. In order to solve such a large number of CMPs we utilize 240 CPU cores split over 10 servers, by making use of GNU-Parallel [23]. Next, we investigate the central tendencies of the clean and corrupt samples separately, as capacity increases (Sect. 5.1) and discuss possible implications of our observations (Sect. 5.2).

5.1 Local Inconsistencies

Figure 2 shows the mean margins as a function of model capacity. We note that expected margins tend to increase along with capacity, in all cases for all types of samples. This is consistent with what we expect when using margins as an indicator of generalization.

However, we see that noise-corrupted and clean data demonstrate different tendencies: we see that margins for both *corrupt:label-corrupted* and *corrupt:input-corrupted* samples tend to be significantly smaller than for the clean samples in the same models at the same capacity. Furthermore, we observe that the *clean:clean* and *clean:input-corrupted* margins are similar, especially at higher capacities, while *clean:label-corrupted* margins are seen to be significantly smaller.

It is interesting that the average margins measured on CIFAR10, Fig. 2 (bottom), are smaller than the MNIST ones. The CIFAR10 features contain 3 072 dimensions and MNIST only 784. We expect Euclidean distance in higher dimensions to be larger. We suspect that the reason for this contradiction is a large

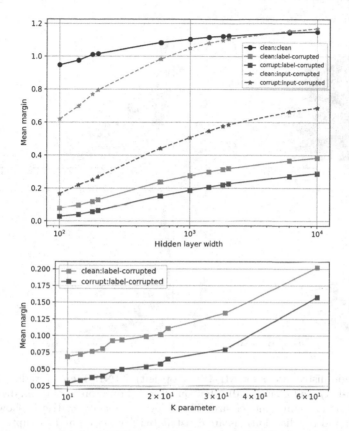

Fig. 2. Mean margins for MNIST models (top) and CIFAR10 models (bottom).

degree of inherent inter-class overlap in CIFAR10 that results in the observed
small margins.

Next, we look at the distributions of margins underlying the means in Fig. 2.
These results are shown in Fig. 3. We construct histograms of the margins measured at each capacity. These histograms share a common set of bins on the
horizontal axes.

We see that most of these distributions are right-skewed distributions with a
long tail containing relatively large margins. This indicates that the mean might
be a slightly overestimated measure of the central tendency of margins. The
margins for the **CIFAR10lc** set show similar trends to that of the **MNISTlc** set,
but the distributions are even more right-skewed, indicating that their central
tendencies are even smaller than the means in Fig. 2 (bottom) suggest.

We also see that the *corrupt:input-corrupted* margin distributions are normally distributed with a relatively low variance, compared to the other distributions. This suggests that all models are constructing similar decision boundaries
around them. There is not much diversity in how far samples tend to be from
their nearest boundary. The *corrupt:label-corrupted* margins for the **MNISTlc**

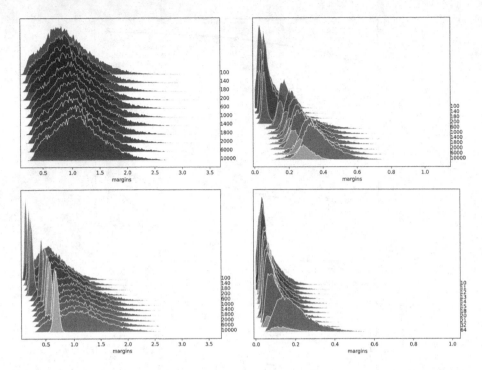

Fig. 3. Margin distributions for **MNIST** (top-left), **MNISTlc** (top-right), **MNIST-gic** (bottom-left), and **CIFAR10lc** (bottom-right). From top to bottom, distributions are ordered by ascending model size. The relevant capacity metric is shown on the right. Green and red distributions are constructed from clean and corrupted samples, respectively. The blue distributions are for the clean models.

set, on the other hand, show much higher diversity. The shape of the distribution changes drastically as capacity increases. At the critically small capacities we see a distribution resembling the *corrupt:input-corrupted* margin distributions and at higher capacities some *corrupt:label-corrupted* samples obtain almost outlying small margins.

5.2 Discussion

We notice a few key inconsistencies regarding the use of a global average margin as an indicator of the extent to which a model separates samples by means of a decision boundary:

1. Samples that are on-manifold, but problematic in terms of their class separability (of which label-corrupted samples are extreme examples) have much lower margins than the global mean would suggest.
2. Samples that are off-manifold and remote (of which the input corrupted samples are extreme examples) also have much smaller margins, even though these samples do not affect the generalization ability of the model.

3. In general, the margin distributions are not normally distributed. A mean might be skewed by the fact that the margin distributions have long tails containing extremely large margins values.

Given these inconsistencies, we can ask whether the global average margin metrics, which are often used to predict or promote generalization, are sound. It seems that the corrupted margins tend to increase in proportion with the clean margins. Assuming the models to be compared fitted exactly the same training samples, this implies that an average margin will only work if the two models for which margins are being compared contain an approximately equal number of samples with these distinct locally inconsistent margin behaviours. If a small set of samples are averaged over, this could become a problem. If the models to be compared have varying training set performance or were not trained on exactly the same training set this could become an even more significant problem.

It can also be argued that margin-based generalization predictors are more sensitive to off-manifold noise than to on-manifold noise. That is because the small *corrupt:label-corrupted* margins rightly indicate the poor generalization of label-corrupted models. However, the small *corrupt:input-corrupted* margins erroneously also indicate poor generalization.

6 A Deeper Look

Now that we have discussed the observed inconsistencies and their possible implications, we explore the origin of these inconsistencies further. We do this by posing three concrete questions based on the results in the previous section. After each question, we propose possible answers, producing additional measurements where these can shed light on the underlying phenomena.

(1) **Why are the *overall:label-corrupted* margins so small?** We note that label corruption is expected to result in many samples that have different targets while being close to each other in the input space. One can think of a sample's minimum distance to another sample with a different target as its absolute maximum possible margin since a boundary needs to be drawn between them, assuming both have been correctly classified during training. We propose that this is the main factor contributing to the small *overall:label-corrupted* margins.

To test this hypothesis, we randomly select 10 000 training samples from the data that the models in the **MNIST1c** and **CIFAR10lc** sets are trained on. We then measure the 'max margin' as the minimum Euclidean distance between each sample (x_1, c_1) and its nearest neighbour (x_2, c_2) (selected from the entire train set) so that:

$$\min_{x_2} |x_1 - x_2|_2, \quad c_1 \neq c_2 \qquad (4)$$

We do this for the same model before and after label corruption. We then construct a scatter plot of these 10 000 training samples with the distance as measured with the original targets on the horizontal axis and the potentially corrupted targets on the vertical axis. The resulting scatter plots are shown in

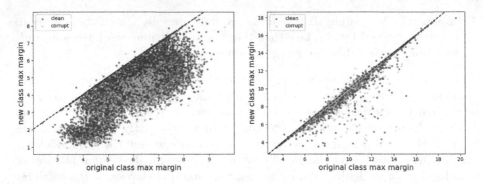

Fig. 4. Maximum margins before vs. after label corruption for **MNISTlc** (left) and **CIFAR10lc** (right). Green points represent clean samples and red points represent corrupt samples. The dashed line indicates $y = x$.

Fig. 4. All samples below the provided identity function line had their max margin reduced due to label corruption. Note that, as expected, the presence of label corruption causes many samples, corrupted and clean, to have drastically reduced upper bounds to their margins.

If our hypothesis is true then the smallest margins should correspond to the *clean:label-corrupted* and *corrupt:label-corrupted* samples that are the closest to each other in the input space. We confirm this by constructing a Euclidean distance matrix comparing the 1 000 *clean:label-corrupted* samples with the smallest margins and the 1 000 *corrupt:label-corrupted* samples with the smallest margins for one of the biggest models from the **MNISTlc** set. This is presented in Fig. 5.

Note that it is the samples that are relatively close to each other that tend to have the smallest margins. From these results we conclude that a significant factor leading to the small margins observed in label-corrupted models, is the proximity (in the input space) of samples with different targets. This also accounts for the observation that *corrupt:label-corrupted* margins tend to be smaller than *clean:label-corrupted* margins. There are more *clean:label-corrupted* samples than *corrupt:label-corrupted* samples. Therefore, fewer clean samples are moved closer to a different target sample. The result is that fewer *clean:label-corrupted* margins are reduced.

(2) Why are the *corrupt:input-corrupted* margins smaller than the *clean:input-corrupted* margins? To determine whether a similar phenomenon is reducing the *corrupt:input-corrupted* margins we generate a similar scatter plot to Fig. 4 but for the **MNISTgic** data. This is seen in Fig. 6. In contrast to the *overall:label-corrupted* samples we see that virtually all *overall:input-corrupted* samples have either increased or unchanged maximum margins. Strikingly, the *corrupt:input-corrupted* samples have extremely high maximum margins.

This is an apparent contradiction. If the proximity of different-target samples reduces the average margin and *corrupt:input-corrupted* samples are extremely distant from any other sample (minimum of 10 in Fig. 6), why are *corrupt:input-*

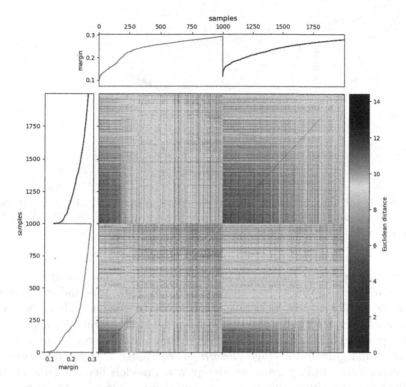

Fig. 5. Euclidean distance between 1 000 *clean:label-corrupted* and 1 000 *corrupt:label-corrupted* samples with the smallest margins for a $1 \times 10\,000$ model **MNISTlc**. Entries in the dissimilarity matrix are ordered by corruption status first, then by margin size. See the corresponding margins in the axis plots for the exact margins. Red curves refer to corrupted margins and green curves refer to clean margins.

corrupted margins small? And why are *clean:input-corrupted* margins slightly smaller than *clean:clean* margins, when they have slightly larger maximum margins?

Regrettably, we do not have as strong an argument for the inconsistency regarding input-corrupted samples as we do for label corruption. We speculate that the relatively small *corrupt:input-corrupted* margins are a result of the remoteness of these samples. They are so far off manifold and far from each other that there is little incentive to increase there respective margins beyond a certain model-specific maximum. The previously-mentioned lack of variance in margins we observe for *corrupt:input-corrupted* margins in Fig. 3 supports this notion.

(3) Why are *clean:input-corrupted* margins smaller than *clean:clean* margins? Without the *corrupt:input-corrupted* samples in close proximity to the *clean:input-corrupted* samples we might expect them to have similar margins to *clean:clean*, across all capacities. However, we see that they tend to be slightly smaller. We hypothesize that this is a result of the capacity it requires to fit

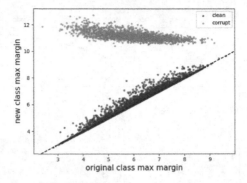

Fig. 6. Maximum margins before vs. after corruption for **MNISTgic**. Green points represent clean samples and red points represent corrupt samples. The dashed line indicates $y = x$.

corrupt:input-corrupted samples. It is known that these samples are fitted later and require more capacity than clean samples [11,24], and that margins tend to increase with capacity. It is then reasonable to conclude that the lack of *clean:input-corrupted* margin size is a result of the lack of available capacity. This idea is supported by Fig. 2, where we observe that the difference between average *clean:clean* margins and average *clean:input-corrupted* margins decreases with added capacity, and disappears completely when models become large enough.

What is the takeaway? To summarize, there seems to be two main mechanisms contributing to the observed local inconsistencies in margin behaviour.

1. Samples that are very close to different-target samples will inevitably have reduced margins. This kind of reduced margin is indicative of poor generalization because it is very likely to pertain to in-distribution and on-manifold regions of feature space that are difficult to model.
2. Samples that are extremely remote, being very distant from any other sample, will also obtain reduced margins, due to a lack of incentive to increase them. This kind of reduced margin is not indicative of poor generalization because it is likely to pertain to out-of-distribution and off-manifold regions of feature space.

7   Conclusion

In this work we show that some training samples are consistently modeled with small margins while affecting generalization in different ways. This is a novel observation of a phenomenon that will require consideration if margins are used in methods to predict generalization of ANNs. We use label and Gaussian input corruption as tools but hypothesize that similar behaviour is possible in natural datasets that contain significant inter-class overlap in the input space, or a large portion of off-manifold samples, respectively. We conclude that a global average margin will be more useful in predicting generalization if it considers these local inconsistencies, or contains an equal proportion of these kinds of samples for all models being compared.

In addition to providing a precise comparison of the way in which different types of margins change with increased capacity, we explore some of the possible reasons for the behaviour observed. These are hypotheses that we will be investigating further in future work. We also plan to extend this work to hidden layers, as hidden layer margins have been shown to similarly relate to generalization behaviour [7,16]. Additionally, a more concrete understanding of the influence (or lack thereof) off-manifold samples have on the margins of on-manifold samples will be useful.

Acknowledgements. We thank and acknowledge the Centre for High Performance Computing (CHPC), South Africa, for providing computational resources to this research project.

References

1. Birgin, E.G., Martinez, J.M.: Improving ultimate convergence of an augmented Lagrangian method. Optim. Methods Softw. **23**(2), 177–195 (2008)
2. Boser, B.E., Guyon, I.M., Vapnik, V.N.: A training algorithm for optimal margin classifiers. In: Proceedings of the Fifth Annual Workshop on Computational Learning Theory, pp. 144–152 (1992)
3. Bousquet, O., Elisseeff, A.: Algorithmic stability and generalization performance. In: Advances in Neural Information Processing Systems, vol. 13 (2000)
4. Elsayed, G., Krishnan, D., Mobahi, H., Regan, K., Bengio, S.: Large margin deep networks for classification. In: Advances in Neural Information Processing Systems, vol. 31 (2018)
5. Guan, S., Loew, M.: Analysis of generalizability of deep neural networks based on the complexity of decision boundary. In: 2020 19th IEEE International Conference on Machine Learning and Applications (ICMLA), pp. 101–106. IEEE (2020)
6. Hochreiter, S., Schmidhuber, J.: Flat minima. Neural Comput. **9**(1), 1–42 (1997)
7. Jiang, Y., Krishnan, D., Mobahi, H., Bengio, S.: Predicting the generalization gap in deep networks with margin distributions. In: International Conference on Learning Representations (ICLR) (2018)
8. Johnson, S.G.: The NLopt nonlinear-optimization package. http://github.com/stevengj/nlopt
9. Karimi, H., Derr, T., Tang, J.: Characterizing the decision boundary of deep neural networks. arXiv preprint arXiv:1912.11460 (2019)
10. Krizhevsky, A.: Learning Multiple Layers of Features from Tiny Images (2012). https://www.cs.toronto.edu/kriz/learning-features-2009-TR.pdf
11. Krueger, D., et al.: Deep nets don't learn via memorization (2017)
12. LeCun, Y., Bottou, L., Bengio, Y., Haffner, P.: Gradient-based learning applied to document recognition. Proc. IEEE **86**(11), 2278–2324 (1998)
13. Lyu, K., Li, J.: Gradient descent maximizes the margin of homogeneous neural networks. In: International Conference on Learning Representations (ICLR) (2019)
14. Moosavi-Dezfooli, S.M., Fawzi, A., Frossard, P.: DeepFool: a simple and accurate method to fool deep neural networks. In: Proceedings of the IEEE Conference on Computer Vision and Pattern Recognition, pp. 2574–2582 (2016)

15. Nakkiran, P., Kaplun, G., Bansal, Y., Yang, T., Barak, B., Sutskever, I.: Deep double descent: where bigger models and more data hurt. J. Stat. Mech: Theory Exp. **2021**(12), 124003 (2021)
16. Natekar, P., Sharma, M.: Representation based complexity measures for predicting generalization in deep learning. arXiv preprint arXiv:2012.02775 (2020)
17. Neyshabur, B., Bhojanapalli, S., McAllester, D., Srebro, N.: Exploring generalization in deep learning. In: Advances in Neural Information Processing Systems, vol. 30 (2017)
18. Northcutt, C., Athalye, A., Mueller, J.: Pervasive label errors in test sets destabilize machine learning benchmarks. In: Vanschoren, J., Yeung, S. (eds.) Proceedings of the Neural Information Processing Systems Track on Datasets and Benchmarks, vol. 1 (2021)
19. Sokolić, J., Giryes, R., Sapiro, G., Rodrigues, M.R.: Robust large margin deep neural networks. IEEE Trans. Signal Process. **65**(16), 4265–4280 (2017)
20. Somepalli, G., et al.: Can neural nets learn the same model twice? Investigating reproducibility and double descent from the decision boundary perspective. In: Proceedings of the IEEE/CVF Conference on Computer Vision and Pattern Recognition, pp. 13699–13708 (2022)
21. Sun, S., Chen, W., Wang, L., Liu, T.Y.: Large margin deep neural networks: theory and algorithms. arXiv preprint arXiv:1506.05232 (2015)
22. Svanberg, K.: A class of globally convergent optimization methods based on conservative convex separable approximations. SIAM J. Optim. **12**(2), 555–573 (2002)
23. Tange, O.: GNU parallel - the command-line power tool. USENIX Mag. **36**(1), 42–47 (2011). http://www.gnu.org/s/parallel
24. Theunissen, M.W., Davel, M.H., Barnard, E.: Benign interpolation of noise in deep learning. South African Comput. J. **32**(2), 80–101 (2020)
25. Tishby, N., Zaslavsky, N.: Deep learning and the information bottleneck principle. In: 2015 IEEE Information Theory Workshop (ITW), pp. 1–5. IEEE (2015)
26. Vapnik, V.N.: An overview of statistical learning theory. IEEE Trans. Neural Netw. **10**(5), 988–999 (1999)
27. Wei, C., Lee, J., Liu, Q., Ma, T.: On the margin theory of feedforward neural networks (2018)
28. Weinberger, K.Q., Saul, L.K.: Distance metric learning for large margin nearest neighbor classification. J. Mach. Learn. Res. (JMLR) **10**(2) (2009)
29. Yousefzadeh, R., O'Leary, D.P.: Deep learning interpretation: flip points and Homotopy methods. In: Lu, J., Ward, R. (eds.) Proceedings of The First Mathematical and Scientific Machine Learning Conference. Proceedings of Machine Learning Research, vol. 107, pp. 1–26. PMLR (2020)
30. Zhang, C., Bengio, S., Hardt, M., Recht, B., Vinyals, O.: Understanding deep learning (still) requires rethinking generalization. Commun. ACM **64**(3), 107–115 (2021)

ST-GNNs for Weather Prediction in South Africa

Mikhail Davidson[1,2]([⊠]) and Deshendran Moodley[1,2]

[1] University of Cape Town, 18 University Avenue, Rondebosch,
Cape Town 7700, South Africa
`m.mikhaildavidson@gmail.com, deshen@cs.uct.ac.za`
[2] Centre for Artificial Intelligence Research, 18 University Avenue, Rondebosch,
Cape Town 7700, South Africa

Abstract. Spatial-temporal graph neural networks (ST-GNN) have been shown to be highly effective for flow prediction in dynamic systems, but are under explored for weather prediction applications. We compare and evaluate Graph WaveNet (GWN) and the Low Rank Weighted Graph Neural Network (WGN) for weather prediction in South Africa. We compare these results to two basic temporal deep neural networks architectures, i.e. the Long Short-Term Memory (LSTM) and the Temporal Convolutional Neural Network (TCN), for maximum temperature prediction across 21 weather stations in South Africa. We also perform rigorous experiments to evaluate the stability and robustness of both ST-GNNs. The results show that the GWN model outperforms the other models across different prediction horizons with an average SMAPE score of 8.30%. We also analyse and compare learnt adjacency matrices of the two ST-GNNs to gain insights into the prominent spatial-temporal dependencies between weather stations.

Keywords: Graph neural networks · Weather forecasting · Spatial-temporal dependencies

1 Introduction

The weather system can be characterised as a highly dynamic and chaotic system that exhibits complex and often latent spatial and temporal dependencies. While Deep Neural Network (DNN) approaches like LSTMs and TCNs [4] can capture both long term and short term temporal patterns, they do not cater for spatial patterns. A novel group of emerging deep neural network architectures, Spatial-Temporal Graph Neural Networks (ST-GNN) have emerged that have been shown to be highly effective for flow prediction in dynamic systems [10,11].

Each variable is typically represented as a node in a graph that captures the local dynamics at that location, whilst the system level or global dynamics are captured by weighted edges that reflect the strength of the dependencies between locations [10]. Thus a variable (node) has strong connections (links) to

A. Pillay et al. (Eds.): SACAIR 2022, CCIS 1734, pp. 93–107, 2022.
https://doi.org/10.1007/978-3-031-22321-1_7

variables that are affected by changes in its values and weaker links to variables that are not affected by changes in its values. In this way, a node captures local dynamics (single location), while the overall graph structure captures the global (inter-location) dynamics of the system. The prevalent application is predicting traffic flow at different points in a city traffic network [10,11]. The inter-variable dependencies for this case are spatial relations between traffic flow at different points in the network. These approaches are referred to as spatial-temporal GNNs (ST-GNNs) [10].

One of the challenges with ST-GNNs is that the spatial dependencies captured in the graph structure can be dynamic. Early ST-GNN approaches [11] used a static graph structure provided as prior knowledge as a fixed adjacency matrix before training. However, spatial dependencies in dynamic systems can change and evolve. Graph WaveNet (GWN) [10] was amongst the first to include a dynamic adjacency matrix that learnt and adapted to evolving spatial dependencies. While these techniques have been applied to other domains like stock market prediction [7], they have not been applied to weather prediction. Despite the emergence of newer approaches like StemGNN [1] and MTGNN [9], which have outperformed Graph WaveNet on traffic flow prediction, GraphWaveNet still provides better performance on some applications like stock market prediction [7].

The use of ST-GNNs for weather prediction is under explored. We found only one ST-GNN architecture that has been proposed specifically for weather prediction, i.e. the Low Rank Weighted Graph Neural Network (WGN) [8]. While the WGN showed good accuracy compared to LSTMs and graph convolutional neural networks, the performance of the WGN was not compared to recent ST-GNN approaches, like GWN emerging from the traffic domain. Furthermore, it was only evaluated on weather prediction in a single region, i.e. the continental United States. The weather system is a highly dynamic and erratic system. DNN models become outdated and must be frequently updated as new data becomes available and to cater for new spatial and temporal patterns that may emerge. However, many studies do not perform adequate model robustness and stability tests for the configured model [6]. The use of a single hold-out set does not provide a sufficiently robust performance measure for datasets that are erratic and non-stationary [10]. Furthermore, DNNs, as a result of random initialisation and possibly other random parameter settings could yield substantially different results when re-run with the same hyperparameter values on the same dataset. It is also crucial that optimal DNN configurations should be stable, i.e. have minimal deviation from the mean test loss across multiple runs.

In this paper, we evaluate and compare two state-of-the-art ST-GNN approaches, i.e. GWN and WGN, for weather prediction in two provinces in South Africa. We formulate the weather prediction problem as a multivariate spatial-temporal graph using six weather features from 21 weather stations and predict the maximum temperature over five time horizons. We perform rigorous experiments to evaluate the stability and robustness of both ST-GNNs over the different time horizons and compare these to an LSTM and TCN. We also analyse and compare the learnt adjacency matrices of the two STGNNs to gain insights into the prominent spatial-temporal dependencies between the 21 weather stations.

In Sect. 2 we review ST-GNN approaches and their applications, then present our experimental design in Sect. 3 and the results in Sect. 4. We provide a discussion and findings in Sect. 5 and describe the limitations and possibilities for future exploration in Sect. 6.

2 Background and Related Work

2.1 Problem Formulation

Consider a graph, $G = (V, E)$, where V is the set of nodes $(v \in V)$ and E is the set of edges, $e \in E$, in the graph. We reformulate the weather prediction problem using a graph representation. Consider a sequence of weather observations at periodic time intervals taken at different weather stations in some region of interest. Then the weather stations are represented as the nodes (V) in graph, and the connections between stations are the edges (E). The adjacency matrix derived from a graph is denoted by $\mathbf{A} \in \mathbb{R}^{N \times N}$. If $v_i, v_j \in V$ and $v_i, v_j \in E$, then $\mathbf{A}_{ij}$ has a non-zero value otherwise it is 0. The adjacency matrix represents the structure and strength of the dependencies between nodes (weather stations) in the graph.

The different weather observations emanating at each station are represented as a multivariate time series X at $|V|$ locations, i.e. $X_1, X_2, ..., X_T$, where each $X_T \in \mathbb{R}^{|V| \times d}$, where T represents the time step. The d features correspond to weather observations i.e. maximum temperature, rain, wind speed and direction, humidity and pressure. Given this sequence of historical weather observations, along with the adjacency matrix $\mathbf{A}$, the goal is to predict the target sequence $y_1, y_1, ..., y_H$, where $y \in \mathbb{R}$ is the target weather variable to be predicted at time t for all $|V|$ weather stations and H is the prediction horizon i.e. the length of the sequence to be predicted. In this research, we explore maximum temperature prediction across five different prediction horizons, i.e. $H = 3, 6, 9, 12,$ and $24\,$h prediction horizons.

2.2 Deep Neural Networks for Weather Prediction

Early deep neural networks (DNNs) for weather prediction consisted of multi-layer perceptrons, radial-basis function neural networks, and ensemble methods. Cifuentes et al. [2] reviewed these early deep learning techniques for air temperature prediction. While these networks showed promising results and were able to capture non-linear patterns within the data, these early DNNs could not fully capture either the temporal or the spatial patterns inherent within the data.

Zahroh et al. [12] explored LSTMs for weather prediction. Since LSTMs provide explicit support for temporal patterns, they showed improved performance over the early DNN approaches. A recent study by Hewage et al. [4]

compared the LSTM and the newer TCN to the established Weather Research and Forecasting (WRF) model and classical machine learning and statistical methods, including Support Vector Regression, Random Forest and Autoregressive Integrated Moving Average. They designed and evaluated both multi-input single-output (MISO) and multi-input multi-output (MIMO) LSTMs and TCNs for short-term forecasting of ten weather features including surface temperature using low resolution (10 km) data from the Global Forecast System (GFS). They found that the LSTM and TCN outperformed other models, including the WRF model, for short term forecasting (3, 6, 9 h). While Hewage et al's work is an important study in the area, the study has some limitations, especially in terms of model robustness and model stability. The models were trained on a single 5 month period (January 2018 to May 2018), with the next two months as the validation (June 2018) and test set (July 2018) respectively. Given the small size of the training data set, the models may well yield different performance if the size of the training data set is increased. The test set is for a single month (July 2018) and it is unclear whether the model will yield similar performance for other months. This brings into question the robustness of the model for predictions over different periods in the data set which may have different patterns. A single figure is given for the performance (MSE) scores, and it is unclear whether the models were retrained over several runs or whether this is the performance of the best model that was found. Finally, the study did not consider ST-GNN models which capture both temporal and spatial dependencies.

2.3   Low Rank Weighted Graph Neural Network (WGN)

The WGN model developed by Wilson et al. [8] was the first ST-GNN model implemented in the weather domain. It is the only STGNN that we found for weather prediction. The WGN model captures both spatial and temporal dependencies by combining an LSTM and graph convolutional layers. The temporal dependencies are captured by the LSTM, whereas the spatial dependencies are captured through a self-adaptive adaptive adjacency matrix. The self-adaptive adjacency matrix is able to dynamically learn the complex spatial dependencies between weather stations through gradient descent. To reduce computational complexity, the WGN model introduces sparsity to the adjacency matrix via a rank parameter which reduces the size of the adjacency matrix.

The WGN model was used to predict temperature and wind speed over two data sets, consisting data from 67 and 332 weather stations respectively. The data sets were split using standard holdout validation with a single test set. Wilson et al. [8] compared a low rank learnt self-adaptive adjacency matrix model to a fixed adjacency matrix model. They found that the low rank model with a sparse adjacency matrix outperformed the fixed adjacency matrix model across both data sets when predicting wind speed and temperature over the 12 h forecasting horizon. Their results also confirmed that utilising a learnt self-adaptive adjacency matrix results in better performance than using a fixed adjacency matrix when predicting temperature and wind speed. The authors also visualised the spatial-temporal dependencies from the adjacency matrix and confirmed that these dependencies aligned with existing domain knowledge of weather systems in the U.S.

The drawback of the WGN model is the use of the LSTM, a comparatively weaker temporal architecture when compared to TCNs. Furthermore, a limitation of the WGN paper was that the model was not effectively evaluated for stability and robustness. Finally, the WGN architecture is not compared to current state of the art ST-GNNs, like Graph WaveNet, emanating from the crowd flow and traffic prediction communities.

2.4   Graph WaveNet (GWN)

Wu et al. [10] developed the Graph WaveNet (GWN) architecture that captures spatial-temporal dependencies via spatial-temporal layers. These layers consisted of TCN layers and graph convolution layers to capture temporal and spatial dependencies simultaneously. The GWN self-adaptive adjacency matrix differs from the WGN matrix. The GWN self-adaptive adjacency matrix is randomly initialised as two node embeddings. These embeddings are then multiplied, this is followed by applying the ReLU function to remove weak connections, and finally, SoftMax is applied to normalise the matrix. This matrix, learnt through gradient descent, is therefore able to capture hidden spatial dependencies without prior knowledge. Furthermore, the GWN graph convolution layer can optionally include prior knowledge that is used in combination with the self-adaptive adjacency matrix.

The GWN was evaluated on two public traffic network data set, for predicting traffic flow across the 15, 30, and 60 min prediction horizons. The results showed that the GWN model outperformed earlier ST-GNN models across all prediction horizons, confirming its ability to capture temporal dependencies and hidden spatial dependencies without prior information.

Pillay et al. [7] explored the GWN architecture for stock price prediction for single and multi-step forecasting. They compared the GWN to newer ST-GNN architectures, i.e. StemGNN [1] and MTGNN [9], and found that the GWN outperformed the newer architectures. Given the prominence of the GWN architecture and its performance in both stock price and traffic flow prediction we decided to explore GWN for weather prediction. One key limitation of both the Wu et al. and Pillay et al. studies was that they used a basic holdout evaluation strategy. There was not interrogation of how their models performed on different parts of the data set which brings into question the robustness of their trained models.

3   Experimental Design

In this section we describe the machine learning pipeline that we used in our experiments. The weather data is pre-processed, this is followed by data partitioning using a sliding window. We then perform hyper-parameter optimisation for selecting the best model configuration and use walk-forward validation for model training and testing.

3.1 Data

The data set used to carry out the experiments performed was provided by the South African Weather Services. It consists of surface weather observations, these are observations made near the earth's surface by automatic weather stations [5]. The original data set comprises surface weather observations across 22 weather stations, shown in Fig. 1, within the Western and Northern provinces of South Africa. The observations are recorded at hourly intervals over a period of 8 years from January 2012 up to December 2019 yielding 70128 time steps. At each time step, all six surface weather observations are recorded at each weather station.

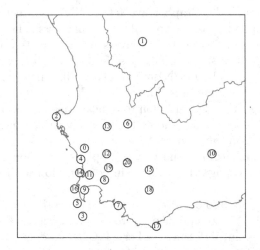

Fig. 1. Figure showing location of weather stations in the Western Cape and Northern Province

3.2 Pre-processing

Following Wilson et al's approach [8] a weather station was only included in the data set if less than 10% of its data is missing. This resulted in a total data set consisting of 21 of a total 22 weather stations. The inverse distance weighting (IDW) method was used to impute missing data in the observations of the 21 stations. This method is computationally easy to implement and showed the strongest performance when imputing meteorological data [3]. The data then underwent normalization so that each of the 6 features of observations fell within the range of [0, 1]. The sliding window technique is then used to form input-output samples from a given time series X. Let x_0 be the current time step, where w is the length of the input sequence window and H the length of the prediction horizon. The input-output pair of the sliding window technique would be $[x_0 - w, ..., x_0]$ and $[x_1, ..., x_1 + H]$ respectively.

3.3 Walk-Forward Validation

Walk-forward validation [6] is a technique where the data set is split into successive train, validation and test sets. This technique is appropriate for time series data as it accounts for the temporal nature of the data. For each split, the train data is used to train the model, the validation data is used for hyper-parameter optimisation, and finally, the test is used to evaluate the models built using the train and validation sets. An expanding walk-forward approach is used which expands the size of the training set with each walk, while the size of the validation and test sets remain the same.

The weather data consists of 8 years worth of surface observations. The first split consists of the 1st year's observations as the training set, the following 3 months as the validation set, and the 3 months after as the test set. The size of the validation and test set are kept the same, whereas the size of the training set expands with each successive split. Using the expanding window technique, this method results in a total of 27 splits over the data set as shown in Table 1. Using walk-forward validation, we were able to use 84.38% of the data set for testing. The predictions on each test set are concatenated into a list. The final metrics presented are calculated between the concatenated list of predictions and true labels.

Table 1. Table showing the walk-forward validation splits

	Surface weather observations
Number of data instances	70128
Chosen test set size	3 Months
Chosen total tests percentage	84.38%
Number of Splits	27
Validation set size	3 Months
Training set size	1 year + 3n months where n = $[0, ..., 26]$

3.4 Baseline TCN and LSTM Models

This section covers how the baseline LSTM and TCN models were built. A TCN and LSTM model were developed for each individual station. We attempted to follow the experiment pipeline used by Hewage et al. [4].

The data was first divided into 27 train, validation, and test splits following walk-forward validation. Hyper-parameter optimisation was then performed over the 24 h prediction horizon using the first 3 splits. For each of the 3 splits, the models were trained on the train set and then evaluated on the validation set. The hyper-parameters that resulted in the lowest validation loss across all 3 splits were selected as the optimal hyper-parameters. Our search space for the LSTM and TCN models both consisted of 8 hyper-parameters each. The search space along with the identified optimal parameters are shown in Table 2 below.

Since HPO was performed for each baseline at each weather station, some of the optimal values are sets consisting of the optimal hyper-parameters across all weather stations.

Table 2. Table showing hyper-parameters explored for the baseline models

Parameter	Search space	LSTM	TCN
# of Epochs	[20, 30, 40]	40	40
Batch Size	[32, 64, 125, 250]	[32, 64, 125]	64
Learning Rate	(0.0001, 0.01)	(0.0003, 0.003)	0.01
Input Window Size (hours)	[12, 24, 48, 72, 168]	[48, 72, 168]	[24, 48]
# of Layers	[1, 2, 3, 4]	1	[1, 2]
Dropout Rate	[0.1, 0.2, 0.3, 0.4]	[0.1, 0.2]	[0.1, ..., 0.5]
# of hidden units	[30, 40, 50, 60]	[30, 40, 50, 60]	–
L1, L2 Regularisers	[0.0001, 0.001, 0.01]	[0.001, 0.01]	–
Layer Normalisation	[True, False]	–	True
# of Filters	[32, 64, 128]	–	[32, 64, 128]

The identified optimal hyper-parameters identified on the 24 h prediction horizon were then used to train and evaluate each baseline model across all 27 splits for each of the 5 prediction horizons (3, 6, 9, 12, 24 h). For evaluation per prediction horizon, the predictions on the test set for each split are concatenated. The SMAPE metrics for each model per prediction horizon are then calculated on the concatenated list of predictions and true labels. Note that the SMAPE scores are not the mean over the 27 splits, but a single score calculated on the concatenated list of predictions and true labels.

3.5   ST-GNNs

The GWN and WGN experimental design closely followed the pipeline used for the LSTM and the TCN. A key difference is that only a single ST-GNN model is developed for all stations, since ST-GNNs are capable of predicting temperature across all weather stations simultaneously.

The WGN and GWN models had additional hyper-parameters relating to the self-adaptive adjacency matrices. The adjacency matrices could either be randomly initialised or initialised by the Euclidean distance between weather stations with a thresholded Gaussian kernel. The WGN ST-GNN had an additional hyper-parameter of rank. A higher rank reduces the number of connections in the adjacency matrix thereby reducing computation complexity at the cost of accuracy. The rank, r, parameter shown in Table 3 reduces the size of the adjacency matrix by reducing it to contain only s/r connections where s is the number of stations. The Graph WaveNet model could utilise just a learnt self-adaptive adjacency matrix that requires no prior knowledge for graph convolution. Alternatively, prior knowledge is given in the form of the thresholded Gaussian kernel values. The prior knowledge is used to derive the forwards and backwards transition matrix that is used in combination with the self-adaptive adjacency matrix in the graph convolution layers. The search space and optimal hyper-parameters for the ST-GNN models are shown below in Table 3.

Table 3. Table showing hyper-parameters of ST-GNN models

Parameter	Search space	WGN	GWN
# of Epochs	[20, 30, 40]	30	30
Batch Size	[32, 64, 125, 250]	32	64
Learning Rate	(0.0001, 0.01)	0.01	0.001
Input Window Size (hours)	[12, 24, 48, 72, 168]	24	12
# of hidden neurons	[10, 20, 30, 40]	20	30
Dropout Rate	[0.1, 0.2, 0.3, 0.4]	0.1	0.3
Adjacency Matrix Init.	[Random, Distance]	Distance	Random
Mask Length	[0, 12, 24, 48]	0	–
# of graph convolution layers	[1, 2, 3, 4]	2	–
Adjacency Matrix Rank (r)	[1, 1.5, 2]	1	–
# of ST layers	[1, 2, 3, 4]	–	1
Adjacency Matrix Type	Self-Adaptive (SA) Forwards-Backwards (FB) SA + FB	–	SA

The optimal hyper-parameters identified for the 24 h prediction horizon were then used to train and evaluate WGN and GWN models across all 27 splits for each of the 5 prediction horizons.

3.6   Implementation

The baseline LSTM and TCN models were implemented in the Keras framework. The WGN and GWN model were implemented using Tensorflow and PyTorch respectively. The models were all trained on an NVIDIA RTX3070 GPU. The models were trained using the Adam optimiser and Mean Squared Error loss function. We evaluate the models performance using both the symmetric mean absolute percentage error (SMAPE) metric and mean square error (MSE). However, we found SMAPE to be more useful than MSE. We calculate SMAPE using the following Eq. 1:

$$SMAPE = \frac{100\%}{n} \sum_{i=1}^{n} \frac{|Y_p - Y_r|}{(|Y_p| + |Y_r|)/2} \qquad (1)$$

where Y_r, Y_p, and n are the true labels, the model's predictions, and the number of samples respectively.

4   Results

This section presents the results of the four deep learning models across the five prediction horizons. We first present a summary performance of the MSE and SMAPE score for the four models across all weather stations and prediction horizons. We found that the TCN outperformed the LSTM. We then provide a deeper analysis of the TCN, WGN and GWN on each station.

4.1   Results Summary

Table 4 shows the average SMAPE and normalized MSE results for each deep learning model and the standard deviation of the results over 10 independent runs. The results are presented across all five prediction horizons. The LSTM model records the worst performance across all prediction horizons with an average accuracy of 16.25% over all prediction horizons. The LSTM model is the least stable model with high standard deviations between SMAPE metrics ranging between 2.52 to 4.12. The TCN model shows a significant accuracy improvement over the LSTM model with an average prediction accuracy of 10.62%. The TCN model performs well, with only a small percentage improvement in accuracy between itself and the GWN and WGN models. The standard deviations across the prediction horizons suggest that the model is also more stable than the LSTM model.

Table 4. Table comparing deep learning models' SMAPE and MSE scores across the prediction horizons

Model		Prediction Horizon					Average
		3	6	9	12	24	
LSTM	$SMAPE$	15.25±2.54	15.87±2.38	16.30±2.28	16.58±2.22	17.27±2.11	16.25%
	MSE	0.005	0.006	0.006	0.007	0.008	0.006
TCN	$SMAPE$	8.41±1.15	9.77±1.22	10.69±1.26	11.35±1.30	12.90±1.39	10.62%
	MSE	0.002	0.002	0.003	0.003	0.003	0.003
WGN	$SMAPE$	6.74±1.63	8.75±2.32	9.99±2.54	9.92±2.48	9.71±2.32	9.02%
	MSE	0.002	0.003	0.004	0.004	0.005	0.004
GWN	$SMAPE$	6.39±1.56	7.82±2.00	8.65±2.20	9.05±2.27	9.60±2.35	8.30%
	MSE	0.002	0.002	0.003	0.003	0.004	0.003

The WGN model outperforms both baseline models, but the performance improvement over the TCN model is not substantial. The WGN model is also slightly outperformed by the GWN model with a performance improvement of 0.72%. The WGN model proves to be as stable as the TCN and GWN model which is surprising considering that the LSTM forms the basis of WGN's architecture. This stability can be attributed to the model reliably capturing spatial dependencies and therefore making consistent, reliable predictions. From the table above it is clear that the GWN model produces the most accurate results on average across all five prediction horizons. The results' standard deviations across the prediction horizons indicate the Graph WaveNet model is the most robust, stable model of four deep learning models implemented in this paper.

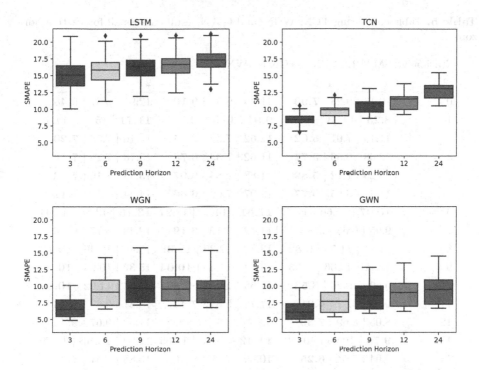

Fig. 2. Box plots showing SMAPE scores of deep learning models over all prediction horizons

Finally, Fig. 2 above shows the distribution of deep learning models' performance across all prediction horizons. The LSTM has the worst performance across the deep learning models, with outliers across 4 of the 5 prediction horizons indicating that is it unstable. The TCN has the smallest distribution of scores of all the deep learning models. While it is the most stable model, its average performance is significantly lower than the ST-GNN models across all prediction horizons. The GWN and WGN clearly outperform both the LSTM and the TCN with the lowest SMAPE scores across all prediction horizons. The addition of spatial information is shown to provide a clear increase in overall prediction performance.

4.2   Performance at Different Weather Stations

We now present the mean SMAPE scores for each station over 10 runs across the 6, 12, and 24 h prediction horizons. We focus on the longer prediction horizons to see the accuracy of these models because predicting further into the future reduces both accuracy and stability.

Table 5. Table comparing TCN, WGN, and GWN results across all forecasting horizons

Stations	SMAPE (TCN | WGN | GWN)		
	6	12	24
0	10.18 | 9.23 | **7.99**	11.49 | 9.51 | **9.10**	12.91 | 9.37 | **9.48**
1	**8.23** | 14.33 | 11.57	**9.91** | 15.57 | 13.51	**11.71** | 15.41 | 14.53
2	11.15 | 7.01 | **6.02**	12.62 | 7.62 | **6.75**	14.16 | 7.50 | **7.20**
3	10.08 | 6.56 | **5.77**	11.62 | 7.33 | **6.78**	13.13 | 7.26 | **7.16**
4	10.25 | 7.22 | **5.89**	11.88 | 7.84 | **6.97**	13.43 | 8.16 | **7.72**
5	11.41 | 6.65 | **5.77**	12.97 | 7.09 | **6.66**	14.44 | 7.19 | **7.11**
6	**10.07** | 13.60 | 11.77	**11.61** | 14.88 | 13.24	**13.16** | 13.88 | 13.94
7	9.95 | 6.65 | **5.38**	11.22 | 7.10 | **6.19**	12.48 | 6.77 | **6.66**
8	11.75 | 11.53 | **9.82**	13.72 | 12.39 | **11.36**	15.49 | 12.46 | **12.12**
9	12.11 | 11.06 | **8.93**	13.83 | 11.48 | **10.04**	15.32 | 10.63 | **10.40**
10	9.69 | 10.26 | **8.53**	11.58 | 11.32 | **9.85**	13.37 | 11.02 | **10.88**
11	9.95 | 6.77 | **6.21**	11.78 | 9.57 | **7.49**	13.81 | 9.67 | **8.16**
12	8.05 | 8.69 | **7.39**	9.86 | 9.57 | **8.80**	11.56 | 9.67 | **9.45**
13	**9.79** | 10.95 | 10.08	**11.42** | 11.43 | 11.65	12.90 | **10.85** | 11.79
14	9.04 | 7.64 | **6.25**	10.00 | 7.89 | **6.99**	10.88 | 7.41 | **7.29**
15	9.93 | 10.37 | **9.04**	11.56 | 11.11 | **10.41**	13.11 | **10.64** | 10.94
16	8.33 | 7.14 | **5.79**	9.63 | 7.60 | **6.75**	10.96 | 7.55 | **7.27**
17	7.88 | 7.91 | **6.08**	9.22 | 8.18 | **7.08**	10.51 | 8.02 | **7.54**
18	10.00 | 11.39 | **9.78**	11.89 | 11.95 | **11.33**	13.56 | **11.37** | 11.73
19	9.48 | 9.11 | **7.71**	11.15 | 9.78 | **9.03**	12.92 | 9.93 | **9.67**
20	**7.88** | 9.71 | 8.53	**9.48** | 10.99 | 10.02	11.03 | 10.30 | **10.52**
Total Top	3 | 0 | **18**	4 | 0 | **17**	2 | 3 | **16**

From Table 5 above, we can see that GWN has the most accurate results across the majority of stations across the 6, 12, and 24 h prediction horizons, with WGN producing the best models at only 3 stations over the 24 h prediction horizon. It is especially interesting that the TCN produced the best models at some further inland stations particularly stations 1 and 6 where it produced the best models across all horizons. The difference in performance between the TCN and GWN over the 24 h horizon for station 1 is substantial, compared to station 6 where it is marginal. One explanation for this is that station 1 is on the north east boundary and there may be other stations outside the study area that may influence it. It may be that the ST-GNNs attempt to force a spatial dependency with stations within the study area when there may be none. This phenomena certainly warrants further exploration.

4.3   Spatial-Temporal Dependency Analysis

The self-adaptive adjacency matrices represent the spatial-temporal dependencies captured by each ST-GNN. The self-adaptive adjacency matrices were extracted from each ST-GNN after training on the final split containing all the training data. The matrices are then normalised. We attempted to follow the method used by Wilson et al. [8] to visualise the spatial-temporal dependencies. We visualized the spatial-temporal dependencies between weather stations by plotting the most important weights within the learnt adjacency matrix. This was done by taking the absolute value of the learned adjacency matrix and finding the largest edge weight in each row of the adjacency matrix. The edge corresponding to each row's largest weight is then plotted along with its direction [8]. Finally, due to the high number of self-loops, when a station's largest edge weight was itself, we included the 2nd largest weight which enabled us to visualise paths within the network.

4.4   Spatial-Temporal Dependencies

First we present the results of the hyper-parameter optimisation that relate to the self-adaptive adjacency matrices. We found that the WGN model performed better when initialised using the thresholded Gaussian kernel values on the Euclidean distance. We also found that a rank of 0 i.e. no sparsity, resulted in the best performance. We found that the Graph WaveNet model required no prior knowledge and was able to utilise only the learnt self-adaptive adjacency matrix for graph convolution to produce the best results.

We visualise the spatio-temporal dependencies overlaid with the geographical map of South Africa. We explored different layout techniques for the network but none resulted in a visualisation where you could easily see the paths in the network. We therefore chose to use the geographical overlay method. The WGN and GWN spatio-temporal dependencies are shown below in Fig. 3 respectively.

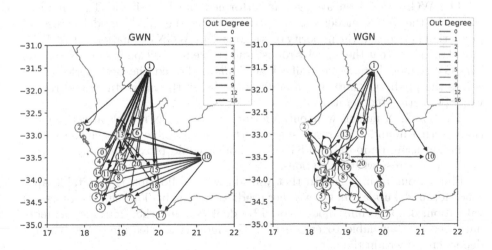

Fig. 3. GWN and WGN spatial-temporal dependency visualisations

The colour of the nodes outline indicates the out degree of the node which relates to the number of nodes that are influenced by the node in question. From the WGN visualisation above, we can see that station one has the highest out degree of 9, with many outgoing arrows to other stations. This is especially interesting considering that station 1 is geographically distant from the stations it is influencing. We can then see stations 0 and 17 as having the 2nd and 3rd highest out degree of the weather stations. Finally, the cluster in the south-east area of the map contains many different paths and nodes with varying out degrees.

The GWN ST dependencies are particularly interesting. We see striking trends of station 1 and 10, the northern and eastern most stations, influencing the most stations with out degrees of 16 and 9 respectively. The WGN and GWN model both learnt that station 1 is highly influential. Furthermore, stations 6 and 13 have the next highest out degrees of 9 and 6 respectively. It is also interesting to note that the remaining stations in the south-west of the map all have an out degree of 0.

5   Discussion and Conclusions

This study evaluated temporal and ST-GNN deep learning architectures for weather prediction. We addressed the prior limitations of evaluating stability and robustness by introducing walk-forward validation. Furthermore, we evaluated Graph WaveNet, a state-of-the-art ST-GNN from the traffic flow domain, for weather prediction.

Our results show that the LSTM model used for weather prediction in the South African region is highly inaccurate, unstable and not robust for weather prediction. Comparatively, the temporal TCN model showed surprising results, outperforming the ST-GNN models at some weather stations over certain prediction horizons. Additionally, the TCN model was shown to be highly stable and robust through walk-forward validation.

The WGN model on average outperformed the LSTM and TCN models. However, the WGN model was outperformed by the GWN model across all prediction horizons at the majority of stations. The WGN evaluation of the WGN model did show that the model is robust and stable in addition to being accurate. The GWN model from the traffic flow domain showed the best performance across the prediction horizons in addition to being the most stable and robust when predicting temperature.

The ST-GNN models were outperformed by the TCN models at certain stations further inland. This result suggests that an ensemble approach to weather prediction using temporal and ST-GNN models could provide reliably accurate predictions across weather stations.

Finally, our results showed that walk-forward validation is a suitable approach for model update in the weather prediction domain. Through walk-forward validation, deep learning models could be deployed and then retrained as more data becomes available to ensure that the models are learning the dynamic dependencies within the data.

6   Limitations and Future Work

The results of this work came with certain limitations and also present opportunities for further investigation. Hyper-parameter optimisation (HPO) was only implemented on the first 3 splits of the full data set. Future work could run HPO over all splits to produce better results. Our models were not compared to the weather prediction models currently in practise, this limits our confidence in the accuracy of our results. Future work could compare ST-GNNs with current approaches for weather prediction. Finally, the window size and its effect on performance of weather prediction models and the ST dependency networks should be investigated further.

References

1. Cao, D., et al.: Spectral temporal graph neural network for multivariate time-series forecasting (2021)
2. Cifuentes Quintero, J., Marulanda, G., Bello, A., Reneses, J.: Air temperature forecasting using machine learning techniques: a review. Energies **13**, 4215 (2020). https://doi.org/10.3390/en13164215
3. Dhevi, A.: Imputing missing values using inverse distance weighted interpolation for time series data, pp. 255–259 (2014). https://doi.org/10.1109/ICoAC.2014.7229721
4. Hewage, P., Trovati, M., Pereira, E., et al.: Deep learning-based effective fine-grained weather forecasting model (2020)
5. Iseh, A.J., Woma, T.Y.: Weather forecasting models, methods and applications. Int. J. Eng. Res. Technol. **2** (2013)
6. Kouassi, K.H., Moodley, D.: An analysis of deep neural networks for predicting trends in time series data. In: SACAIR 2021. CCIS, vol. 1342, pp. 119–140. Springer, Cham (2020). https://doi.org/10.1007/978-3-030-66151-9_8
7. Pillay, K., Moodley, D.: Exploring graph neural networks for stock market prediction on the JSE. In: Jembere, E., Gerber, A.J., Viriri, S., Pillay, A. (eds.) SACAIR 2021. CCIS, vol. 1551, pp. 95–110. Springer, Cham (2022). https://doi.org/10.1007/978-3-030-95070-5_7
8. Wilson, T., Tan, P.N., Luo, L.: A low rank weighted graph convolutional approach to weather prediction (2018). https://doi.org/10.1109/ICDM.2018.00078
9. Wu, Z., Pan, S., Long, G., Jiang, J., Chang, X., Zhang, C.: Connecting the dots: multivariate time series forecasting with graph neural networks. In: Proceedings of the 26th ACM SIGKDD International Conference on Knowledge Discovery & Data Mining, pp. 753–763 (2020)
10. Wu, Z., Pan, S., Long, G., Jiang, J., Zhang, C.: Graph WaveNet for deep spatial-temporal graph modeling (2019). https://doi.org/10.24963/ijcai.2019/264
11. Yu, B., Yin, H., Zhu, Z.: Spatio-temporal graph convolutional networks: a deep learning framework for traffic forecasting. In: Proceedings of the Twenty-Seventh International Joint Conference on Artificial Intelligence (2018). http://dx.doi.org/10.24963/ijcai.2018/505
12. Zahroh, S., Hidayat, Y., Pontoh, R., Santoso, A., Bon, A., Firman, S.: Modeling and forecasting daily temperature in Bandung (2019)

Multi-modal Recommendation System with Auxiliary Information

Mufhumudzi Muthivhi[1]([✉]) [iD], Terence van Zyl[2] [iD], and Hairong Wang[1] [iD]

[1] University of the Witwatersrand, Johannesburg 2000, GT, South Africa
1599695@students.wits.ac.za
[2] University of Johannesburg, Johannesburg 2092, GT, South Africa
tvanzyl@uj.ac.za

Abstract. Context-aware recommendation systems improve upon classical recommender systems by including, in the modelling, a user's behaviour. Research into context-aware recommendation systems has previously only considered the sequential ordering of items as contextual information. However, there is a wealth of unexploited additional multi-modal information available in auxiliary knowledge related to items. This study extends the existing research by evaluating a multi-modal recommendation system that exploits the inclusion of comprehensive auxiliary knowledge related to an item. The empirical results explore extracting vector representations (embeddings) from unstructured and structured data using data2vec. The fused embeddings are then used to train several state-of-the-art transformer architectures for sequential user-item representations. The analysis of the experimental results shows a statistically significant improvement in prediction accuracy, which confirms the effectiveness of including auxiliary information in a context-aware recommendation system. We report a 4% and 11% increase in the NDCG score for long and short user sequence datasets, respectively.

Keywords: Recommendation systems · Multi modal · Auxiliary information · Context aware · Transformer · Data2vec

1 Introduction

Web 2.0 gave rise to an information overload. In response, the industry has introduced Recommendation Systems to suggest a relevant item to users based on their preferences [6]. Recommending the most relevant item to a user requires an understanding of item characteristics and user preferences [2]. Recently, researchers have proposed sequential recommendation to predict the next item that a user would prefer. Sequential recommendation models the historical consumption of items made by a user. Some state-of-the-art recommendation models only use item identifiers to model user behaviour [12,14,25]. Item identifiers are unique numbers assigned to each item. Furthermore, by exploiting the various forms of auxiliary information, we can enhance the modelling of user behaviour [30].

A. Pillay et al. (Eds.): SACAIR 2022, CCIS 1734, pp. 108–122, 2022.
https://doi.org/10.1007/978-3-031-22321-1_8

One area of research has explored the use of auxiliary information through tabular data. They consider the keyword descriptions of items. Keyword descriptions are attribute data in tabular form usually transformed into one-hot encoded embeddings [3,10,28]. However, their work neglects the use of unstructured data for the sequential recommendation task. This study explores an enhanced user-item representation learning framework. We use multi-modal auxiliary information whilst evaluating state-of-the-art transformer architectures that exploit the sequential dependencies between items.

The results demonstrate that transformer architectures with multi-modal auxiliary information improved the prediction results. Furthermore, SASRec, a unidirectional transformer architecture, displayed improvement over the bidirectional implementations. Finally, our analysis shows that the presented model considers each user's context and can make recommendations of items across the different categories.

We contribute to the existing literature by:

- using multi-modal auxiliary information, including both text and image data to enhance the sequential recommendation task;
- using an embedding of continuous values instead of a multi-hot encoding of categorical data.
- using XGBoost to extract embeddings from tabular data
- showing that a unidirectional transformer, SASRec, yields the greatest improvement over the bidirectional transformers, BERT4Rec; and
- conducting an ablation study to analyse the effects of the various forms of modalities and their combinations.

2   Background and Related Work

Recommendation datasets consist of user feedback data. User feedback data provides information on whether a user has consumed an item or not. It may come in the form of a rating, purchase or webpage view of an item by a user. Users rate a small percentage of the item collection. Researchers refer to the existence of finite ratings between user-item pairs as the *data sparsity* problem [21]. As a result, recommendation models have to predict the next item a user will consume from limited historical data.

Early work employed Matrix Factorization (MF) methods to find the latent factor space that encodes the user-item interaction matrix [17]. Koren *et al.* found that a combination of implicit and explicit feedback alleviates the data sparsity problem and increases prediction accuracy [16]. It became increasingly evident that understanding item characteristics and user preferences would aid in recommending the most relevant item to a user [2]. Koren's findings [16] motivated researchers to expand their experiments towards auxiliary information in the form of keyword descriptions.

Subsequently, research in recommendation systems focused on finding a recommendation dataset's latent factor space. Hence, recommendation systems have predominately become a representation learning problem. The representation learning task led to using neural networks, which learn non-linear patterns from user-item data [30]. Simultaneously, within computer vision and natural language processing (NLP), researchers found that effective data representations significantly improve existing models [8,13,15]. Inspired by the developments in NLP, computer vision and other fields, the authors believe that the relevance of an item recommended to a user depends on the quality of an item's representation [9,18,26,27]. By exploiting comprehensive knowledge from item and user auxiliary information, we can design a rich contextualized embedding space of users, items, and their relationships. Auxiliary knowledge exists in a variety of formats. Item descriptions and user profile information may be tabular or text/image/audio format (unstructured). In addition, we may also have the time at which an item is consumed, known as sequential data. The combination of these various forms of data allows us to model user behaviour effectively [30].

Researchers have employed auxiliary information in recommendation systems to understand user behaviour. They model the sequential patterns within a user's consumption history and make predictions of the following preferred item. Recently, several novel transformer architectures have emerged for the sequential recommendation task. Kang *et el.* proposed a two-layer Transformer decoder that models users' sequential behaviour, known as SASRec [14]. However, Sun *et el.* argue that unidirectional models, like SASRec, do not sufficiently learn optimal representations for user behaviour sequences [8]. Inspired by the success of BERT [8], the authors proposed a bidirectional model called BERT4Rec [25]. Then, Fischer *et al.* improved on BERT4Rec by integrating auxiliary information in the form of keyword embedding vectors [10].

The current trend in research is to append keyword descriptions of items to item identifiers [10,31]. Keyword descriptions are attribute data in tabular form. In most cases, the data is categorical and transformed into a one-hot encoded embedding. This paper argues that comprehensive knowledge about users and items exists in different modalities. Extracting this knowledge should result in better embeddings of users and items. We aim to reflect on the significance of auxiliary knowledge by exploiting various forms of data, including tabular, unstructured and sequential data.

We contribute to the field by exploring the design of an enhanced modelling process of user behaviour. We explore extracting embeddings from tabular data using gradient boosting and from unstructured data using data2vec. We fuse both structured and unstructured embeddings and learn the sequential behaviour of a user's preference through a unidirectional and bidirectional transformer architecture, SASRec and BERT4Rec respectively [14,25].

3   Experimental Methodology

3.1   Problem Statement

The problem addressed in this paper is formulated as follows. Given a set of users $\mathcal{U} = \{u_1, u_2, \dots, u_{|\mathcal{U}|}\}$, a set of items $\mathcal{V} = \{v_1, v_2, \dots, v_{|\mathcal{V}|}\}$, each item's

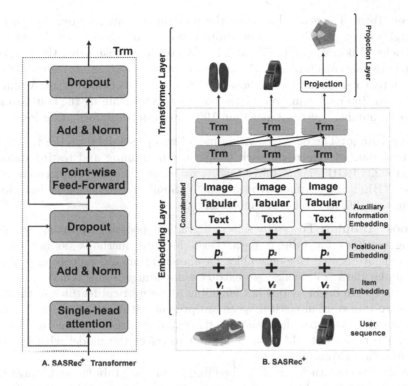

Fig. 1. The SASRec$^+$ architecture with the inclusion of multi-modal auxiliary information embedding

auxiliary information embedding $\mathcal{K} = \{\mathbf{k}_1, \mathbf{k}_2, \ldots, \mathbf{k}_{|\mathcal{V}|}\}$ and the sequential item data $\mathcal{S}_u = \{v_1^u, v_2^u, \ldots, v_{n_u}^u\}$, predict the next item $v_{n_u+1}^u$ that is most preferred by user u, where n_u is the number of items rated by user u, v_t^u ($1 \leq t \leq n_u$) represents the relative time t when user u interacted with item v, and $\mathbf{k}_v$ is a joint vector embedding of text, tabular and image data for item v.

3.2 Multi-modal Auxiliary Information

Auxiliary information may exist as tabular or unstructured data. Some research is focused on tabular data, usually categorical data [3,10,28]. Practitioners transform the categorical data by one-hot encoding the embeddings. In contrast, this study uses vector representations transformed into embeddings from tabular and unstructured data.

The study uses data2vec to extract embeddings from unstructured data [4]. Baevski *et al.* proposed a novel framework that uses the same learning method for speech, natural language and images by predicting latent representations of the input data. Mainly, data2vec is a general self-supervised learning method that predicts the latent representations of the full input data using a standard transformer architecture.

dat2vec Text: The study builds on the pre-trained data2vec text model from HuggingFace [1]. Data2vec was pre-trained on five datasets consisting of texts from books, Wikipedia and news articles. We begin by tokenizing the text data using a byte-pair encoding [22]. Then we pass the array of tokens of size 196 to the data2vec text model and use the last hidden state as its embeddings. A number greater than 196 results in overly large parameters for training the transformers. Whereas a number much smaller than 196 may result in information loss.

dat2vec Image: The data2vec vision model was pre-trained on the ImageNet-1k fashion dataset. We resize each image in our training and testing datasets into 224×224 RGB images. Furthermore, we follow the same feature extraction process as BEiT and normalize the images [5]. Similarly, we retain the last hidden state, of size 196, as the image embedding.

XGBoost Tabular: The study selects a gradient-boosting approach over deep learning methods for tabular data. Deep learning methods do not outperform gradient-boosted tree ensembles for classification and regression problems with tabular data [23]. Hence, we extract embeddings from tabular data using XGBoost [7]. We begin by one-hot encoding all categorical features in the table and ensure that only numerical types are present within the dataset. We train XGBoost on 70% of the data and set the training target variable as the user ratings. We perform 4-fold cross-validation to select the model with the least Mean Absolute Error.

Finally, we construct a joint embedding $\|_v$ of text, tabular and image data for item v. The embedding is either the concatenations or the summation of the text, tabular and image representations. The paper uses the concatenation approach to compare our proposed approach to baselines.

3.3   Embedding Layer

To integrate auxiliary information consisting of both tabular and unstructured data into a transformer architecture, we take an embedding approach inspired by KeBert4Rec [10]. The embedding layer for KeBert4Rec consists of three different learned embeddings: (i) an item identifier embedding $E_\mathcal{V} \in \mathbb{R}^{|\mathcal{V}| \times d}$, (ii) a positional embedding $E_\mathcal{P} \in \mathbb{R}^{|N| \times d}$ and (iii) an auxiliary information embedding $\mathcal{K}$. The positional embedding encodes the positions of the items within a user's sequential item data $\mathcal{S}_u$. N is the configurable maximum input sequence length. The auxiliary information embedding encodes categorical data into a multi-hot encoded vector. The vector is scaled to the embedding size d using a linear layer. For each item in the sequence $\mathcal{S}_u$, we have the item embedding $e_{v_t} = E_\mathcal{V} v_t$, the positional embedding $p_{v_t} = E_\mathcal{P} t$ and auxiliary information embedding $\mathbf{k}_{v_t}$. Where t is the time at which the user consumed the item. The sum $h_t^0 = e_{v_t} + p_{v_t} + k_{v_t}$ is used as the input into the transformer.

The critical difference in our implementation, compared to KeBert4Rec, is that we do not use a multi-hot encoding of categorical data. Instead, we make use of all multi-modal auxiliary information available and transform it into an embedding of continuous values. The study investigates whether concatenated or summation (point-wise addition of vector items) produces the best embeddings.

Finally, we sum the item, positional and auxiliary information embeddings and pass it as input to the transformer layer. Figure 1B depicts the embedding layer for the unidirectional model SASRec. We use the same embedding layer for the bidirectional model BERT4Rec.

3.4 Transformers

The study adopts uni- and bi-directional transformer architecture. The state-of-the-art uni- and bi-directional transformers for sequential recommendation are SASRec and BERT4Rec, respectively [14,25]. Given an input sequence, both transformers iteratively compute hidden representations h_i^l at each layer l for each position i, simultaneously. The scaled dot-product attention is [29]:

$$\text{Attention}(\mathbf{Q}, \mathbf{K}, \mathbf{V}) = \text{softmax}\left(\frac{\mathbf{Q}\mathbf{K}^T}{M}\mathbf{V}\right) \tag{1}$$

where $\mathbf{Q}$ represents the queries, $\mathbf{K}$ the keys and $\mathbf{V}$ the values (each row represents an item). M is a scale factor used to avoid overly large inner product values.

Self-attention. SASRec uses a single-head self-attention mechanism, whereas BERT4Rec uses a multi-head self-attention mechanism. SASRec attends to information from left to right, whereas BERT4Rec attends to information from different representation subspaces at different positions. Take the input embedding $\mathbf{H}^l$; we convert it into three matrices through linear projections, for SASRec:

$$\text{SA}(\mathbf{H}^l) = \text{Attention}(\mathbf{H}^l\mathbf{W}^Q, \mathbf{H}^l\mathbf{W}^K, \mathbf{H}^l\mathbf{W}^V) \tag{2}$$

and h subspaces for BERT4Rec

$$\text{MH}(\mathbf{H}^l) = [\text{head}_1, \text{head}_2, \ldots, \text{head}_h]\mathbf{W}^O \tag{3}$$

$$\text{head}_i(\mathbf{H}^l) = \text{Attention}(\mathbf{H}^l\mathbf{W}_i^Q, \mathbf{H}^l\mathbf{W}_i^K, \mathbf{H}^l\mathbf{W}_i^V) \tag{4}$$

where $\mathbf{W}_i^Q \in \mathbb{R}^{d\times d/h}$, $\mathbf{W}_i^K \in \mathbb{R}^{d\times d/h}$ and $\mathbf{W}_i^V \in \mathbb{R}^{d\times d/h}$ are the projection matrices with learnable parameters. $h = 1$ for SASRec.

Feed-Forward Network. To endow the model with non-linearity and consider interactions between different dimensions, SASRec adopts a point-wise Feed-Forward Network

$$\text{FFN}_S(\mathbf{h}_i^l) = \text{ReLU}(\mathbf{h}_i^l\mathbf{W}^{(1)} + \mathbf{B}^{(1)})\mathbf{W}^{(2)} + \mathbf{B}^{(2)} \tag{5}$$

and Bert4Rec adopts a position-wise Feed-Forward Network

$$\text{PFFN}_B(\mathbf{h}_i^l) = [\text{FFN}_B(\mathbf{h}_1^l)^T, \text{FFN}_B(\mathbf{h}_2^l)^T, \ldots, \text{FFN}_B(\mathbf{h}_t^l)^T] \tag{6}$$

$$\text{FFN}_B(\mathbf{h}_i^l) = \text{GELU}(\mathbf{h}_i^l \mathbf{W}^{(1)} + \mathbf{B}^{(1)})\mathbf{W}^{(2)} + \mathbf{B}^{(2)} \qquad (7)$$

with a Gaussian Error Linear Unit (GELU) activation. $\mathbf{W}^{(1)}, \mathbf{W}^{(2)} \in \mathbb{R}^{d \times d}$ and $\mathbf{d}^{(1)}, \mathbf{d}^{(2)} \in \mathbb{R}^d$ are learnable parameters.

Figure 1 depicts the structure of the proposed model SASRec$^+$, which is a slight modification of SASRec for multi-model data.

Proposed Method: The study adopts a similar design to the KeBERT4Rec transformer architecture [10]. KeBERT4Rec integrates auxiliary information into the bidirectional transformer BERT4Rec [25], along with the sequential item data. Denote the sequential item data for user u as $\mathcal{S}_u = \{v_1^u, v_2^u, ..., v_{n_u}^u\}$, where n_u is the number of items rated by user u and v_t^u ($1 \leq t \leq n_u$) represents the relative time t when user u interacted with item v. We consider for each item $v \in \mathcal{V}$ an embedding of structured and unstructured data $\mathbf{k}_v$. Given the history $\mathcal{S}_u$ of user u and the auxiliary information $\mathcal{K}_u = \{\mathbf{k}_{v_t^u}, 0 \leq t \leq n_u\}$ that corresponds to $\mathcal{S}_u$, the transformer must predict $v_{n_u+1}^u$ from the sequence of user's interaction and its corresponding auxiliary information. It can be formalized as modelling the probability over all possible items for user u at relative timestamp $n_u + 1$:

$$p(v_{n_u+1}^u) = (v|\mathcal{S}_u, \mathcal{K}_u) \qquad (8)$$

We refer to the uni- and bi-directional model that includes multi-model data as SASRec$^+$ and BERT4Rec$^+$.

3.5   Datasets

The study explores the Amazon Fashion dataset and ML-20M. The Amazon Fashion dataset contains a moderate amount of interactions with a relatively sparse interaction matrix [19]. ML-20M is a benchmark dataset widely used in the literature, and it has more users than items [11]. However, the interactions in ML-20M are usually more evenly distributed across users than items. Each dataset consists of tabular, text and image data and the time a user consumed an item.

3.6   Baselines

The study uses the following three groups of recommendation baselines.

1. General recommendation methods that only consider user feedback without modelling sequential behaviour. The methods considered in this group consist of the following:
   - **PopRec** which ranks items according to their popularity based on the number of interactions [24];
   - **Bayesian Personalized Ranking (BPR)** optimizes matrix factorisation with implicit feedback using pairwise Bayesian Personalized Ranking [20].

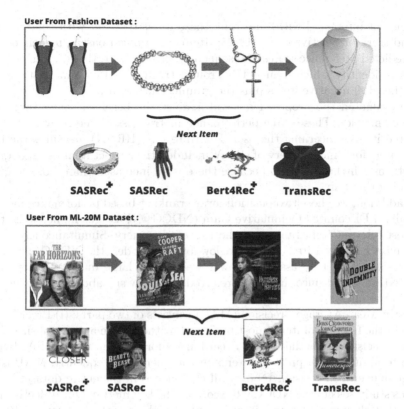

Fig. 2. A sample of two users' historical consumption of items and the predicted next item by each model

2. Sequential recommendation methods based on first-order Markov chains. **Translation-Based Recommendation (TransRec)** is a state-of-the-art approach that models each user as a translation vector to capture item transitions [12].
3. Selected state-of-the-art deep learning-based sequential recommendation systems, namely, **SASRec**, **BERT4Rec** and **KeBERT4Rec**.

Figure 2 gives an example of a user from each dataset, and their sequential consumption of items. The diagram also depicts the predicted next item by each sequential recommendation model.

3.7 Evaluation

Following previous research, we applied the *leave-one-out* strategy to recommend the next item [14,25,31]. For each user interaction sequence, we used the last item as the test data and the item prior to it as the validation data. The remaining items are used for training. The entire item set is usually too large to rank all candidate items. Hence, we randomly selected 100 negative items (the user has not interacted with) and rank these items together with the test item (ground truth or positive item) [14,25]. Items are ranked based on their respective predicted preference scores. Subsequently, given a user, the recommendation task is to identify which item is the ground truth next item amongst the 101 items (i.e., 100 negative items plus the ground truth item).

We evaluated the proposed method against the baselines using three performance metrics. These three performance metrics assess the relevancy of the predicted items concerning the user. Hit Rate @N (HR@N) measures the fraction of users for which a query of size N included the correct item. A large query size returns a higher hit ratio because there is an increased chance of the query containing the item.

In addition, we chose two position-aware ranking-based performance metrics: Normalized Discounted Cumulative Gain (NDCG@N) assigns higher weights to the most relevant item ranked at the top of the query. Simultaneously, if the user's most preferred item is ranked low by the model, the NDCG algorithm penalises the model by assigning lower weights. A large query size returns a higher NDCG score since the most preferred item now sits above more irrelevant items in the query.

Lastly, Mean Average Precision (MAP) consists of two parts; (i) Precision@N measures the fraction of items most preferred by the user in a query of size N (ii) average Precision@N is the sum of Precision@N for different values of N divided by the total number of preferred items in the query. Consequently, MAP is the average of average Precision@N over all the queries in the entire dataset.

This study selects the NDCG@10 score as the key metric. Our objective is to ensure that the most preferred item is ranked highest within a query of 10 randomly selected items. NDCG@10 is an accuracy-based metric that emphasises the correct positioning of items in a query. Hence, a model that achieves the highest NDCG@10 score better conceptualises the user's behaviour and preferences.

4   Results

Table 1 presents the performance of all models on the two datasets. BERT4Rec[+] and SASRec[+] are our proposed implementations. The + refers to the inclusion of multi-modal auxiliary information. At first glance, SASRec[+] scores the highest against each baseline across all metrics. With SASRec achieving the second-highest performance. The results show that a unidirectional transformer architecture is far superior in modelling context data. In addition, we observe that BERT4Rec only experiences marginal gains from including multi-modal data. The NDCG@10 score only improves by 0.04% and 0.09% for the ML-20M and Fashion dataset, respectively. As expected, non-context-aware models, PopRec and BPR, perform worse than the other models across all metrics. Proving that information about a user's sequential consumption of items is essential in understanding a user's preference.

Table 1. Performance of each recommendation model. Top two models in grey. Best model in bold[a].

Dataset Metrics	Pop-Rec	BPR	Trans-Rec	BERT-4Rec	Ke-BERT-4Rec	SAS-Rec	BERT-4Rec+	SAS-Rec+	Improve[b]
ML-20M									
HR@1	.014	.202	.217	.227	.229	.429	.229	**.431**	0.4 %
HR@5	.054	.489	.563	.569	.568	.802	.570	**.818**	2.0 %
HR@10	.081	.630	.755	.740	.746	.904	.746	**.920**	1.7 %
NDCG@5	.034	.350	.394	.403	.404	.629	.405	**.638**	1.4 %
NDCG@10	.043	.396	.457	.445	.461	.679	.462	**.706**	4.0 %
MAP	.046	.338	.379	.387	.387	.387	.388	**.597**	53.9 %
Fashion									
HR@1	.001	.049	.057	.049	.069	.437	.075	**.525**	16.9 %
HR@5	.001	.208	.214	.199	.244	.760	.244	**.883**	16.3 %
HR@10	.020	.229	.266	.284	.266	.870	.267	**.955**	9.7 %
NDCG@5	.001	.130	.133	.122	.162	.608	.166	**.720**	18.4 %
NDCG@10	.007	.136	.151	.161	.169	.643	.176	**.713**	11.0 %
MAP	.015	.114	.126	.120	.149	.580	.154	**.677**	16.8 %

[a] **BERT4Rec+** and **SASRec+** are the proposed methods.
[b] percentage improvement achieved by **SASRec+** against the best baseline.

Table 1 shows that the highest performance gains were achieved on the MAP score by SASRec+. Precisely, our model is 54% and 17% more capable of uncovering the most preferred items above a model without auxiliary information, for the ML-20M and Fashion dataset, respectively. Therefore, multi-modal auxiliary

Table 2. Paired t-test of NDCG@10 scores with and without multi-modal auxiliary information

Dataset		BERT4Rec	BERT4Rec+	SASRec	SASRec+
ML-20M	Mean	.4450	.4615	.6787	.7060
	Standard deviation	.0009	.0004	.0187	.0026
	Observations		10		10
	degrees of freedom		9		9
	t statistic		64.2321		4.0780
	p-value		2.7e-13		.0028
Fashion	Mean	.1605	.1757	.6428	.7134
	Standard deviation	.0128	.0089	.0117	.0326
	Observations		10		10
	degrees of freedom		9		9
	t statistic		3.7576		5.5814
	p-value		.0045		.0003

information adds further context that aids in understanding the item's characteristics. Notably, all three BERT4Rec implementations perform only slightly better than TransRec and substantially below all two SASRec variants. Presumably, the bidirectional structure of BERT4Rec is trained on both past and future items consumed by a user. Whereas a unidirectional model like SASRec is only trained on the past values consumed by a user. The poor performance of the bidirectional transformer proves that a user does not consume an item based on future consumption patterns. However, for each uni- and bi-directional model, including auxiliary information, adds to the performance of each model. Table 2 reports on the statistical significance of adding auxiliary information into the two transformer architectures. We observe a p-value less than 0.01 for BERT4Rec$^+$ and SASRec$^+$. Hence, multi-modal auxiliary information provides a better context of a user and item's relationship.

4.1   Ablation Study

Table 3. Ablation analysis (NDCG@10) on two datasets

| Dataset | ML-20M | | Fashion | |
Transformers	BERT4Rec$^+$	SASRec$^+$	BERT4Rec$^+$	SASRec$^+$
(1) Text	.46181	.70260	.18384	**.72807**
(2) Image	.46219	.68590	**.20737**	.58797
(3) Tabular	.46182	.68815	.19242	.59929
Concatenated				
(4) Text + Image + Tabular	.46146	.70600	.17567	.71342
(5) w/o Tabular	.46203	.70676	.18828	.69398
(6) w/o Image	.46198	.70837	.17734	.72643
(7) w/o Text	.46182	**.71004**	.15517	.62355
Summation				
(8) Text + Image + Tabular	.46218	.67628	.19266	.43453
(9) w/o Tabular	.46211	.68225	.19236	.53307
(10) w/o Image	.46198	.65881	.19469	.65882
(11) w/o Text	**.46226**	.68386	.17060	.41108

Top two best performing variants along the column are in grey. Best model in Bold

We performed an ablation study of the impact of each modality on both the uni- and bi-directional transformer. We recorded the results of a combined embedding of modalities and investigated whether a concatenated or summation approach works best. Table 3 reports the NDCG@10 scores achieved for each variant over SASRec$^+$ and BERT4Rec$^+$. The essential characteristic of the ML-20M dataset is that it is a large dataset with long item sequence vectors, an average length of 65. In contrast, the Fashion dataset is relatively sparse with concise item sequence vectors, an average length of six.

We observe that short sequence datasets benefit most from a single modality embedding (see rows (1) – (3) in Table 3), whereas longer sequence datasets produce the best results from a joint embedding. Long sequence datasets produce more complex relationships between different users' behaviour. Hence, we can uncover similarities between items within the embedding space by extending our knowledge of the item. Often the bi-directional transformer yields the highest results from an embedding that includes image data (see rows (2), (6) and (10)), whereas the uni-directional model requires text data(see rows (1) and (11)). In addition, a concatenated joint embedding works best for a uni-directional model, whereas a bi-directional model can exploit the summation approach (see rows (4) and (8), respectively). Bi-directional models worsen with longer text descriptions or concatenated joint embeddings. Similarly, uni-directional models worsen with image data and the summation of embeddings. Presumably, both models experience some information loss. The multi-head self-attention mechanism deteriorates from longer and more detailed data formats, such as text descriptions and concatenated embeddings. The uni-directional model deteriorates from less descriptive data formats such as images and summation of embeddings.

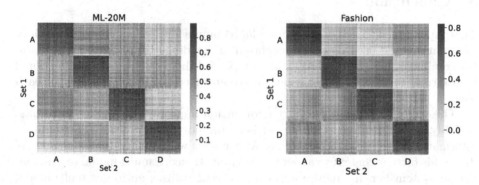

Fig. 3. Heatmaps of the similarity scores between two sets of user's average attention weights

4.2   Visualizing Attention Weights

Finally, we show that our model is user context-aware. Our objective is to demonstrate that we can learn the context of each user by looking at sequential and multi-modal data. We do not want a scenario where a recommendation is being made by clustering similar items. Instead, we want to group users by their behaviour and recommend items from other users that share similar behaviour. The paper investigates if our model is user context-aware by visualizing the attention weights from the best performing model, i.e. SASRec$^+$.

First, we use the K-means algorithm to group users into clusters according to their behaviour. Then we use the naive elbow method to select the optimal number of clusters that separates the users at a more significant level. We had seven and five optimal clusters for the ML-20M and Fashion datasets, respectively. We selected four clusters for visualization purposes. Each user sequence vector belongs to one of the four clusters, denoted by A, B, C or D for brevity. Then, we randomly select two disjoint sets from each cluster. Each set contains 100 users; therefore, we have 200 users per cluster. We pass each of the 200 user's sequence vectors and auxiliary information into SASRec$^+$ and obtain their average attention weights.

Lastly, we calculate the cosine similarity of the average attention weights between sets one and two for each cluster. Our goal is to observe the highest similarity between sets one and two that belong to the same cluster. Figure 3 depicts two heatmaps of the similarity scores of the average attention weights between the two sets, for ML-20M and Fashion datasets, respectively. Both heatmaps are approximately block diagonal matrices, meaning that SASRec$^+$ can identify users that share similar behaviour and tend to assign larger weights between them. This adds evidence to our model being user context-aware.

5   Conclusion

This paper aimed to design a multi-modal recommendation system with auxiliary information. We used data2vec to obtain embeddings from multi-modal data. We combined these embeddings and then passed them to a uni- or bi-directional transformer to model the behaviour of sequential consumption of items for each user.

The study found that auxiliary information improves the prediction results. Notably, a unidirectional transformer benefits most from multi-modal auxiliary information than a bidirectional one. As a result, we conclude that multi-model data adds further context to a user's behaviour. Hence, a multi-modal recommendation system benefits more from a model that utilizes an item's multi-modal data together in the embedding space.

This study used a general learning framework called data2vec to obtain embeddings. However, we did not assess the quality of the image and text embeddings obtained from data2vec. Though from the ablation study, we observed inconsistent results when combining different modalities, we conjecture that the embeddings obtained from the text and image data, belonging to the same item, are similar. We encourage future research to investigate the similarity of embeddings belonging to the same item but exist in different modalities.

References

1. HuggingFace Data2Vec. https://huggingface.co/docs/transformers/v4.21.2/en/model_doc/data2vec
2. Adomavicius, G., Tuzhilin, A.: Toward the next generation of recommender systems: a survey of the state-of-the-art and possible extensions. IEEE Trans. Knowl. Data Eng. **17**(6), 734–749 (2005)
3. Arun, B., Gupta, U., Soeny, K., Jain, S., Mondal, R., Agarwal, G.: Metatransformer4rec-sequential recommendation using meta information and transformers

4. Baevski, A., Hsu, W.N., Xu, Q., Babu, A., Gu, J., Auli, M.: Data2Vec: a general framework for self-supervised learning in speech, vision and language. arXiv preprint arXiv:2202.03555 (2022)
5. Bao, H., Dong, L., Wei, F.: BEiT: BERT pre-training of image transformers. arXiv preprint arXiv:2106.08254 (2021)
6. Billsus, D., Pazzani, M.J., et al.: Learning collaborative information filters. In: ICML, vol. 98, pp. 46–54 (1998)
7. Chen, T., et al.: XGBoost: extreme gradient boosting. R package version 0.4-2 1(4), 1–4 (2015)
8. Devlin, J., Chang, M.W., Lee, K., Toutanova, K.: BERT: pre-training of deep bidirectional transformers for language understanding. arXiv preprint arXiv:1810.04805 (2018)
9. Dlamini, N., van Zyl, T.L.: Author identification from handwritten characters using Siamese CNN. In: 2019 International Multidisciplinary Information Technology and Engineering Conference (IMITEC), pp. 1–6. IEEE (2019)
10. Fischer, E., Zoller, D., Dallmann, A., Hotho, A.: Integrating keywords into BERT4Rec for sequential recommendation. In: Schmid, U., Klügl, F., Wolter, D. (eds.) KI 2020. LNCS (LNAI), vol. 12325, pp. 275–282. Springer, Cham (2020). https://doi.org/10.1007/978-3-030-58285-2_23
11. Harper, F.M., Konstan, J.A.: The MovieLens datasets: History and context. ACM Trans. Interact. Intell. Syst. (TIIS) 5(4), 1–19 (2015)
12. He, R., Kang, W.C., McAuley, J.: Translation-based recommendation. In: Proceedings of the Eleventh ACM Conference on Recommender Systems, pp. 161–169 (2017)
13. Hjelm, R.D., et al.: Learning deep representations by mutual information estimation and maximization. arXiv preprint arXiv:1808.06670 (2018)
14. Kang, W.C., McAuley, J.: Self-attentive sequential recommendation. In: 2018 IEEE International Conference on Data Mining (ICDM), pp. 197–206. IEEE (2018)
15. Kong, L., de Masson d'Autume, C., Ling, W., Yu, L., Dai, Z., Yogatama, D.: A mutual information maximization perspective of language representation learning. arXiv preprint arXiv:1910.08350 (2019)
16. Koren, Y.: Factor in the neighbors: scalable and accurate collaborative filtering. ACM Trans. Knowl. Discov. Data (TKDD) 4(1), 1–24 (2010)
17. Koren, Y., Bell, R., Volinsky, C.: Matrix factorization techniques for recommender systems. Computer 42(8), 30–37 (2009)
18. Manack, H., Van Zyl, T.L.: Deep similarity learning for soccer team ranking. In: 2020 IEEE 23rd International Conference on Information Fusion (FUSION), pp. 1–7. IEEE (2020)
19. Ni, J., Li, J., McAuley, J.: Justifying recommendations using distantly-labeled reviews and fine-grained aspects. In: Proceedings of the 2019 Conference on Empirical Methods in Natural Language Processing and the 9th International Joint Conference on Natural Language Processing (EMNLP-IJCNLP), pp. 188–197 (2019)
20. Rendle, S., Freudenthaler, C., Gantner, Z., Schmidt-Thieme, L.: BPR: Bayesian personalized ranking from implicit feedback. arXiv preprint arXiv:1205.2618 (2012)
21. Sarwar, B., Karypis, G., Konstan, J., Riedl, J.: Application of dimensionality reduction in recommender system-a case study. Technical report, Minnesota University of Minneapolis, Department of Computer Science (2000)
22. Sennrich, R., Haddow, B., Birch, A.: Neural machine translation of rare words with subword units. arXiv preprint arXiv:1508.07909 (2015)
23. Shwartz-Ziv, R., Armon, A.: Tabular data: deep learning is not all you need. Inf. Fusion 81, 84–90 (2022)

24. Steck, H.: Item popularity and recommendation accuracy. In: Proceedings of the Fifth ACM Conference on Recommender Systems, pp. 125–132 (2011)
25. Sun, F., et al.: BERT4Rec: sequential recommendation with bidirectional encoder representations from transformer. In: Proceedings of the 28th ACM International Conference on Information and Knowledge Management, pp. 1441–1450 (2019)
26. Van Zyl, T.L., Woolway, M., Engelbrecht, B.: Unique animal identification using deep transfer learning for data fusion in Siamese networks. In: 2020 IEEE 23rd International Conference on Information Fusion (FUSION), pp. 1–6. IEEE (2020)
27. Variawa, M.Z., Van Zyl, T.L., Woolway, M.: Transfer learning and deep metric learning for automated galaxy morphology representation. IEEE Access **10**, 19539–19550 (2022)
28. Vasile, F., Smirnova, E., Conneau, A.: Meta-Prod2Vec: product embeddings using side-information for recommendation. In: Proceedings of the 10th ACM Conference on Recommender Systems, pp. 225–232 (2016)
29. Vaswani, A., et al.: Attention is all you need. In: Advances in Neural Information Processing Systems, vol. 30 (2017)
30. Zhang, S., Yao, L., Sun, A., Tay, Y.: Deep learning based recommender system: a survey and new perspectives. ACM Comput. Surv. (CSUR) **52**(1), 1–38 (2019)
31. Zhou, K., et al.: S3-Rec: self-supervised learning for sequential recommendation with mutual information maximization. In: Proceedings of the 29th ACM International Conference on Information & Knowledge Management, pp. 1893–1902 (2020)

Cauchy Loss Function: Robustness Under Gaussian and Cauchy Noise

Thamsanqa Mlotshwa, Heinrich van Deventer⬤,
and Anna Sergeevna Bosman(✉)⬤

Department of Computer Science, University of Pretoria, Pretoria, South Africa
anna.bosman@up.ac.za

Abstract. In supervised machine learning, the choice of loss function implicitly assumes a particular noise distribution over the data. For example, the frequently used mean squared error (MSE) loss assumes a Gaussian noise distribution. The choice of loss function during training and testing affects the performance of artificial neural networks (ANNs). It is known that MSE may yield substandard performance in the presence of outliers. The Cauchy loss function (CLF) assumes a Cauchy noise distribution, and is therefore potentially better suited for data with outliers. This papers aims to determine the extent of robustness and generalisability of the CLF as compared to MSE. CLF and MSE are assessed on a few handcrafted regression problems, and a real-world regression problem with artificially simulated outliers, in the context of ANN training. CLF yielded results that were either comparable to or better than the results yielded by MSE, with a few notable exceptions.

Keywords: Mean squared error (MSE) · Cauchy loss function (CLF) · Loss functions · Generalisation · Outliers

1 Introduction

Use of the mean squared error (MSE) loss function in the context of regression problems has been and remains fairly common owing to convention and its computational simplicity [1]. MSE depends on the assumption that noise in the data has a Gaussian distribution [2]. But in practise, it is unlikely that the true noise distribution can be determined without a thorough understanding of the underlying data generating process [3].

MSE tends to perform poorly in the presence of residuals that deviate significantly in size from the trend of the data [3]. Large residuals can exceed the expected bounds that the assumption of Gaussianity imposes. This is often seen in real world problems, where particularly large residuals or outliers account for 1% to 10% or even more of typical datasets [3]. There exist a number of mechanisms for handling outlying points in limited numbers; however, when there are

Supported by the NRF Thuthuka Grant Number 13819413.

too many outliers, especially in large highly structured or multivariate datasets, detection and handling of outliers becomes difficult or even unfeasible [3]. Using a loss function robust to outliers could mitigate the need to clean training data.

Hence, alternative loss functions for regression have been used in attempts to achieve robustness to outliers and non-Gaussian noise profiles. Particularly, M-estimators based on the maximum likelihood for the density functions of an assumed distribution are common candidates [4]. The Cauchy loss function (CLF) is one popular choice that assumes a Cauchy noise profile [1]. The Cauchy distribution has statistical properties that align closely with the intuition of what an outlier might be, such as its undefined mean and infinite variance, accounting for more extreme residuals than would be expected under a Gaussian assumption [5]. The most convincing argument for the CLF is that the influence that residuals have on the estimator is bounded and tends towards zero (illustrated in Fig. 1) [6].

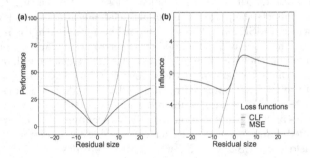

Fig. 1. (a) Quantification of performance according to MSE and CLF varying with residual size, (b) Influence on performance quantification varying with residual size.

Despite some theoretical and empirical evidence of the advantages of using CLF over MSE [3], there is a lack of corresponding systematic studies that would highlight the behaviour of the two loss functions under different parameter settings. Thus, the research objectives of this study are summarised as follows:

- To establish whether the use of CLF consistently leads to similar or better performance of the artificial neural network (ANN) models compared to MSE, applied to regression problems with varying degrees of Gaussian and Cauchy noise.
- To elucidate whether the use of CLF in place of MSE yields ANN models that are more robust in the presence of deterministic noise, arising from complexities in a regression dataset that a model has limited flexibility to capture otherwise.
- To contrast the behaviour and robustness of ANN models trained with CLF to those trained with MSE in the presence of outliers in real world data.

The rest of the paper is structured as follows: Sect. 2 covers the relevant background to contextualise the research. Section 3 outlines the methodology that is used in addressing the research goals. Section 4 then presents and discusses the results of the research, and conclusions are drawn in Sect. 5.

2 Background

This section provides background into MSE, CLF, and their statistical considerations. Furthermore, a brief discussion is offered on deterministic noise, and an argument is made for the validity of the inferences drawn from the results of the experiments.

2.1 Consequences of the Gaussian Assumption

The normal or Gaussian distribution is arguably the most extensively used distribution in most fields of science [7]. In particular, the noise found in empirical data for many scientific domains is modelled as or assumed to be Gaussian [7], based on the statistical central limit theorem and the law of large numbers, which assume a number of regularity conditions [5]. Furthermore, a number of advantageous properties regarding unbiased parameter estimates and least squares methods come under the assumption of Gaussianity [8]. Namely, the maximum likelihood estimator is equivalent to the least squares estimator for data containing Gaussian noise [4].

As it applies to ANNs, the minimisation of the MSE loss can be shown to be equivalent to the selection of parameters that maximise the likelihood for a Gaussian model, i.e. its maximum likelihood estimator [9]. Hence, the convention of Gaussian noise is subsumed in optimisation of the MSE loss for regression problems [2,10].

However, this conventional assumption of Gaussianity is often unfounded in practice, leading to significant degradation in results in the presence of adversarial, impulsive noise components, and particularly in the case of non-Gaussian noise [11]. A fundamental error is often made in the use of the central limit theorem, where finite variances are assumed although this may not be appropriate [4]. In these cases, the MSE loss breaks down, and the estimator vector may shift significantly from the true regression plane to minimise the errors that are inflated by being squared [10]. Moreover, the influence of these outliers on the overall estimate increases proportionally with their magnitude in an unbounded manner [6,12]. These characteristics have led practitioners to seek performance criteria that are more robust to outliers and non-Gaussian noise profiles.

2.2 Stable Distributions

Although there exist a number of distributions accounting for large residuals (i.e. having heavier tails) than the Gaussian distribution, Borak et al. [5] express a theoretical reasoning for assessing the use of the family of stable distributions in particular. A generalisation of the central limit theorem states that stable distributions "are the only possible limit distributions for properly normalised and centred sums of independent, identically distributed random variables" [5]. This statement of the central limit theorem drops the regularity conditions (such as finite variance) necessary to converge to a Gaussian, and even subsumes the convergence to a Gaussian as a member of the family of stable distributions. Hence, the use of a heavy-tailed stable distribution offers a correction to the frequent misapplication of the central limit theorem that accompanies the conventional assumption of Gaussianity. Although there exist an infinite number of stable distributions, there exist only two closed form formulas for symmetric stable distributions: the Gaussian and Cauchy distributions [5].

2.3   Cauchy Distribution and Cauchy Loss Function

The choice of the Cauchy distribution is more intuitive for modelling practical occurrences of noise and outliers than the Gaussian. Its close-form density function also makes Cauchy distribution a candidate for maximum likelihood estimation [5], and hence its use as a loss function based on a heavy-tailed stable distribution. This paper considered CLF scaled by a constant as presented in [1] and shown as necessary to achieve the theoretical behaviour of CLF in preliminary testing:

$$L_{CLF} = \frac{c^2}{2} \log \left(1 + \left(\frac{y - \hat{y}}{c} \right)^2 \right)$$

where $c \in (0, \infty)$ is the CLF constant, and y and $\hat{y}$ refer to the target and actual output of the model, respectively, their difference constituting the residual. The influence function associated with the CLF indicates any residual's influence on the estimate has an upper bound, and that for arbitrarily big residuals, their influence tends towards zero [6]. This is as a result of the Cauchy distribution's power-law behaviour [5], and qualifies as robust to noise when contrasted to MSE [6].

2.4   Deterministic Noise

Although the influence of additive noise, whether Gaussian or Cauchy, has been the primary grounds for discussion until this point, a discussion on the tolerance of the MSE and CLF performance criteria would be incomplete without consideration for the internal complexities of a dataset. In the field of Bayesian statistical models, a given signal from a random process can be decomposed into a structured deterministic component (explained by the model) and a random stochastic component (attributed to noise) [13].

An analogue is drawn between the Bayesian model and a supervised learning algorithm by Abu-Mostafa et al. [14], who assert that a well-performing model not only accounts for random stochastic noise (in the case of this paper assumed to be either Gaussian or Cauchy), but also for complexity in the structured deterministic component. Complexity that is not modelled is treated as deterministic noise, which a model may fail to fit, even in the absence of stochastic noise [14].

Ultimately, the performance of models using MSE or CLF are subject not only to the noise present in the data, but also whether or not they offer the degrees of freedom to capture the deterministic complexities of the target function.

2.5   On the Validity of Inferring from the Results

The performance of statistical inference based on the samples generated from the experiments of this study warrants careful consideration, particularly the results generated in the presence of Cauchy noise. The assumption of Cauchy noise implies indeterminate moments [5]. In particular, Cauchy population mean is undefined, and its variance is infinite [5]. Consequently, inference that depends on any assumption of convergence to a finite population variance or determinate population mean can be thought to be limited in usefulness. Thus, inferences based on the root mean squared error (RMSE) are likely to be unreliable in the presence of Cauchy noise.

An empirical study conducted by Balkema and Embrechts [15] aimed to assess the performance of various estimators in a simple linear regression problem with deterministic and stochastic components, both modelled as potentially heavy tailed random variables. Balkema and Embrechts showed that under the condition that the deterministic component is of finite variance itself, the least absolute deviation (i.e. the minimisation of the mean absolute error (MAE)) is typically a good estimator for the linear regression parameters, even if the stochastic component has infinite variance [15]. In the context of stochastic gradient descent, the minimisation of MAE is equivalent to the least absolute deviation. Hence, by assuming finite variance of the multivariate random process giving rise to the data set, the use of MAE is considered to be fairly reliable [15].

The inferential tools that were chosen in this study (the Kruskal-Wallis [16] and Wilcoxon Rank Sum [17] tests) are rank-based tests to avoid mean-based testing. They are hence equivalent to median-based testing when the various inferential population distributions (i.e., the actual performance of the various models) are assumed to be of similar shapes [18].

2.6   Related Work

The CLF has been applied and achieved satisfactory results in ANNs as well as other machine learning domains such as subspace clustering [3,6,12]. It is suggested by Li et al. [6] that the nature of the CLF's influence function means that it is generally insensitive to the distribution of the noise; however, developing a better theoretical understanding is still worthwhile. Furthermore, El-Melegy et al. [3] suggest that CLF may only be robust to a certain extent. As such, a comparison of how CLF and MSE perform in the presence of varying degrees of both Gaussian and non-Gaussian stable noise is necessary in order to evaluate CLF's ability to generalise, and its robustness to the implied Cauchy noise profile.

Additionally, a gap in the literature was found regarding the impact of deterministic noise on the performance of CLF models, and how this compares to the performance of MSE models. Lastly, none of the literature surveyed exhibited a statistically sensitive approach to the drawing of conclusions.

3   Methodology

Three experiments were conducted for this study. The first two used handcrafted datasets of varied complexity to assess the responses of MSE and CLF to various Gaussian and Cauchy noise profiles. The third experiment made use of a real-world dataset to assess the response of the two loss functions to simulated outliers. This section describes each of the datasets, the structural and implementational considerations of each experiment, and the corresponding experimentation procedures.

3.1   Handcrafted Experiments

Two regression datasets were generated according to mathematical functions. The first dataset is in two variables, given by:

$$y_1 = e^{x_1} - \sin x_2 \tag{1}$$

for $x_1 \in [-6, 2]$ and $x_2 \in [-3, 9]$.

The second dataset is generated from a function of eight variables, characterised by quickly oscillating behaviour. It is given by:

$$y_2 = 0.03 \Big(\sin^2(x_1)(x_2 - 2)(x_3 - 8)(x_4 - 11) \\ + \cos^2(x_5)(x_6 - 6)(x_7 - 6)(x_8 + 5)^2 \Big) \tag{2}$$

for $x_1 \in [-6, 17]$, $x_2 \in [-7, 20]$, $x_3 \in [-2, 17]$, $x_4 \in [-6, 10]$, $x_5 \in [-10, 16]$, $x_6 \in [-5, 10]$, $x_7 \in [-15, 9]$, and $x_8 \in [-1, 14]$.

Each dataset consisted of 5000 randomly sampled datapoints, where the target values followed the given mathematical function y_1 or y_2.

In the experimentation, noise was introduced into each dataset for training purposes, and the integrity of the data was maintained for testing. For the Gaussian configuration of each experiment, additive noise ϵ was injected into y_i, where $\epsilon \sim N(0, \sigma^2)$, given that the standard deviation $\sigma \in \{1, 10, 50, 100\}$.

Likewise, for the Cauchy configuration, additive noise ϵ was injected into y_i where $\epsilon \sim Cau(x_0, \tau)$, given that the location parameter $x_0 = 0$ and the scale parameter $\tau \in \{1, 10, 50, 100\}$.

3.2   Seoul Bike Sharing Demand Experiment

The dataset is a compilation of a year's worth of data from the Seoul Public Data Park website. Each datapoint describes the number of bikes rented in the city for a single hour of a particular day, along with a number of meteorological and date-specific details that may have correlated with the demand for bikes [19]. This dataset is available on the UC Irvine Machine Learning Repository.

In the experimentation, noise was injected according to a uniform distribution with a range that is 500 times the range of the actual data. This was done in order to simulate outliers. Noise was randomly injected into variable proportions of the training data targets according to the experiment configuration (i.e. 2.5%, 5%, 7.5%, 10% of the training data). The testing set was left uncorrupted.

It is noted by Qi et al. [20] that data from the UC Irvine Machine Learning Repository is typically clean, so inconsistencies are introduced here in the form of simulated outliers. This noise is essentially impulsive, as the uniform distribution acts as a thick-tailed distribution along its range, and was picked as an outlier generator on this basis.

3.3   General Setup

For each experiment, there were two model types used. Each type is topographically analogous: a feed-forward ANN with the Adam optimiser [21] and the ReLU activation function [22] applied in the hidden layers. For the handcrafted experiments, the ANN architecture featured a single hidden layer of ten nodes, and a single output node. For the Seoul bike sharing demand experiment, a bigger ANN architecture was used, featuring two hidden layers of fourteen nodes each, and a single output node. The ANN architecture parameters were not optimised for a specific loss function to avoid bias.

For each experiment, the models differed only in the loss function employed. The first model type makes use of MSE, and the second makes use of CLF_c, where c is the CLF constant. For the handcrafted experiments, five CLF_c models were considered for $c \in \{0.1, 1, 10, 20, 100\}$. For the Seoul bike sharing demand experiment, six CLF_c models were considered for $c \in \{1, 10, 100, 200, 1000, 10000\}$.

All experiments were conducted on Google Colab, using the Python 3 Google Compute Engine backend. The setup features 12.78 GB memory and two Intel® Xeon® 2.30 GHz CPUs. All code was written in Python, using algorithms provided by the open-source *Keras*, *TensorFlow* and *scikit-learn* libraries. As no CLF implementation exists in any of the used libraries, a custom implementation was written for the purposes of the research.

3.4   General Procedure

In all experiments, 10-fold cross-validation was used to split the datasets into training and testing sets. As a negative control and benchmark, training and testing were initially performed on uncorrupted data using both the MSE model and all CLF_c models for each of the datasets.

Subsequently, training was performed on the dataset in question with corrupted data. In the handcrafted experiments, this means corrupting the data with either Gaussian or Cauchy noise. Four extents of Gaussian noise were assessed, where the noise profile had a standard deviation $\sigma \in \{1, 10, 50, 100\}$.

Likewise, four extents of Cauchy noise were assessed, where the noise profile had a scale parameter $\tau \in \{1, 10, 50, 100\}$. The training was performed for the MSE model and all CLF_c models, which were then tested on uncorrupted data.

In the Seoul bike sharing demand experiment, training data was corrupted with simulated outliers, generated according to a uniform distribution. The proportion of the dataset to be corrupted was set to one of four levels: 2.5%, 5%, 7.5% or 10%. The training was performed for the MSE model and all CLF_c models, which were then tested on uncorrupted data.

To score the models' performance on the test data, two different metrics were used for all experiments: RMSE and MAE. Five values of each score were generated from five independent training-testing runs for each model, i.e., each value is the average of the 10 scores obtained from each cross-validation process.

4   Discussion

The results from the experimental procedures are presented and discussed in this section. The 2-variable and 8-variable handcrafted experiments are presented first, followed by the results from the Seoul bike sharing demand experiment.

4.1   2-Variable Handcrafted Experiment

Tables 1 and 2 summarise the MSE and CLF_c performance under various extents of Gaussian and Cauchy noise, respectively.

The negative control configuration (refer to Table 1) saw no additive noise in the data, and the performances of the six models were comparable, with no conclusive evidence suggesting significant differences between them. This behaviour was as expected, indicating that CLF is at the least not inferior to MSE.

The experiment configuration that used Gaussian noise profiles also exhibited expected behaviour. Increasing the standard deviation of the Gaussian noise resulted in a fairly steady decline in performance in all models (see Fig. 2). However, the CLF models did not consistently perform equivalently to the MSE model. From the point where $\sigma = 10$, WRS tests revealed evidence ($p < 0.01$ for both scores) of notable difference between the other models and the CLF_{20} model. By the point where $\sigma = 50$, all CLF models have performed worse than the MSE model according to KW and WRS testing ($p < 0.05$ for both scores). The superior performance of the MSE model on its loss function's assumed noise profile is to be expected, and ultimately the rates of degradation of CLF models' performance are comparable for limited extents of Gaussian noise.

However, beyond $\sigma = 10$, the CLF_1 model was shown to perform better than the worst performing CLF models in each case, according to evidence from WRS testing ($p < 0.05$ in both scores). This is illustrated in Fig. 2, and may speak to the optimisation obtained from choosing the appropriate CLF constant, noting that worse performance was seen for both smaller and larger CLF constants.

For the Cauchy noise configuration, the results were as expected once more (refer to Table 2). Noting that inference can only be thought to be statistically

Table 1. Handcrafted experiments: negative control and Gaussian noise. Lowest errors for each set-up are indicated in red.

Var	Model	Negative		$\sigma = 1$		$\sigma = 10$	
		MAE	RMSE	MAE	RMSE	MAE	RMSE
2	$CLF_{0.1}$	0.485 (0.013)	0.67 (0.006)	0.492 (0.021)	0.675 (0.031)	0.617 (0.021)	0.8 (0.026)
	$CLF_{1.0}$	0.476 (0.012)	0.663 (0.008)	0.491 (0.006)	0.666 (0.012)	0.614 (0.013)	0.799 (0.019)
	$CLF_{10.0}$	0.479 (0.009)	0.657 (0.008)	0.487 (0.009)	0.664 (0.014)	0.61 (0.016)	0.79 (0.015)
	$CLF_{20.0}$	0.482 (0.01)	0.668 (0.007)	0.495 (0.008)	0.672 (0.006)	0.604 (0.007)	0.795 (0.005)
	$CLF_{100.0}$	0.48 (0.012)	0.659 (0.012)	0.494 (0.008)	0.674 (0.008)	0.653 (0.019)	0.841 (0.01)
	MSE	0.485 (0.01)	0.677 (0.013)	0.496 (0.009)	0.678 (0.008)	0.604 (0.01)	0.785 (0.01)
8	$CLF_{0.1}$	39.261 (0.059)	55.384 (0.155)	39.383 (0.188)	55.607 (0.359)	39.075 (0.033)	54.881 (0.06)
	$CLF_{1.0}$	39.573 (0.485)	55.91 (0.839)	39.195 (0.053)	55.255 (0.134)	39.059 (0.041)	54.859 (0.074)
	$CLF_{10.0}$	39.351 (0.234)	55.533 (0.439)	39.424 (0.318)	55.703 (0.568)	39.073 (0.035)	54.885 (0.085)
	$CLF_{20.0}$	39.445 (0.152)	55.709 (0.244)	39.306 (0.134)	55.478 (0.303)	39.043 (0.05)	54.851 (0.084)
	$CLF_{100.0}$	39.394 (0.17)	55.649 (0.314)	39.572 (0.457)	55.917 (0.815)	39.063 (0.04)	54.848 (0.068)
	MSE	41.81 (0.073)	52.425 (0.028)	41.749 (0.074)	52.398 (0.042)	41.822 (0.085)	52.426 (0.024)

Var	Model	$\sigma = 50$		$\sigma = 100$	
		MAE	RMSE	MAE	RMSE
2	$CLF_{0.1}$	1.189 (0.03)	1.531 (0.043)	1.873 (0.502)	2.34 (0.51)
	$CLF_{1.0}$	1.106 (0.132)	1.454 (0.151)	1.655 (0.181)	2.056 (0.164)
	$CLF_{10.0}$	1.217 (0.082)	1.555 (0.125)	1.865 (0.373)	2.341 (0.368)
	$CLF_{20.0}$	1.216 (0.124)	1.535 (0.108)	1.803 (0.385)	2.266 (0.431)
	$CLF_{100.0}$	1.401 (0.176)	1.784 (0.228)	1.732 (0.441)	2.173 (0.514)
	MSE	0.994 (0.156)	1.296 (0.18)	1.376 (0.174)	1.729 (0.171)
8	$CLF_{0.1}$	40.113 (0.179)	53.442 (0.142)	41.686 (0.651)	55.161 (0.714)
	$CLF_{1.0}$	40.305 (0.26)	53.454 (0.192)	41.975 (0.526)	54.97 (0.856)
	$CLF_{10.0}$	40.033 (0.114)	53.359 (0.084)	42.051 (0.508)	54.837 (0.597)
	$CLF_{20.0}$	40.45 (0.39)	53.721 (0.352)	42.185 (0.73)	54.642 (0.581)
	$CLF_{100.0}$	40.023 (0.032)	53.497 (0.221)	41.669 (0.411)	54.18 (0.44)
	MSE	41.803 (0.087)	52.438 (0.038)	41.84 (0.196)	52.504 (0.036)

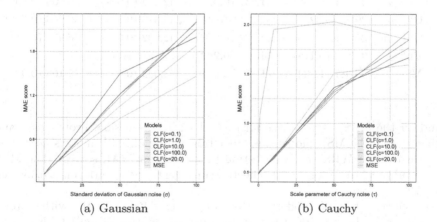

(a) Gaussian   (b) Cauchy

Fig. 2. MAE scores for models varying with standard deviation σ of Gaussian noise and scale parameter τ of Cauchy noise (2-variable handcrafted experiment)

Table 2. Handcrafted experiments: Cauchy noise. Lowest errors for each set-up are indicated in red.

Var	Model	$\tau = 1$		$\tau = 10$	
		MAE	RMSE	MAE	RMSE
2	$CLF_{0.1}$	0.503 (0.011)	0.68 (0.014)	0.655 (0.025)	0.867 (0.029)
	$CLF_{1.0}$	0.506 (0.009)	0.685 (0.005)	0.64 (0.018)	0.847 (0.025)
	$CLF_{10.0}$	0.517 (0.009)	0.693 (0.006)	0.635 (0.01)	0.836 (0.015)
	$CLF_{20.0}$	0.503 (0.006)	0.69 (0.011)	0.645 (0.03)	0.85 (0.027)
	$CLF_{100.0}$	0.5 (0.011)	0.689 (0.016)	0.659 (0.017)	0.861 (0.036)
	MSE	1.088 (0.11)	1.377 (0.111)	1.952 (0.22)	2.426 (0.223)
8	$CLF_{0.1}$	39.289 (0.125)	55.446 (0.323)	39.028 (0.016)	54.786 (0.093)
	$CLF_{1.0}$	39.347 (0.113)	55.511 (0.23)	38.99 (0.038)	54.669 (0.084)
	$CLF_{10.0}$	39.507 (0.459)	55.848 (0.916)	39.118 (0.164)	54.939 (0.318)
	$CLF_{20.0}$	39.194 (0.089)	55.236 (0.191)	39.179 (0.139)	55.066 (0.237)
	$CLF_{100.0}$	39.206 (0.061)	55.251 (0.13)	39.017 (0.033)	54.726 (0.071)
	MSE	41.772 (0.099)	52.531 (0.127)	45.717 (2.667)	60.302 (4.179)
		$\tau = 50$		$\tau = 100$	
2	$CLF_{0.1}$	1.289 (0.147)	1.654 (0.146)	1.937 (0.254)	2.392 (0.266)
	$CLF_{1.0}$	1.512 (0.131)	1.931 (0.167)	1.595 (0.228)	2.007 (0.222)
	$CLF_{10.0}$	1.319 (0.209)	1.669 (0.239)	1.769 (0.334)	2.282 (0.384)
	$CLF_{20.0}$	1.365 (0.181)	1.772 (0.168)	1.669 (0.32)	2.082 (0.352)
	$CLF_{100.0}$	1.337 (0.118)	1.749 (0.12)	1.847 (0.318)	2.255 (0.272)
	MSE	2.031 (0.436)	2.494 (0.533)	1.85 (0.318)	2.305 (0.318)
8	$CLF_{0.1}$	40.085 (0.176)	54.368 (0.338)	42.112 (0.44)	56.261 (0.502)
	$CLF_{1.0}$	39.941 (0.195)	54.237 (0.385)	42.515 (0.697)	56.806 (0.859)
	$CLF_{10.0}$	39.95 (0.327)	53.966 (0.313)	42.373 (0.474)	56.966 (0.788)
	$CLF_{20.0}$	39.866 (0.244)	54.171 (0.268)	42.092 (0.547)	56.938 (0.549)
	$CLF_{100.0}$	39.888 (0.118)	54.194 (0.247)	41.853 (0.653)	55.916 (0.516)
	MSE	50.987 (1.734)	71.354 (2.23)	53.683 (0.664)	75.103 (0.987)

sound for the MAE scores, these form the basis of discussion. From $\tau = 1$ and until $\tau = 100$, the MSE model was outperformed by all CLF models according to evidence from WRS and KW testing ($p < 0.005$ for MAE scores). The MSE model had test MAE scores that reflected consistently poor performance, as seen in Fig. 2.

In contrast, the CLF models' performances degraded gradually with increasing τ. However, by $\tau = 100$, the performance degradation of most CLF models led to insufficient evidence of their being different from the MSE model, according to KW testing.

Furthermore, WRS testing confirmed ($p < 0.05$ in both scores) that CLF models performed differently from each other. Thus, the results showed that in the presence of all the considered extents of additive noise, at least one CLF model outperformed the MSE model. This indicates that there may be some correlation between the scale of Cauchy noise profiles and optimal value of the CLF constant.

4.2   8-Variable Handcrafted Experiment

The negative control configuration saw no additive noise in the data (see Table 1). The results obtained from this benchmark present a seeming contradiction in the scores. According to the MAE score, KW and WRS testing reflected significant evidence ($p < 0.005$) of the CLF models outperforming the MSE model. However, according to the RMSE score, KW and WRS testing reported equally significant evidence ($p < 0.005$) of the MSE model performing better than any of the CLF models. It is likely that this apparent contradiction is a consequence of deterministic noise. Nonetheless, WRS testing asserted further that there was evidence ($p < 0.05$ in both scores) of different performances between the CLF models, indicating that the performance of CLF is dependent on the CLF constant choice. An example of this is in the $CLF_{0.1}$ model performing better than the CLF_{20} model.

As was the case for the negative control, there was an apparent contradiction in the scores for the various models under Gaussian noise (see Table 1). The MSE model could be shown to perform better than the CLF models in the RMSE scores for all considered values of σ (refer to Fig. 3). Similarly, the CLF models could be shown to perform better in the MAE scores (as seen in Fig. 3). Each contradicting case of better performance was supported by statistical evidence ($p < 0.05$) through KW and WRS testing.

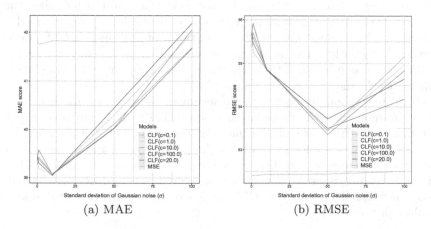

(a) MAE   (b) RMSE

Fig. 3. MAE and RMSE scores for models varying with standard deviation σ of Gaussian noise (8-variable handcrafted experiment)

As a consequence of the contradiction in results, the reliability of the said results must be discussed. A case can be formed against the RMSE as a least squares score, owing to its dependence on the assumption of Gaussianity and the conceptual link to the MSE model in that way. The inability of the CLF models to perform comparatively to the MSE model for even extreme values of σ may reflect a bias in least squares scoring towards the way in which the MSE facilitates learning. In contrast, the MSE and CLF models could be shown to perform similarly for larger extents of Gaussian noise in MAE scoring. This may speak to less bias towards the MSE model as a consequence of being based on different statistical assumptions.

The performance of the MSE model seemed to be almost completely invariant to changes in the value of σ (refer to Table 1). Its performance remaining constant in the absence and presence of all considered extents of Gaussian noise reflects a potentially unforeseen consequence of introducing deterministic noise into the dataset. The additive noise introduced may have been insufficient to alter the performance seen in light of the deterministic noise. Comparatively, the CLF models were more responsive to the extent of Gaussian noise present in the data.

Nonetheless, a behaviour of the CLF models that was identified in both scores is that their performances collectively improved with σ until a certain threshold, after which their performances degraded. This was seen in the MAE scores, where their best performances were reported for $\sigma = 10$ (see Fig. 3), and for the RMSE scores, where their best performances were reported for $\sigma = 50$ (see Fig. 3). This was likely owing to the limited extents of Gaussian noise having a 'smoothing' effect on the oscillations of the data. This ultimately made the data easier for the CLF models to learn, somewhat countering the effects of the deterministic noise.

The KW and WRS tests provided evidence ($p < 0.05$ for all scores) of differing performances between the CLF models themselves for varying additive Gaussian noise. Moreover, the models that performed best before the general performance turning point were not necessarily the models that performed best afterwards. This speaks again to the importance of appropriate CLF constant selection, but also how the most appropriate CLF constant may change based on the noise present in the data. In this case, the appropriate CLF constant may be linked to the data smoothing taking place.

The Cauchy configuration of the experiment behaved in a much more predictable fashion than the Gaussian one (see Table 2). It is worth noting once more that the MAE score will be the only one used as evidence for inference, but the RMSE scores align closely with them. For $\tau = 10$ and above, the MSE model's performance degraded significantly in both scores (refer to Fig. 4). KW and WRS testing asserted that there was significant evidence ($p < 0.005$ in MAE score) to indicate that the CLF models outperformed the MSE model for all values of τ above 10.

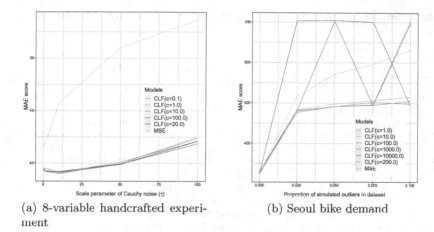

(a) 8-variable handcrafted experiment

(b) Seoul bike demand

Fig. 4. MAE scores for models varying with scale parameter τ of Cauchy noise (8-variable handcrafted experiment), and with varying proportion of simulated outliers in Seoul bike demand dataset.

Looking specifically at the CLF models, we see a similar behaviour to the Gaussian configuration; the performance of CLF models increased with τ until a certain threshold, after which their performance degraded, as shown in Fig. 4. This behaviour was likely due to some degree of smoothing owing to the additive noise that made the oscillations in the data easier to learn. This ultimately counteracted the deterministic noise, until the additive noise itself began to dominate the signal. At this point, performance degraded; but unlike with the MSE model, the CLF models' performances degraded more gradually. There was suggestive evidence ($p < 0.1$ in MAE) of differences between the CLF models at all values of τ, according to WRS testing. This suggests that the combination of additive and deterministic noise makes it harder for a single CLF constant to be considered optimal.

4.3   Seoul Bike Sharing Demand Experiment

Results obtained by corrupting various proportions of the training data with simulated outliers were largely in line with expectations (see Table 3). As a negative control, the initial configuration saw all seven models trained and tested on uncorrupted data, and neither KW nor WRS testing revealed significant evidence to suggest different performances between CLF and MSE for any of the scores.

Table 3. Seoul bike sharing demand experiment. Smallest error value for each setup is shown in red.

Model	Negative (0.000)		0.025		0.05	
	MAE	RMSE	MAE	RMSE	MAE	RMSE
CLF 1.0	329.01 (12.76)	469.19 (18.32)	482.02 (5.07)	618.75 (5.104)	491.03 (7.25)	635.72 (1.20)
CLF 10.0	322.85 (1.75)	459.64 (1.19)	483.63 (9.11)	621.61 (4.85)	497.75 (9.91)	635.74 (6.49)
CLF 100.0	323.69 (1.82)	460.42 (1.55)	476.8 (3.79)	614.59 (6.23)	491.14 (3.93)	635.59 (5.65)
CLF 200.0	322.75 (1.98)	460.43 (0.81)	480.47 (4.14)	616.84 (1.43)	491.57 (4.85)	630.43 (5.36)
CLF 1,000.0	326.67 (2.19)	462.21 (1.23)	480.21 (5.24)	615.29 (3.45)	702.99 (2.97)	953.56 (2.65)
CLF 10,000.0	324.25 (2.60)	460.39 (1.75)	703.57 (1.94)	953.99 (1.74)	704.49 (0.12)	954.85 (0.07)
MSE	323.16 (1.71)	461.45 (1.65)	515.71 (15.61)	674.29 (16.76)	571.89 (16.64)	751.85 (17.49)

Model	0.075		0.100			
	MAE	RMSE	MAE	RMSE		
CLF 1.0	502.19 (9.57)	658.37 (9.12)	497.72 (5.33)	651.63 (7.97)		
CLF 10.0	506.11 (10.49)	649.22 (9.31)	515.16 (15.55)	658.05 (14.93)		
CLF 100.0	494.88 (9.58)	642.99 (8.36)	504.47 (4.61)	652.59 (2.88)		
CLF 200.0	495.97 (3.19)	640.2 (4.95)	704.55 (0.04)	954.94 (0.06)		
CLF 1,000.0	494.16 (6.81)	639.24 (8.83)	697.36 (8.06)	947.05 (9.26)		
CLF 10,000.0	700.89 (5.02)	951.44 (4.7)	494.21 (5.75)	642.88 (5.42)		
MSE	595.93 (25.18)	783.75 (22.75)	630.86 (47.43)	815.79 (57.71)		

However, as the proportion of outliers increased, the performance of the models began to shift. Firstly, the MSE model's performance degraded noticeably in comparison to the performance of the CLF models. From the 2.5% proportion of noise onward, significant evidence ($p < 0.001$ for all scores) was found using the WRS test suggesting that the MSE model was outperformed by the CLF models (refer to Fig. 4). The rate of performance degradation as outliers increased was higher for the MSE model than for the high-performing CLF models.

The models trained with the Cauchy loss function with reasonable CLF constants (less than or equal to 100) consistently performed better than the MSE model. This supports the hypothesis that CLF works well in the presence of outliers. It should be noted that the outliers were generated with a distribution that is very different from the Cauchy distribution. The models trained with extreme CLF constants (greater than or equal to 200) exhibited odd behaviour, sometimes performing better and sometimes performing worse than the MSE model. This highlights the sensitivity to the choice of CLF constants, hyper-parameters (such as the learning rate), and the optimiser.

5   Conclusion

Examining the underlying statistical assumptions in machine learning provides insight and possible improvements for practical applications. This study illustarted the shortcomings of the mean squared error (MSE) that arise owing to the implicit assumption of Gaussian noise. This assumption is prone to poor performance when the distribution of random noise in a dataset can be modelled by heavier-tailed profiles than the Gaussian, or when the data has a large number of outliers. The Cauchy loss function (CLF) was studied as an alternative, and was shown to be capable of performing well for its own assumed noise profile, as well as generalising better to the Gaussian noise profiles than the MSE could generalise to Cauchy noise profiles. CLF was also shown to be more robust to the combination of deterministic and additive noise in this study; promoting the learning of data in spite of, and even because of, the additive noise through smoothing. This is in contrast with MSE, which is heavily influenced by the deterministic noise. The CLF was also shown to perform better in the presence of outliers in a real dataset.

The choice of the CLF constant significantly affects the performance of CLF. The means by which to choose an appropriate CLF constant, learning rate, optimiser and hyperparameter optimisation fall outside of the scope of this paper, and is suggested for future research. Other recommendations for future research include a recreation of similar tests, but with more powerful inferential tools and more data.

References

1. Zahra, M.M., Essai, M.H., Ellah, A.: Performance functions alternatives of MSE for neural networks learning. Int. J. Eng. Res. Technol. (IJERT) **3**(1), 967–970 (2014)
2. Heravi, A.R., Hodtani, G.A.: Where does minimum error entropy outperform minimum mean square error? a new and closer look. IEEE Access **6**(1), 5856 5864 (2018)
3. El-Melegy, M.T., Essai, M.H., Ali, A.A.: Robust training of artificial feedforward neural networks. In: Hassanien, A.E., Abraham, A., Vasilakos, A.V., Pedrycz, W. (eds.) Foundations of Computational. Studies in Computational Intelligence, vol. 201, pp. 217–242. Springer, Heidelberg (2009). https://doi.org/10.1007/978-3-642-01082-8_9
4. Brunet, F.: Contributions to parametric image registration and 3D surface reconstruction. PhD thesis, University of Auvergne, Auvergne, France (2010)
5. Borak, S., Härdle, W., Weron, R.: Stable distributions. In: Čížek, P., Weron, R., Härdle, W. (eds.) Statistical Tools for Finance and Insurance, pp. 21–44. Springer, Heidelberg (2005). https://doi.org/10.1007/3-540-27395-6_1
6. Li, X., Lu, Q., Dong, Y., Tao, D.: Robust subspace clustering by Cauchy loss function. IEEE Trans. Neural Netw. Learn. Syst. **30**(7), 2067–2078 (2019)
7. Park, S., Serpedin, E., Qaraqe, K.: Gaussian assumption: the least favorable but the most useful. IEEE Signal Process. Mag. **30**(3), 183–186 (2013)
8. Pearson, R.K.: Control Systems, Identification, pp. 687–707. Academic Press, California (2003)
9. Chambers, R.L., Steel, Wang, D.G., Welsh, A.: Maximum Likelihood Estimation for Sample Surveys. Chapman and Hall/CRC (2012)
10. Chen, R., Paschalidis, I.C.: A robust learning approach for regression models based on distributionally robust optimization. J. Mach. Learn. Res. **19**, 517–564 (2018)
11. Tsakalides, P., Nikias, C.L.: Maximum likelihood localization of sources in noise modeled as a Cauchy process. In: Proceedings of MILCOM 1994, vol. 2, pp. 613–617 (1994)
12. Barron, J.T.: A general and adaptive robust loss function. In: Proceedings of IEEE/CVF Conference on Computer Vision and Pattern Recognition (CVPR), pp. 4326–4334 (2019)

13. Huang, H.-C., Cressie, N.: Deterministic/stochastic wavelet decomposition for recovery of signal from noisy data. Technometrics **42**(3), 262–276 (2000)
14. Abu-Mostafa, Y.S., Magdon-Ismail, M., Lin, H.-T.: Learning from data : a short course. AMLbook.com, USA (2012)
15. Balkema, G., Embrechts, P.: Linear regression for heavy tails. Risks **6**, 93 (2018)
16. Fan, C., Zhang, D., Zhang, C.-H.: On sample size of the Kruskal-Wallis test with application to a mouse peritoneal cavity study. Biometrics **67**, 213–24 (2010)
17. Brcich, R., Iskander, D., Zoubir, A.: The stability test for symmetric alpha-stable distributions. IEEE Trans. Signal Process. **53**(3), 977–986 (2005)
18. Hart, A.: Mann-Whitney test is not just a test of medians: differences in spread can be important. BMJ **323**(7309), 391–393 (2001)
19. Sathishkumar, V.E., Park, J., Cho, Y.: Using data mining techniques for bike sharing demand prediction in metropolitan city. Comput. Commun. **153**, 353–366 (2020)
20. Qi, Z., Wang, H.: Dirty-data impacts on regression models: an experimental evaluation. In: Jensen, C.S., et al. (eds.) DASFAA 2021. LNCS, vol. 12681, pp. 88–95. Springer, Cham (2021). https://doi.org/10.1007/978-3-030-73194-6_6
21. Zhang, Z.: Improved Adam optimizer for deep neural networks. In: 2018 IEEE/ACM 26th International Symposium on Quality of Service (IWQoS), pp. 1–2 (2018)
22. Banerjee, C., Mukherjee, T., Pasiliao, E.L.: An empirical study on generalizations of the relu activation function. In: Proceedings of the 2019 ACM Southeast Conference (2019)

CASA: Cricket Action Similarity Assessment in Video Footage Using Deep Metric Learning

Tevin Moodley[ID] and Dustin van der Haar[✉][ID]

University of Johannesburg, Kingsway Avenue and, University Road, Auckland Park,
Johannesburg 2092, South Africa
{tevin,dvanderhaar}@uj.ac.za

Abstract. Cricket batters will often measure their performance through comparisons against successful batters or feedback provided by experts. Action Similarity Assessment is the task of comparing the similarity or dissimilarity of an action between two actors to determine how similar the actions they perform are to one another. This research paper proposes the use of a Siamese Convolution Neural Network to compute the similarity distances between different batters using video footage. Due to the limited research surrounding action similarity, a new dataset is proposed to help foster future works pertaining to action similarity. Three architectures are proposed to determine which architecture is best suited for the domain: *a custom CNN, Inception Resnet V2*, and *Xception*. From the results obtained, it can be concluded that the best solution for the *action similarity assessment* task within cricket video footage is a Siamese *Xception* architecture, achieving a model accuracy of 98%.

Keywords: Cricket action similarity · Xception · Inception Resnet V2 · Siamese network

1 Introduction

Research in computer vision has significantly grown over recent years due to digital video becoming readily available in large amounts, consequently increasing the demand for video processing and analysis. Human action recognition is a key task in computer vision, used in various applications, including video surveillance, health services, robotics, and many more [12]. A *human action* can be defined as the physical movements of the head, hands, legs or the body to convey meaningful information or interact with the environment [14]. Early works focused on action recognition, but more recent works are venturing out beyond the recognition task.

Cricket is a bat and ball game played by two teams, each consisting of eleven players on a team governed by specific game mechanics, as shown in Fig. 1 [3]. During the game, the bowling team must attempt to dismiss the batting team

A. Pillay et al. (Eds.): SACAIR 2022, CCIS 1734, pp. 139–153, 2022.
https://doi.org/10.1007/978-3-031-22321-1_10

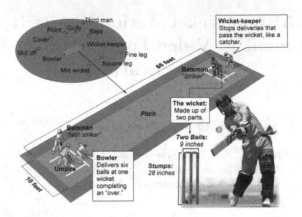

Fig. 1. A figure, which represents the game of cricket and the mechanics, taken from Chicago Tribune Articles [2].

by delivering the ball to the batter on strike. The batting team has two batters on the field at all times until the bowling team dismisses 10 out of the eleven batters. The team that accumulates the most number of runs by the end of the game is declared the winner [3]. A significant contributing factor to the outcome of a game is the batter's ability to score runs which is achieved by a batter striking the ball to different parts of the field.

Previous research focuses on recognising different batter strokes using machine learning and deep learning methods [15]. However, recognising cricket strokes have limited use to coaches during training. Athletes will often measure their performance through comparisons or feedback provided by experts [10]. This work defines the *Action Similarity Assessment (ASA)* task to derive the similarity of an action between two different actors.

This research aims to perform Action Similarity Assessment for a given stroke (specifically a drive in this study) performed between different batters. The study has the following contributions. **1)** A modified Siamese network architecture that achieves similarity comparison using deep metric learning is proposed. **2)** The ideal Convolutional Neural Network (CNN) based backbone, which is most applicable to the research domain, is determined. Finally, **3)** A new dataset is produced and will be disseminated upon request to foster future work for the cricket similarity assessment task.

This research paper performs Action Similarity Assessment between different cricket batters using deep metric learning. *Section 2* will unpack ASA by describing the core problems surrounding similarity between different actions and by describing related works. *Section 3* will describe the dataset used for this research and the collection process involved. The algorithms and methods used to achieve action similarity assessment will be unpacked, followed by the results in *Sect. 4*. The paper will be concluded with a detailed discussion of the results and findings.

2   Problem Background

Understanding the batting stroke is necessary to determine the key phases that are inspected when performing ASA between different batters. When a coach analyses similarities between batters, they look at a stroke in three phases; pre-delivery, execution, and post-delivery. Each phase inspects different criteria, and each criterion analyses the batter's stance before, during, and after the ball is bowled, along with the movement of the hands, feet, head, and body during and after the execution of a cricket stroke. The different phases will be unpacked in Sect. 3 to illustrate the mechanisms involved and needed to execute a cricket stroke correctly.

Action similarity involves computing the similarity scores between two actors to determine how similar or dissimilar the actions are and is considered an important facet of action learning and understanding actions [16]. Action similarity can be seen as a complex task within a realistic setting and can cause class ambiguity in multi-class action recognition [16]. To address the ambiguity, action similarity labelling was introduced by *Kliper-Gross, Hassner & Wolf*. Action similarity labelling aims to determine if actors in two different video sequences perform the same action or not, dubbed as the "same/not same" or as a binary classification task. The labelling algorithm relies primarily on creating a suitable metric for the differences between the actions from extracted kinematic features [12].

According to *Kliper-Gross*, action similarity addresses several noteworthy problems within the action recognition domain. Action similarity labelling has the following advantages over multi-class action labelling [12]: 1) It relaxes the problem of ambiguous classes as it is easier to label pairs as "same/not same" rather than pick one class out of hundreds. 2) Pair matching has interesting applications in its own right, such as data collection based on similar actions. Action similarity labelling has limited research directed towards focused applications.

Metric learning is an approach that aims at automatically constructing task-specific distance metrics from (weakly) supervised data in a machine learning manner to establish similarity or dissimilarity between objects. Metric learning aims to decrease the distance between similar objects and increase the distance between dissimilar objects [11]. Metric learning uses distance metrics to measure the similarities between different samples [11]. Deep metric learning has been used in the past to achieve similarity comparisons and is a continuation of *metric learning*. Current research shows that existing metric learning studies are based on Siamese and Triplet Networks, which will be unpacked in the sections to follow.

2.1   Related Works

It has already been mentioned that the field of action similarity is vastly under-studied, and limited works are available. However, image similarity is a field that has received more attention in the past, which can serve as inspiration for the action similarity task. Image similarity involves obtaining a similar image based

on a referenced image [1]. Authors *Appalaraju & Chaoji* developed a solution
for image similarity that makes use of a deep Siamese network that is trained on
positive and negative image pairs using an online pair mining strategy inspired
by Curriculum learning [1]. Their solution uses a multi-scale convolutional neural network in a Siamese network that learns a joint image embedding of the
top and lower layers. Their model is said to outperform traditional tasks for
image similarity. The proposed model was able to improve on the baseline and
traditional image similarity models with a 92.6% accuracy.

A similar task to action similarity is action quality assessment, where action
scoring is modelled as a regression task. Authors *Jain, Harit, & Sharma* use
deep metric learning with a custom CNN-based Siamese Network to perform
Action Quality Assessment (AQA) [10]. Their research noted that automated
vision score estimation models had been used to judge how well an action was
performed to avoid a biased judgement [10]. Actions performed by athletes are
usually scored based on estimation models by regressing the video into ground
truth scores provided by judges. However, the authors noted that these methods
are unbiased and cannot be explained. To enhance the scoring mechanism, *Jain
et al.* proposed a solution to make the scores more concrete by comparing an
action with a referenced video that would compare temporal variations of the
two videos and map the variations to a final score [10]. Their research proposes
an action scoring system, which comprises of a deep metric learning method that
learns the similarity between two referenced videos and a score estimation model
that derives a similarity score based on the two referenced videos. In Fig. 2 the
deep metric learning model demonstrates how the similarity is learnt between
two videos in a pair. Each backbone CNN takes in one video sequence as an
input, producing an output vector concatenated and fed into a fully-connected
dense layer. These dense layer outputs are fed to a sigmoid unit to map the two
videos' combined features to a binary similar/non-similar classification output
[10].

The Action Similarity Assessment is the task of computing the similarity distances between different actors to determine how similar an action is, whereas
Action Quality Assessment aims to determine the quality of an action by predicting an action score aligned to expert-provided priors to make recommendations
on the action. Action Similarity Assessment, Action Quality Assessment, and
Action Similarity Labelling are all based on the foundation of computing the
similarity between actions.

Other research achieves Action Similarity Assessment using skeletal human
keypoint detection. Researchers *Wai Feng, Lei, Liu* et al. propose a method that
makes use of OpenPose and byte pair encoding algorithms. These algorithms are
used to analyse and compare fitness movements based on human pose estimation, and movement similarity assessment that can provide real-time movement
comparison analysis and evaluation results [20]. The proposed model extracts the
action keypoints to generate the corresponding action vector. The action vector obtained from each corresponding video is then compared to determine the
similarity between the videos. It was noted that both algorithms require a great

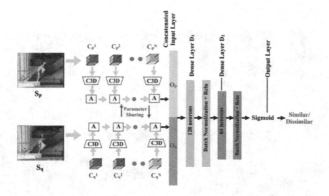

Fig. 2. The Siamese deep metric learning architecture proposed by *Jain, Harit, & Sharma*, which demonstrates the manner in which the two referenced videos are compared [10].

deal of computational resources, which may result in the delay of instantaneous feedback [20].

This study will attempt to take different cricket batters and deduce whether the stroke performed is the "same/not same" as a reference batter, which will allow batters to gauge their performance through comparison.

3   Experiment Setup

In Sect. 2, the different phases of the cricket stroke were outlined. It is necessary to understand these phases as it will justify why the dataset was created in its specific manner. The phases that determine how a stroke is performed are pre-delivery, execution, and post-delivery. Each phase will involve a complex array of movements involving the head, hands, feet, bat, and general bodily positioning. The pre-delivery determines how a batter will set up to face a delivery, as seen in *a* of Fig. 3. The pre-delivery phase occurs between the bowler's run-up and before the ball's release. The execution phase can be seen in Fig. 4, which occurs between the ball's release and impact on the bat. The type of delivery usually determines the execution. The execution phase is critical, determining where the ball will end. If the execution phase is not performed correctly, it can lead to the batter's dismissal. The final phase can be seen in Figure *b* of 3, which is the post-delivery phase, often analysed by many coaches as it allows one to determine whether the execution phase was executed correctly. Often coaches will look at the batter's head position after the execution to ensure it remains still and over the ball. Each of these highlighted phases determines how a stroke is performed. Therefore, the dataset complied will be made up of snippets that involve all three phases. Furthermore, it is important to note that when performing action similarity for a cricket stroke, a single image cannot be inspected as a stroke involves an array of complex movements.

(a) The pre-delivery phase of the stroke, which illustrates the manner in which the batter prepares to face the delivery [9]

(b) The post-delivery phase of the stroke, which illustrates the manner in which the batter performs a series of movements after the ball has been stroked [9].

Fig. 3. A figure illustrating the pre-delivery (a) and post-delivery (b) phase, phase a is the initial frame selected where frame b serves as the last phase considered for each video clip in the dataset

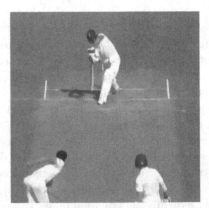

(a) The initial frame of the batter performing the execution phase

(b) The last frame before the post-delivery phase illustrating the batter performing the execution phase

Fig. 4. The execution (middle) phase of the stroke, which illustrates the manner in which the batter strokes the ball after the ball has been released by the bowler, the post-delivery phase proceeds this phase [9].

To perform the task of ASA using a Siamese network, two classes were created called; *reference* and *sample*. The reference class consisted of retired international batsman *AB de Villiers*, who has an International Test batting average of 50.66 [8], demonstrating his success within the sport. There are several candidates that have seen more success in the game and could be considered. However, AB de Villers was selected due to the availability of data and the batter serves as a good baseline measure for initial studies. It is well-known fact, that all batters exhibit different batting techniques and therefore research must be cognisant of the context in which strokes are performed for future works. The second class, called sample, comprises of different current International Test batters. Comparing current batters to a successful retired batter may yield benefits in the way of coaching and improving the technique of current batters in future works. Since there are several types of batting strokes where each stroke entails different movements, the batting stroke selected and analysed for this study is the drive stroke. The drive stroke was chosen due to the amount of video footage available on YouTube for various international test batters. The drive stroke consisted of two classes; reference and sample. Each class contained 100 video clips collected from YouTube that consists of the three key phases as identified earlier. Since batter AB de Villers was selected, there was a limited amount of data available, which subsequently restricted the number of samples available for the dataset. During the video collection phase, International Test Match highlights were collected, and each video was fed into video editing software, OpenShot. The video was manually edited, where specific clips of the batsman performing the drive stroke were captured and saved for later processing.

Once the classes were created, data pre-processing was applied to reduce complexity and ensure the data was in the correct format for analysis. Each pixel was normalised using (0, 1), and a select number of frames were chosen. Through testing and validation, the videos were separated using an 85:15 split. The ablation study discussed later shows that the middle 20 frames extracted from each video yield the best performance. It was noted that the selected frames covered all three phases required to perform a stroke and removed any unnecessary frames within the video.

The following process within the pipeline was followed to create the image pairs. Once the data was correctly prepared, the image pairs were created, using positive and negative pairs. The positive and negative image pairs were created by looping over every frame, randomly selecting an image with the referenced/sample label index, and pairing it up with the same label.

3.1 Methods

The model proposed is a Siamese Convolutional Neural Network. Siamese networks were first introduced in the 90s by *Bromly, Guyon, LeCun, Sackinger & Shah*. The Siamese network consists of two identical sub-networks joined at their outputs. During the training phase, the two sub-networks extract features from samples, while the joining neuron measures the distance between the two feature vectors [4]. The Siamese network uses twin networks to realise non-linear embedding features from its input domain. The non-linear embedding is accomplished using feature extraction followed by learning the discriminative embedding space [17]. In their original form, the sister networks are identical, where the embedding part identifies a metric in the resulting space to increase the discriminatory power. These characteristics of the Siamese networks have many key properties, such as; guaranteed consistency of predictions, as the weight sharing ensures that similar samples will not map to different parts of the embedded space. Another key property lies within its symmetry, which allows the input pair to be fed into the network in any manner [17].

In order to successfully implement the Siamese network, an appropriate distance measure must be selected. Traditionally, many distance measures are used to determine the similarity between a set of specified features. Some popular distance measures known within the research domain are Euclidean, Manhattan, and Hausdorff. Through testing and validation, it was noted that the Euclidean distance measure yielded the best performance. Euclidean distance is defined as the matrices of the squared distances between two points [7]. The Euclidean Distance is also denoted by L2 distance; if $u = (x1, y1)$ and $v = (x2, y2)$, then the Euclidean distance between u and v are given by [13].

$$D(u, v) = \sqrt{(x1 - x2)^2 + (y1 - y2)^2} \tag{1}$$

Various backbone architectures were selected to determine which architecture is best suited for the research domain. The architectures proposed for this study are; a *custom CNN*, *Inception Resnet V2*, and the *Xception* architecture. A typical CNN consists of two components; a feature extractor and a classifier [5]. The feature extractor is used to filter images into feature maps that represent a variety of features of the image. These features may include corners, lines, and edges, which are relatively invariant to position shifting or distortions. The output created from the feature vector is a low-dimensional vector composed of the features extracted. The vector is then fed into a classifier based on a traditional artificial neural network [5]. In Fig. 5 an illustration of the proposed solution is highlighted that implements a Siamese CNN using euclidean distance.

A *custom CNN* was created to serve as a baseline for comparison against more well-established CNNs. The *custom CNN* was built as follows. The input shape used was 128×128, with 3 channels. Two-layer sets were defined, where each convolution layer learns a total of $64 \times 2 \times 2$ filters, the layers both use the *relu* activation function and apply 2×2 max pooling. Two-dimensional global average pooling is applied as it is more native to the convolution structure by enforcing correspondences between feature maps and categories. As will be shown in the ablation study, various embedding dimensions were used to determine the optimal value. It can be seen that the image embedding dimension of 64 proved to be the optimal selection.

Once the baseline performance was computed, more well-known architectures were tested and validated to improve the performance. Based on the top-1 accu-

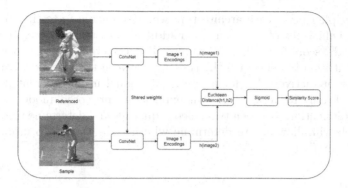

Fig. 5. The proposed Siamese CNN using euclidean distance to determine the similarity between a referenced and sample video sequence.

racy and top-5 accuracy [6,18], it was determined that *Inception Resnet V2* and *Xception* architectures would be used.

The *Inception Resnet V2* is a convolutional neural network trained on more than a million images from the *ImageNet* database that implements concatenation in each multi-branch architecture [18]. Using the filter expansion layer and residual modules after each inception block, the dimensionality of the filter bank is scaled up to compensate for the dimensionality reduction induced by the inception block [18].

Research conducted by *Chollet* illustrated the manner in which the *Xception* architecture was able to outperform the *Inception V2 Resnet* architecture on the *ImageNet* dataset [6]. The *Xception* architecture comprises 36 convolutional layers forming the feature extraction base of the network, and the layers can decouple the mapping between cross-channel and spatial correlations. The 36 layers are structured into 14 modules. All 14 modules have linear residual connections around them, apart from the first and last modules [6]. The *Xception* architecture is a linear stack of depthwise separable convolution layers with residual connections.

Once the architectures were implemented, the left input and right input images had to be created in order to compare the similarity between the referenced video and the sample video. The left and right image pairs were created by passing the input shape through the proposed model. Once the features were extracted, the euclidean distance measure was computed to determine the distance between the features. Once the distance measure was obtained, an output layer was created with the sigmoid activation function to match the number of specified classes. Based on experimentation (that is also discussed in the ablation study), the loss function selected was binary cross-entropy. *Adam* learning optimiser was selected, and the number of epochs for each architecture was set to 150 with early stopping.

Each architecture's performance will be evaluated on the model's accuracy, confusion matrix, precision, recall, and f1-score metrics. During the implementation, multiple runs of each architecture were completed to testify to the effectiveness and efficiency of the results. An additional metric used was the Average Precision (AP) score from the precision-recall Receiver Operating Curve (ROC), which was utilised to determine the tradeoff between the true positive rate and the positive predictive value for a predictive model using different probability thresholds. From the results, we can expect the proposed model to successfully draw distinctions between two video sequences and determine the similarity score, which will allow one to determine whether the two video inputs are the "same/not same".

4   Results

The results highlight the ability to successfully determine whether different video sequences are the "same/not same". However, the results further highlight the improvements needed to gain better performance. The test set consists of 15 video sequences, where each video consists of 20 frames. The sample class struggled with 8 video sequences incorrectly labelled as referenced for the *custom CNN*. The *Xception* architecture had 4 incorrectly labelled video sequences for each class and performed more consistently, which is further supported using the f1-score found in Table 1.

The accuracy can be seen in Table 1. The *Xception* model performs best with an accuracy score of 98%. The *custom CNN* also provides promising results with an accuracy of 90%. The *Inception Resnet V2* does not yield the expected performance scoring a low accuracy score of 64%. The discussion Section will unpack these results to determine why each architecture achieved its respective accuracy and loss scores.

Table 1. The accuracy and loss scores for the proposed architectures on 150 epochs.

Architecture	Accuracy	Precision	Recall	F1-score
Custom CNN	90%	76%	70%	68%
Inception Resnet V2	64%	59%	57%	54%
Xception	98%	73%	73%	73%

4.1   Ablation Study

To understand the performance of each component of the system, an ablation study was conducted. The components assessed during the ablation study were: the number of frames, the loss function, and the embedding dimension size for the *custom CNN*. To determine what the ideal number of frames was for the study, two variations were considered, 20 and 15 frames. It was noted that some video clips had a varied number of frames. To mitigate this issue, the middle frames were selected to ensure the three key phases were all included, and unnecessary frames were subsequently removed. From Table 1, the different accuracy and average precision scores are computed for each variation, and the optimal components for each CNN are realised. The *Xception* architecture is optimal using 20 frames with a binary cross-entropy function, with an accuracy of 98% and an average precision of 88%, respectively. Traditionally in similarity problems, the contrastive loss function is said to yield better performance [19], but it can be seen that binary cross-entropy performs better within the problem domain (Table 2).

Table 2. The ablation study where the components considered were the number of frames and loss function.

Architecture	Loss function	No. frames	Accuracy	Average precision
Custom CNN	Binary	20	90%	80%
Custom CNN	Contrastive	20	40%	33%
Inception Resnet V2	Binary	20	64%	64%
Inception Resnet V2	Contrastive	20	35%	30%
Xception	Binary	20	98%	88%
Xception	Contrastive	20	64%	65%
Custom CNN	Binary	15	84%	78%
Custom CNN	Contrastive	15	52%	41%
Inception Resnet V2	Binary	15	54%	60%
Inception Resnet V2	Contrastive	15	28%	20%
Xception	Binary	15	92%	81%
Xception	Contrastive	15	54%	51%

Table 3. The embedding sizes; 32, 64, and 128 used to determine the optimal embedding size for the *custom CNN*

Architecture	Embedding	No. frames	Accuracy	Average precision
Custom CNN	32	20	85%	74%
Custom CNN	64	20	90%	80%
Custom CNN	128	20	86%	76%

For the *custom CNN*, the number of embedding dimensions was altered to determine the optimal embedding size for the use case. The embedding dimensions used were 32, 64, and 128. It was noted that the embedding dimension with 64 yielded the optimal performance. The accuracy of 90% and an average precision score of 80% are achieved.

The precision-recall graphs were mapped for each architecture for the respective positive labels. The *custom CNN* achieved an Average Precision (AP) score of 80%. The *Xception* architecture yielded the best performance with an AP score of 83%. The *Inception Resnet V2* architecture achieved an AP score of 66%.

(a) Custom CNN ROC with an AP of 80%.

(b) *Inception Resnet V2* ROC with an AP of 66%.

(c) *Xception* ROC with an AP of 84%.

Fig. 6. An illustration of the ROC on the positive label to determine the similarity threshold that should be used for each architecture.

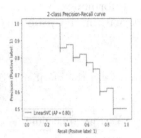

(a) A figure illustrating AB de Villers from the reference class performing the drive stroke

(b) A figure illustrating a different batter from the sample class performing the drive stroke

Fig. 7. A comparison between two different batters illustrating the similarity score of 0.47 obtained using the Siamese CNN.

Since the ASA task tries to determine how similar an action is, the distance measures between each video are computed to understand how the proposed model performs. The execution phase frames for two different videos are shown in Fig. 7.

5   Discussion of Results

This study aims to perform ASA on video sequences to determine if two different batters perform the "same/not same" stroke. The proposed *Siamese Xception model* successfully achieves the aim of this study based on the results highlighted in Sect. 4. The Xception architecture misclassifies a total of 8 video samples. In order for two different videos to be considered the same, they need to have a similarity score above 0.5, which was determined by the AP score found in *c* of Fig. 6. A similarity score that is closer to 1 (greater than 0.5) is considered the same, whereas a similarity score that is further away from 1 (less than 0.5) is considered to be not the same. When analysing the misclassifications for the *Xception* architecture, it is noted that the similarity score for each misclassification falls between 0.4 and 0.5, which falls under the required threshold, which can be further supported using Fig. 7. The analysis of these videos, which fall under the required threshold, could serve as a possible reason for the incorrect classifications.

A comparison of the proposed architectures was made as there was little research to indicate which architecture would best suit the problem domain. From the results, it can be concluded that the best solution for the similarity task within cricket video footage is a *Siamese Xception architecture*.

The AP score for the *Xception* is 84%. The AP score combines precision and recall in a single visualisation. The higher the y-axis, the better the performance of the model. The precision is mapped for the positive label to determine how well the positive label performs during the mapping stage. It can be seen from Fig. 7 and in the ablation study that the ideal threshold is 0.5. Hence, the threshold score of 0.5 is used to determine whether a video is positive or negative.

From Table 1, the architecture accuracy and loss scores were highlighted. The *Xception* architecture had an accuracy score of 98%. It has been demonstrated that the *Xception* architecture in both a highly specialised context, as seen in this study, and a generalised context, outperforms the *Inception Resnet V2* architecture as well as the *custom CNN*, thus highlighting its efficiency within the given context.

Using Fig. 7, the similarity scores for the video sequences can be compared, and one can determine how similar a drive stroke performed between different batters is. It is noted that when the bat is in a similar position, the similarity score tends to increase, thus emphasising the bat's positioning. However, the bat position is also determined by the pitch of the ball and can vary based on the type of delivery. Future works must cater for the type of delivery to improve the proposed model's performance.

In a study by *Jain, Harit, & Sharma* the authors create a Siamese Network to perform Action Quality Assessment (AQA) [10]. In the study, each of the two Long-short Term Memory (LSTM) networks take in one video sequence as an input and produces an output vector concatenated and fed into a fully-connected dense layer. Their performance of the feedback module on the original and misaligned data when using the Reference Guided Regression yielded a precision score of 58%, a recall score of 95%, and an f1-score of 72%. *Jain et al's* works, and this study both addresses action similarity within video footage but analyse different domains. When comparing the metrics across both studies, it is noted that the Siamese *Xception* model is a valid selection for this research study and future Action Similarity Assessment tasks. Future works will consider using an LSTM for Action Similarity Assessment to determine whether a correlation between ASA and AQA can be made.

6 Conclusion

This research article performed Action Similarity Assessment by comparing two video sequences using a Siamese *Xception* model to determine the similarity between different cricket batters performing the drive stroke. The findings highlight the advantages of transfer learning and identify the *Xception* architecture as the best performing architecture for the similarity task, which is a largely unexplored area and creates a new dataset within the domain. The proposed solution provides meaningful contributions in that a Siamese network is created with varying CNN backbones. Using the individual similarity scores between different batters, it can be deduced that the bat's positioning is a key outlier for the similarity, and future works must factor in the type of delivery to make further improvements. Future improvements can be made by factoring in more temporal aspects, such as finer-grain gestures. Additional work should also evaluate the generalisation ability of similar models in a match setting and analyse players of varied demographics, including age, gender, skill level and format type.

References

1. Appalaraju, S., Chaoji, V.: Image similarity using deep CNN and curriculum learning. arXiv preprint arXiv:1709.08761 (2017)
2. Articles, C.T.: Cricket basics explaination. http://www.chicagotribune.com/chi-cricket-basics-explanation-gfx-20150215-htmlstory.html
3. Auerbach, Michael P.M.: Cricket (sport). Salem Press Encyclopedia (2022)
4. Bromley, J., Guyon, I., LeCun, Y., Säckinger, E., Shah, R.: Signature verification using a "siamese" time delay neural network. Adv. Neural Inf. Process. Syst. **6** (1993)
5. Chang, J., Sha, J.: An efficient implementation of 2D convolution in CNN. IEICE Electron. Express **14**, 20161134 (2016)
6. Chollet, F.: Xception: deep learning with depthwise separable convolutions. In: Proceedings of the IEEE Conference on Computer Vision and Pattern Recognition, pp. 1251–1258 (2017)
7. Dokmanic, I., Parhizkar, R., Ranieri, J., Vetterli, M.: Euclidean distance matrices: essential theory, algorithms, and applications. IEEE Signal Process. Mag. **32**(6), 12–30 (2015)
8. ESPNcricinfo: abdevilliers (n.d.). https://www.espncricinfo.com/player/ab-de-villiers-44936. Accessed 04 July 2022
9. Goel, D.: Ab Devilliers 126 Vs Aus in Port Elizabeth (2021). https://www.youtube.com/watch?v=CCRS-z31g60&t=32s. Accessed 23 June 2022

10. Jain, H., Harit, G., Sharma, A.: Action quality assessment using Siamese network-based deep metric learning. IEEE Trans. Circ. Syst. Video Technol. **31**(6), 2260–2273 (2020)
11. Kaya, M., Bilge, H.Ş: Deep metric learning: a survey. Symmetry **11**(9), 1066 (2019)
12. Kliper-Gross, O., Hassner, T., Wolf, L.: The action similarity labeling challenge. IEEE Trans. Pattern Anal. Mach. Intell. **34**(3), 615–621 (2011)
13. Malkauthekar, M.: Analysis of euclidean distance and manhattan distance measure in face recognition. In: Third International Conference on Computational Intelligence and Information Technology (CIIT 2013), pp. 503–507. IET (2013)
14. Mitra, S., Acharya, T.: Gesture recognition: a survey. IEEE Trans. Syst. Man Cybern. Part C (Appl. Rev.) **37**(3), 311–324 (2007)
15. Moodley, T., van der Haar, D.: Cricket stroke recognition using computer vision methods. In: Kim, K.J., Kim, H.-Y. (eds.) Information Science and Applications. LNEE, vol. 621, pp. 171–181. Springer, Singapore (2020). https://doi.org/10.1007/978-981-15-1465-4_18
16. Nair, V., et al.: Action similarity judgment based on kinematic primitives. In: 2020 Joint IEEE 10th International Conference on Development and Learning and Epigenetic Robotics (ICDL-EpiRob), pp. 1–8. IEEE (2020)
17. Roy, S.K., Harandi, M., Nock, R., Hartley, R.: Siamese networks: the tale of two manifolds. In: Proceedings of the IEEE/CVF International Conference on Computer Vision, pp. 3046–3055 (2019)
18. Szegedy, C., Ioffe, S., Vanhoucke, V., Alemi, A.A.: Inception-v4, inception-ResNet and the impact of residual connections on learning. In: Thirty-first AAAI Conference on Artificial Intelligence (2017)
19. Wang, F., Liu, H.: Understanding the behaviour of contrastive loss. In: Proceedings of the IEEE/CVF Conference on Computer Vision and Pattern Recognition, pp. 2495–2504 (2021)
20. Zhou, J., et al.: Skeleton-based human keypoints detection and action similarity assessment for fitness assistance. In: 2021 IEEE 6th International Conference on Signal and Image Processing (ICSIP), pp. 304–310. IEEE (2021)

From GNNs to Sparse Transformers: Graph-Based Architectures for Multi-hop Question Answering

Shane Acton[ID] and Jan Buys[✉][ID]

Department of Computer Science, University of Cape Town, Cape Town, South Africa
ACTSHA001@myuct.ac.za, jbuys@cs.uct.ac.za

Abstract. Sparse Transformers have surpassed Graph Neural Networks (GNNs) as the state-of-the-art architecture for multi-hop question answering (MHQA). Noting that the Transformer is a particular message passing GNN, in this paper we perform an architectural analysis and evaluation to investigate why the Transformer outperforms other GNNs on MHQA. We simplify existing GNN-based MHQA models and leverage this system to compare GNN architectures in a lower compute setting than token-level models. Our results support the superiority of the Transformer architecture as a GNN in MHQA. We also investigate the role of graph sparsity, graph structure, and edge features in our GNNs. We find that task-specific graph structuring rules outperform the random connections used in Sparse Transformers. We also show that utilising edge type information alleviates performance losses introduced by sparsity.

Keywords: Transformers · Graph neural networks · Question answering

1 Introduction

Multihop question answering (MHQA) is a challenging language understanding task which involves reasoning over multiple facts often found across multiple documents [17,19]. Standard sequence models such as LSTM-based models have struggled on this task due to the need to model long-distance dependencies and to potentially perform multiple reasoning steps to answer a question.

Facts that are required to answer a question are often linked via common entities in what is called a reasoning chain [5]. Various MHQA works [7,9,14] demonstrate that MHQA input can be represented as an entity graph, and that Graph Neural Networks (GNNs) can be used to encode the graphs and learn to do reasoning to answer multihop questions. On tasks that involve long sequences, the Transformer [15] has been shown to exhibit higher task performance than previous approaches. However, as the Transformer processes information by performing full self-attention between all pairs of input tokens, its compute requirements scale quadratically with the number of tokens. This makes modeling very long token sequences with full self-attention Transformers intractable.

Sparse Transformers such as the Longformer [2] and Big Bird [20] perform self-attention between a subset of input token pairs. This allows them to process longer token sequences, at the cost of theoretical model expressiveness [20]. Sparse Transformers use rules to construct adjacency matrices that define which token pairs should communicate. This is conceptually identical to the function of a GNN, which makes use of an adjacency matrix to communicate node state information around a graph. GNNs, however, are a much more general class of models than Transformers. Sparse Transformer models are currently state-of-the-art on the WikiHop MHQA dataset [17], although a GNN-based model [9] is competitive with Sparse Transformers on the HotpotQA dataset [19].

In this paper, we aim to improve the performance of entity graph-based GNN models on the WikiHop dataset by leveraging the observation that the Transformer is a specific instance of a GNN in the message-passing framework. Typically, Transformer models operate at the token level only. However, we show that the GNN architecture can be substituted with a Transformer without changing any other part of the MHQA pipeline. The GNN-based MHQA pipeline involves encoding documents independently before aggregating token representations into course grained nodes. These nodes correspond to token spans in the document and can represent entities, sentences, and even whole documents. Finally, these node representations are encoded by a GNN [9,14], before being used to predict an answer. Despite the success of the Transformer, to the best of our knowledge, it has not previously been evaluated in such a GNN-based MHQA setting.

We consider a set of architectural design choices in GNN-based MHQA models, incorporating best practices established in the Transformer literature. We focus on three aspects in particular. First, we compare additive attention and scaled dot product (SDP) attention as used by Graph Attention Neural Networks (GAT) [16] and the Transformer respectively. We find that while both forms of attention perform favourably when compared to a non-attention baseline, SDP attention results in our best task performance.

Second, we evaluate the use of the Transformer's residual connections and position-wise Feed Forward Neural Networks (FFNN) in-between self-attention layers, which we refer to as the Transformer Update Function (TUF). We compare this to the position-wise gating mechanism common in GNNs [7,14]. We find that while gating has mixed effects on task performance, the TUF improves performance in all tested settings. Furthermore, we find that SDP attention pairs especially well with the TUF.

Third, we investigate graph structure and edge information in our GNN-based MHQA model. Full self-attention outperforms task-specific sparse attention when no edge information is available, supporting the idea that sparsity degrades model expressiveness [20]. However, when edge information is available, our task-specific sparse attention model outperforms the full self-attention model, indicating that edge information alleviates some disadvantages of sparse models.

2   Background

Graph Neural Networks (GNNs) were introduced to generalise Convolutional Neural Networks (CNNs) to non-Euclidean data [18]. GNNs have been applied widely due to the generality of the data they can model, and their natural modeling of sparsity. In this work we consider GNNs that can be defined in the message passing framework, which are able to operate on graphs of arbitrary topology [18]. After introducing this framework, we use it to define the Graph Attention Neural Network (GAT) [16], and the Transformer [15].

2.1   Message Passing GNNs

A graph can be represented by a set of n nodes and an edge matrix $E \in \mathbb{Z}^{n \times n}$ whose entries represent the edge types between any two nodes, with $E_{ij} = 0$ indicating that nodes i and j are unconnected. We use the notation given in the following equations to describe the general form of updating a single node state ($\boldsymbol{h}_i \in \mathbb{R}^f$) in the kth message passing GNN (MP-GNN) layer.

1. Message:
$$M_{ij}^k = \phi^k(\boldsymbol{h}_i^k, \boldsymbol{h}_j^k, E_{ij}), \tag{1}$$

2. Aggregate:
$$A_i^k = \sum_{j \in N(i)} \omega^k(M_{ij}^k, \boldsymbol{h}_i^k, \boldsymbol{h}_j^k), \tag{2}$$

3. Update:
$$\boldsymbol{h}_i^{k+1} = \gamma^k(\boldsymbol{h}_i^k, A_i^k). \tag{3}$$

Here ϕ^k is the message function at layer k and M_{ij}^k is its output which represents the message vector being passed to node i from node j along a single edge E_{ij}. ω^k is the aggregate function which prepares the message M_{ij}^k to be summed with all messages bound for node i. This sum forms a single aggregated message A_i^k to be passed to node i, which represents information from $N(i)$, the set of node i's neighbour nodes as defined by the edge matrix E, represented by their node indices. γ^k is the update function which combines the aggregated message A_i^k with the nodes current state $\boldsymbol{h}_i^k$ to create the node's state at the next layer, $\boldsymbol{h}_i^{k+1}$. The full algorithm for using these equations to encode a set of node states is given as Algorithm 1.

2.2   Attention

In order to define the GAT and Transformer as MP-GNNs we first define their attention mechanisms, restricted to the case of a single head for clarity. Additive attention involves the use of a multilayer perceptron (MLP) to predict *compatibility scores* between pairs of vectors:

$$\text{MLP}(\boldsymbol{h}_i, \boldsymbol{h}_j) = \sigma(\text{Concat}_f(\text{W}_v \boldsymbol{h}_i; \text{W}_v \boldsymbol{h}_j)\text{W}_a), \tag{4}$$

Algorithm 1. MP-GNN Node Encoding Procedure

Require:
  Initial node features $\{\boldsymbol{h}_1^1, .., \boldsymbol{h}_n^1\}$
  Edge matrix $E \in \mathbb{Z}^{n \times n}$

1: **for all** $k \in \{1, .., L\}$ **do**
2:     **for all** $i \in \{1, .., n\}$ **do**
3:         $A_i^k \leftarrow Zeros(f) \in \mathbb{R}^f$
4:         **for all** $j \in N(i)$ **do**
5:             $M_{ij}^k \leftarrow \phi^k(\boldsymbol{h}_i^k, \boldsymbol{h}_j^k, E_{ij})$
6:             $A_i^k \leftarrow A_i^k + \omega^k(M_{ij}^k, \boldsymbol{h}_i^k, \boldsymbol{h}_j^k)$
7:         **end for**
8:         $\boldsymbol{h}_i^{k+1} = \gamma^k(\boldsymbol{h}_i^k, A_i^k)$
9:     **end for**
10: **end for**
11: **return** $\{\boldsymbol{h}_0^L, .., \boldsymbol{h}_n^L\}$

---

where W_a and W_v are model weights and Concat_f denotes concatenation along the feature dimension. Finally, σ is a non-linear activation function.

Scaled dot product (SDP) attention calculates compatability scores as:

$$\text{Dot}(\boldsymbol{h}_i, \boldsymbol{h}_j) = \frac{(\boldsymbol{h}_i W_q)(\boldsymbol{h}_j W_k)^T}{\sqrt{f}}, \tag{5}$$

where W_q and W_k are model weights for queries and keys respectively.

The compatibility scores S_{ij} are computed using either MLP(.) or Dot(.) for every pair of nodes i and j. The softmax function is used to normalise the compatibility scores and produce *attention scores*:

$$\alpha(\boldsymbol{h}_i, \boldsymbol{h}_j) = \text{softmax}_j(S_{ij}) = \frac{exp(S_{ij})}{\sum_{k=1}^{l} exp(S_{ik})}. \tag{6}$$

To distinguish the use of additive or SDP attention in our models, we denote the GAT's additive attention as α_g and the Transformer's SDP attention as α_t.

2.3 GAT

We define the message passing equations for the GAT, in the case of a single attention head.

1. Message:

$$\phi^k(\boldsymbol{h}_i^k, \boldsymbol{h}_j^k, E_{ij}) = W_v \boldsymbol{h}_j, \tag{7}$$

2. Aggregate:

$$\omega^k(M_{ij}^k, \boldsymbol{h}_i^k, \boldsymbol{h}_j^k) = \alpha_g(\boldsymbol{h}_i^k, \boldsymbol{h}_j^k)M_{ij}^k, \tag{8}$$

3. Update:

$$\gamma^k(\boldsymbol{h}_i^k, A_i^k) = \sigma(A_i^k). \tag{9}$$

Here, $W_v \in \mathbb{R}^{f \times f}$ is a learned linear transformation. The additive attention function α_g is used in the aggregate step.

2.4 Transformer

While not typically thought of as a GNN, the Transformer [15] in fact performs message passing over the elements inputted to it. These elements are typically referred to as tokens, however they can naturally be considered as nodes. The standard Transformer performs full self-attention, meaning all nodes are connected to all nodes. Using the Transformer to model sparse graphs is an implementation detail, but typically would involve masking out the attention matrix with an adjacency matrix. This allows the Transformer to operate over arbitrary graph topology. The Transformer is described using the message passing notation using the equations that follow.

1. **Message:**
$$\phi^k(\boldsymbol{h}_i^k, \boldsymbol{h}_j^k, E_{ij}) = W_v \boldsymbol{h}_j, \tag{10}$$

2. **Aggregate:**
$$\omega^k(M_{ij}^k, \boldsymbol{h}_i^k, \boldsymbol{h}_j^k) = \alpha_t(\boldsymbol{h}_i^k, \boldsymbol{h}_j^k) M_{ij}^k, \tag{11}$$

3. **Update:**
$$\gamma^k(\boldsymbol{h}_i^k, A_i^k) = \text{TUF}(\boldsymbol{h}_i^k, A_i^k). \tag{12}$$

Here the SDP attention function α_t is used. The other difference is that the Transformer uses the TUF update function introduced in the original Transformer paper [15]. The Transformer Update Function (TUF) consists of residual connections, Layernorms [1], and a Feed Forward Neural Network (FFNN), formally given by equations:

$$\text{TUF}(\boldsymbol{x}_1, \boldsymbol{x}_2) = \text{Norm}(\boldsymbol{x}' + \text{FFNN}(\boldsymbol{x}')), \tag{13}$$

$$\boldsymbol{x}' = \text{Norm}(\boldsymbol{x}_1 + \boldsymbol{x}_2). \tag{14}$$

Norm is the LayerNorm [1] function, and FFNN is an MLP with a single hidden layer.

Both the Transformer and GAT make use of multi-headed attention. While the details differ slightly between the two models, the concept is the same. Multi-headed attention involves computing multiple attention scores for every pair of nodes using distinctly parameterised functions (heads). These multiple attention scores are then used to perform multiple weighted sums which are then combined. Making use of multiple attention heads increases the theoretical expressiveness of the attention mechanism [15], possibly by allowing different heads to focus on different criteria when comparing vectors [6]. Multi-headed attention is compatible with the message passing framework, but for brevity we omit its definitions. When the Transformer is viewed as an MP-GNN, the multi-headed attention block is the GNN's message and aggregate functions, while the rest of the operations all fall under the GNN's update function.

2.5   Gating and Over-Smoothing

MHQA GNN's commonly [7,14] employ some form of a gating [13] mechanism. Gating can be used in a GNN's update function to modulate how neighbour node information updates node states. We consider a particular gating function common in GNN-based MHQA models [7,14], defined as:

$$G(\boldsymbol{x}_1, \boldsymbol{x}_2) = \tanh(\boldsymbol{u}) \odot \boldsymbol{g} + \boldsymbol{x}_1 \odot (1 - \boldsymbol{g}), \tag{15}$$

$$\boldsymbol{u} = \boldsymbol{x}_1 W_u + \boldsymbol{x}_2, \tag{16}$$

$$\boldsymbol{g} = \sigma(\text{Concat}_f(\boldsymbol{x}_1, \boldsymbol{u})W_g), \tag{17}$$

where $\odot$ denotes element-wise multiplication. $W_g \in \mathbb{R}^{2f \times f}$ and $W_u \in \mathbb{R}^{f \times f}$ are learned matrices. Using gating this way in a GNN's update function is believed to reduce the over-smoothing problem [14]. Over-smoothing prevents the effective use of deep GNNs. It is thought to be caused by the loss of node identities as node representations become more similar to each other every layer [4].

Although the Transformer is an MP-GNN, it does not suffer from the over-smoothing problem, with extremely deep Transformers like GPT3 [3] (96 layers) producing state-of-the-art results. This may be due to the use of residual connections in the TUF, which are not commonly used in GNN-based MHQA models. Thus, in this work we investigate whether the TUF and gating are complimentary, or if one is superior to the other.

3   Model

We introduce our Graph Neural Network (GNN)-based Multihop Question Answering (MHQA) model. The model is trained to answer multiple-choice questions in the form of a question q, a set of text passages S_q containing information needed to answer the question, a set of answer candidates C_q, and the answer $a_q \in C_q$. For WikiHop, the question comes in the form of the semantic triple (Subject, Relationship, ?) where Subject is an entity and "?" the answer to be predicted. We use heuristic rules proposed by previous work [14] to construct graphs based on the question inputs. The model uses a pre-trained token embedder such as GloVe [11] or BERT [8], alongside a parameterised Transformer-encoder to create graph node embeddings. The graphs are then encoded by the GNN, and the graph encodings are used to predict the answer to the question.

3.1   Graph Construction

We follow the Heterogeneous Document Entity model (HDE) [14] to construct a graph composed of entities, answer candidates, and documents for a given multiple-choice question. These graphs are heterogeneous in that they contain multiple node and edge types. They also include information at multiple levels of granularity, namely the phrase-level entity and candidate nodes, and the course-grained document nodes. An example is shown in Fig. 1.

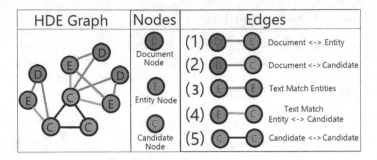

Fig. 1. An example graph used in our model, similar to HDE graphs [14], shown here together with node and edge types.

Entity nodes represent token spans in containing documents. The entities considered by our model are only those which are either a string match with any of the answer candidates, or the subject entity found in the question. There is an entity node for each extracted entity mention in each document. Therefore there may be multiple entity nodes representing identical text from different locations: these are referred to as co-mentions. The other node types are: a document node for each supporting document, and a candidate node for each answer candidate.

A set of heuristic rules connect nodes with the following edge types:

1. Document nodes connect to entity nodes extracted from the document.
2. Document nodes are also connected to candidate nodes if their text is found in the document.
3. *Comention Edge*: Entity comentions are connected. This edge type enables connecting mentions of the same entity across long distances in the input.
4. Entity nodes are connected to the candidate node whose text it matches.
5. All candidate nodes are connected to each other.

3.2   Graph Node Embedding

The first step in constructing embeddings for each of the graph nodes is to generate token embedding sequences for each document S_q^i, candidate C_q^j, and the query q. The token embeddings are obtained by use of a pre-trained token embedder such as GloVe or BERT. This involves first breaking the text up into tokens recognised by the embedder's vocabulary, and then using the embedder to produce a vector for each token. This yields *token matrices* X_q, X_s^i, and X_c^j for the query, i^{th} document, and j^{th} candidate respectively. Beyond this point, the model architecture is not affected by the choice of token embedder.

Each node represents a token span, i.e., it corresponds to a subsequence of one of the token matrices. We refer to these subsequences of token matrices as *node matrices*, and denote the node matrix for a node with index t as $X_t \in \mathbb{R}^{l \times f}$ for a node with l tokens within its span.

We concatenate the query token matrix X_q with each node matrix X_t, and encode the result with a Transformer encoder Trans_c:

$$X_{qt} = \text{Trans}_c(\text{Concat}_s(X_q, X_t)), \tag{18}$$

where Concat_s denotes concatenation along the sequence dimension. The output of the Transformer is a query-aware matrix X_{qt}. We use the vector corresponding to the first token in the sequence X_{qt} as our initial node embedding $\mathbf{h}_t^0$.

3.3   GNN Encoding

We use message passing GNNs with L-layers to encode the graphs. We consider several GNN variations: We compare additive attention, SDP attention and a GNN without attention. We also consider using the Transformer Update Function (TUF), gating, both, or neither.

GNN Cores with Edge Types. We introduce the term *GNN Core* to refer to the choice of message and aggregate functions in a GNN. All our GNNs include edge type embeddings. Given the edge matrix $E \in \mathbb{Z}^{n \times n}$ such that E_{ij} represents the edge type between nodes i and j, the edge embeddings are defined as

$$V_{ij} = \text{EdgeTypeEmb}(E_{ij}) \in \mathbb{R}^f, \tag{19}$$

where EdgeTypeEmb maps each unique edge type to a learned f-dimensional vector.

We define three GNN cores:

1. SDP-Att Core. The SDP attention mechanism with edge type embeddings as described by the Relative Positional Embedding Transformer [12].
2. MLP-Att Core. A version of the GAT modified to use edge type embeddings with additive attention.
3. Mean Core. A version of a GNN without attention similar to previous GNN-based MHQA models, modified to use edge type embeddings.

All of our GNN Cores share the same message function, which applies a linear transformation to the node embedding and adds the edge type embedding:

$$\phi(\mathbf{h}_i^k, \mathbf{h}_j^k, E) = \mathbf{W}_v \mathbf{h}_j^k + V_{ij}. \tag{20}$$

Next, we define the aggregate function for each of the GNN Cores.

i. Mean Core Aggregate:

$$\omega^k(M_{ij}^k, \mathbf{h}_i, \mathbf{h}_j) = \frac{M_{ij}^k}{|N(i)|}. \tag{21}$$

ii. SDP-Att Core Aggregate:

$$\omega^k(M_{ij}^k, \mathbf{h}_i, \mathbf{h}_j) = \alpha_t(\mathbf{h}_i, \mathbf{h}_j) M_{ij}^k. \tag{22}$$

iii. MLP-Att Core Aggregate:

$$\omega^k(M_{ij}^k, \mathbf{h}_i, \mathbf{h}_j) = \alpha_g(\mathbf{h}_i, \mathbf{h}_j) M_{ij}^k, \tag{23}$$

Table 1. Model hyperparameters for all experiments.

Hyper parameter	Value
Training epochs	30
Learning rate (LR)	0.01
LR schedule exponential decay	0.9
Dropout	0.1
Batch size	1
Num GNN layers	9
Model dimension	**300** for GloVe **512** for BERT
Num GNN heads	**4** if attention-based

Update Functions. As a GNN update function we consider the gating function G described in Sect. 2.5, the TUF described in Sect. 2.4, or composing them together as follows:

$$\gamma(\boldsymbol{h}_i^k, A_i^k) = \mathrm{G}(\boldsymbol{h}_i^k, \mathrm{TUF}(\boldsymbol{h}_i^k, A_i^k)). \tag{24}$$

It is also possible to use neither, as the canonical GAT [16] does.

3.4 Output Model

Finally, our model outputs a probability distribution over answer candidates C_q. The output of the Lth GNN layer is a set of encoded node states denoted $\mathbf{h}_i^L$ for the ith node. For each candidate $c \in C_q$ the model extracts the set of all entity node vectors which correspond to candidate c: $E_c = \{\mathbf{h}_i^L \; \forall i \mid i$ is a mention of $c\}$.

Candidate score c_s is based on both entity and candidate node states:

$$c_s = \mathrm{MLP}_c(\mathbf{h}_c^L) + \max_{\mathbf{e} \in E_c}(\mathrm{MLP}_\epsilon(\mathbf{e})), \tag{25}$$

where MLP_c and MLP_ϵ are MLP's to score candidates and entities respectively. The final probability distribution over all candidates is obtained by performing a softmax over the candidate scores c_s.

4   Experimental Setup

We train and evaluate our models on the WikiHop MHQA dataset [17]. WikiHop follows the multiple choice MHQA structure described in Sect. 3. Table 1 shows the hyperparameters used in all of our experiments. No model-specific hyperparameter tuning was performed for any of the results. The model (hidden state) dimension is determined by the dimensionality of the token embedder, i.e., 300 for GloVe-based models and 512 for BERT-based models. We use an LR scheduler with exponential decay [10]. In preliminary experiments mini-batching (using a batch size greater than 1) did not improve performance.

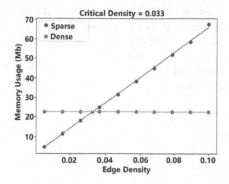

Fig. 2. Edge density vs. memory used for dense and sparse attention implementations. Tested with 500 graph nodes, 100 features per node, and 10 attention heads.

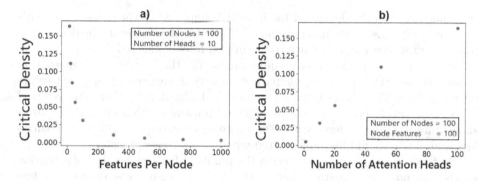

Fig. 3. Plots showing how the critical edge density changes while varying (a) the number of features per node, and (b) the number of attention heads.

4.1   Implementation

Our models are implemented in Pytorch. We use Pytorch Geometric[1] to implement our GAT and MLP-based GNNs. For our SDP GNNs we use the native Pytorch implementation of the Transformer's multi-headed attention. Tokenisation is either word-based to be compatible with the GloVe vocabulary, or performed using Huggingface's subword tokenisation utilities for our BERT-based models.[2]

Sparse and Dense GNN Implementations. Pytorch Geometric is designed to be optimal for sparse graphs (the *sparse approach*). The Transformer implementation (the *dense approach*) is optimised for fully connected graphs, although it can also model sparse graphs. The sparse approach's memory requirements

---

[1] https://pytorch-geometric.readthedocs.io/en/latest/.
[2] https://huggingface.co/.

Table 2. MHQA accuracies on the Wiki-Hop development set for our best GNN configurations using GloVe or BERT embeddings, compared to HDE (the best previous GNN-based model) and BigBird (the best-performing Transformer).

Model	Dev accuracy
GNN (GloVe)	66.0
GNN (BERT)	71.4
HDE [14]	68.1
BigBird [20]	**75.9**

Table 3. WikiHop development set accuracies of our GloVe-based GNN models with different architectural choices.

GNN core	Gating	✗	✓	✗	✓
	TUF	✗	✗	✓	✓
		Dev accuracy			
Mean		52.7	58.5	62.5	60.2
MLP-Att		29.0	62.9	60.4	64.0
SDP-Att		13.0	61.4	**66.0**	64.7

scale linearly with the density of the inputted graph, while the dense approach's memory requirements are independent of graph density. However, the dense approach requires less memory when operating on fully connected graphs.

We analyse the memory requirements of these two implementation approaches for SDP attention, comparing memory usage across graphs of varying edge densities. This allows us to find the critical edge density (hereafter referred to as the *critical density*) where the optimal implementation changes. Figure 2 shows that the critical density is 0.033, meaning only graphs with edge densities less than 0.033 should be implemented with the sparse approach.

We further investigate how varying the number of nodes, number of attention heads, and number of features per node affects the memory scaling of the two implementation approaches. We found that the number of nodes has no significant relationship with Critical Density. Figure 3 (a and b) shows that when the number of features is very small, or the number of heads is very large, the sparse approach may be preferable, but only for very sparse graphs. Therefore, in all but the most extreme cases the memory requirements of the dense approach will be lower. The PyTorch Geometric implementation is therefore suboptimal for tasks such as multi-hop question answering.

5   Results

All results presented in this section are accuracies on the WikiHop development set.[3] Table 2 shows the results of our best performing models when using GloVe and BERT as token embedders. These two models make use of the SDP-Att Core and the TUF without gating. The performance of our model with GloVe embeddings is lower than that of HDE [14]. HDE uses character n-gram embedding embeddings in addition to GloVe and a custom attention mechanism for getting node embeddings from token embeddings, which we substituted for a

---

[3] Evaluation on the hidden test set was not possible due to incompatible software versions on the evaluation portal.

Table 4. WikiHop development set accuracies of our GloVe-based GNNs, comparing sparse vs fully connected graphs, with or without edge embeddings.

Sparsity	Edge embeddings	Dev accuracy
✓	✓	**64.7**
✗	✓	64.4
✓	✗	60.2
✗	✗	61.5

Table 5. Task-specific vs random graph structure, with or without edge embeddings on the WikiHop development set, sing BERT-based GNNs

Structure	Edge embeddings	Dev accuracy
Task-Specific	✓	**68.2**
Task-Specific	✗	64.4
Random	✓	63.8
Random	✗	64.5

Transformer encoder. Using BERT as token embedder however, our model is more accurate than HDE, which is to date the highest performing GNN-based MHQA model on the WikiHop dataset. Our best performing model falls short of the performance of the BigBird Sparse Transformer [20]. BigBird encodes token sequences directly, and the full model is pre-trained with Masked Language Modeling (MLM), while in our approach only the BERT token encoder is pre-trained. BigBird makes use of random connections between tokens, in combination with some rule-based connections. However, it does not include edge type information to distinguish these distinct edges.

5.1 GNN Architecture

Table 3 shows the performance of our GNN-based MHQA model with various GNN Cores as well as different update functions (Gating and/or TUF). The results demonstrate a number of key findings. First, the attention-based GNN Cores (MLP-Att and SDP-Att) outperform the GNN without attention (Mean Core) in most settings. The attention-based GNNs also achieve better maximum task performances. Second, when neither gating nor the TUF is included, all evaluated GNN variants perform worse. TUF boosts task performance in all settings where it is included. Finally, the attention-based GNNs are especially reliant on the TUF and/or gating compared to the non attention-based GNN Core.

The Mean Core and SDP-Att Core models perform best when the TUF is used without gating. On the other hand, the combination of gating and the TUF is beneficial for MLP-Att Core. Thus we can draw no clear conclusion about the use of gating in our GNNs as these results may be due to noise.

5.2 Graph Structure and Edge Embeddings

Table 4 compares models which use sparse graphs against variants which use fully connected graphs (similar to the vanilla Transformer). The GNNs used in this experiment all use SDP-Att Core with gating and the TUF. Here sparse graphs are the rules-based task-specific graphs described in Sect. 3. We construct fully connected graphs by starting with our sparse graphs, and adding in a new edge type *unconnected* which connects all unconnected nodes. Thus, there is no loss in edge information when comparing our sparse and fully connected graphs, although we also evaluate without using edge embeddings. The results demonstrate that without edge information, using fully connected graphs performs better than sparse graphs. However, when including edge information, sparse graphs can produce similar performance.

Table 5 compares the use of task-specific graph structure based on rules developed in the GNN-based MHQA literature [14] against random sparse structure. In this experiment we use GNNs with SDP-Att Core including gating and the TUF, using BERT as token encoder. Here, the random structure involves replacing edges with random edges, defined by randomly selecting two nodes to connect. We do not replace (Candidate-Candidate) edges with random edges. This ensures that candidates remain fully connected to each other. This random shuffling preserves the number of edges.

In random graph structure with edge embeddings, the edge type is simply the tuple of the node types which it connects (e.g., Candidate-Entity). The results clearly show that task-specific structure in combination with edge information is important to model performance. Neither task-specific structure nor edge information alone significantly boosts model performance. This serves to validate the specific graph structuring rules which have been developed in the GNN-based MHQA literature [14].

6   Discussion

There are several limitations to how much can be claimed about the generalizability of our experimental results. Due to high computational requirements we only trained a single model per configuration, without model-specific hyperparameter tuning. We also only evaluated on a single dataset for which the test set was unavailable. However, we believe our results are nonetheless valuable and may spur further research into the connection between GNNs and Sparse Transformers.

Bigbird, the Sparse Transformer, [20] makes use of random connectivity to decrease the average number of connections between each node. Our results indicate that random connectivity should be replaced by problem-specific structure where possible. Our results also motivate the use of edge type information alongside the problem specific structure. Finally, our results demonstrate the value of using SDP attention in combination with the TUF, exactly as is done in the Transformer model [15]. Given the popularity of the GAT GNN, there may be many models which could benefit from (1) switching from additive attention to SDP attention, and (2) including the TUF in the GNN's update function. Essentially, this is a recommendation to replace existing attention-based GNNs with the Transformer. Our empirical analysis on memory requirements in Sect. 4.1 also indicate that for graphs used in realistic Natural Language Processing scenarios, using a vanilla Transformer implementation is preferable to a message passing implementation which is optimised for sparse graphs.

Many token-level NLP models make use of end-to-end pre-training to prepare the model for the final task [8,20]. GNN-based NLP models, including ours, do not make use of end-to-end pre-training, instead relying only on pre-trained token embedders [9,14]. This motivates the future development of pre-training strategies for GNN-based NLP models. A graph-based model which performs on par with Sparse Transformers would reduce the memory requirements compared to token-level models.

7 Conclusion

We implemented and evaluated a simplified version of the GNN-based Heterogeneous Document Entity (HDE) MHQA model. We used the WikiHop dataset to evaluate two primary differences between the GAT and the Transformer, namely (1) the type of attention mechanism, and (2) the use of the Transformer Update Function (TUF) as an update function. Our results serve as a case study motivating the use of scaled dot product (SDP) attention and the TUF—essentially the canonical Transformer. We also investigate the role of graph sparsity, graph structure, and edge information in our MHQA model. Our results demonstrate the value in task-specific graph structure rules over random connectivity and fully connected graphs with an emphasis on the use of problem specific edge information. The results further indicate that without edge information, task-specific connection rules may not yield performance gains over random sparse connections or fully connected graphs. These insights may provide a path to improving token-level Sparse Transformer performance. Finally, our results show that there is room for further research to close the performance gap between token-level models and graph-level models with coarse-grained nodes.

References

1. Ba, J.L., Kiros, J.R., Hinton, G.E.: Layer normalization. stat. **1050**, 21 (2016)
2. Beltagy, I., Peters, M.E., Cohan, A.: Longformer: the long-document transformer. CoRR abs/2004.05150 (2020). https://arxiv.org/abs/2004.05150
3. Brown, T., et al.: Language models are few-shot learners. In: Advances in Neural Information Processing Systems, vol. 33, pp. 1877–1901 (2020)
4. Chen, D., Lin, Y., Li, W., Li, P., Zhou, J., Sun, X.: Measuring and relieving the over-smoothing problem for graph neural networks from the topological view. In: Proceedings of the AAAI Conference on Artificial Intelligence, vol. 34, pp. 3438–3445 (2020)
5. Chen, J., Lin, S., Durrett, G.: Multi-hop question answering via reasoning chains. CoRR abs/1910.02610 (2019). http://arxiv.org/abs/1910.02610
6. d'Ascoli, S., Touvron, H., Leavitt, M.L., Morcos, A.S., Biroli, G., Sagun, L.: ConViT: improving vision transformers with soft convolutional inductive biases. In: Proceedings of the 38th International Conference on Machine Learning, ICML 2021, 18–24 July 2021, Virtual Event, vol. 139, pp. 2286–2296. PMLR (2021). http://proceedings.mlr.press/v139/d-ascoli21a.html

7. De Cao, N., Aziz, W., Titov, I.: Question answering by reasoning across documents with graph convolutional networks. In: Proceedings of the 2019 Conference of the North American Chapter of the Association for Computational Linguistics: Human Language Technologies, Volume 1 (Long and Short Papers), pp. 2306–2317 (2019)
8. Devlin, J., Chang, M.W., Lee, K., Toutanova, K.: Bert: pre-training of deep bidirectional transformers for language understanding. In: NAACL-HLT (1) (2019)
9. Fang, Y., Sun, S., Gan, Z., Pillai, R., Wang, S., Liu, J.: Hierarchical graph network for multi-hop question answering. In: Proceedings of the 2020 Conference on Empirical Methods in Natural Language Processing (EMNLP), pp. 8823–8838 (2020)
10. George, A.P., Powell, W.B.: Adaptive stepsizes for recursive estimation with applications in approximate dynamic programming. Mach. Learn. **65**(1), 167–198 (2006). https://doi.org/10.1007/s10994-006-8365-9
11. Pennington, J., Socher, R., Manning, C.D.: GloVe: global vectors for word representation. In: Proceedings of the 2014 Conference on Empirical Methods in Natural Language Processing, EMNLP 2014, 25–29 Oct 2014, Doha, Qatar, A meeting of SIGDAT, A Special Interest Group of the ACL, pp. 1532–1543. ACL (2014). https://doi.org/10.3115/v1/d14-1162
12. Shaw, P., Uszkoreit, J., Vaswani, A.: Self-attention with relative position representations. In: Proceedings of the 2018 Conference of the North American Chapter of the Association for Computational Linguistics: Human Language Technologies, Volume 2 (Short Papers), pp. 464–468 (2018)
13. Sigaud, O., Masson, C., Filliat, D., Stulp, F.: Gated networks: an inventory. CoRR abs/1512.03201 (2015). http://arxiv.org/abs/1512.03201
14. Tu, M., Wang, G., Huang, J., Tang, Y., He, X., Zhou, B.: Multi-hop reading comprehension across multiple documents by reasoning over heterogeneous graphs. In: Proceedings of the 57th Annual Meeting of the Association for Computational Linguistics, pp. 2704–2713 (2019)
15. Vaswani, A., et al.: Attention is all you need. In: Advances in Neural Information Processing Systems, pp. 5998–6008 (2017)
16. Veličković, P., Cucurull, G., Casanova, A., Romero, A., Liò, P., Bengio, Y.: Graph attention networks. In: International Conference on Learning Representations (2018)
17. Welbl, J., Stenetorp, P., Riedel, S.: Constructing datasets for multi-hop reading comprehension across documents. Trans. Assoc. Comput. Linguis. **6**, 287–302 (2018)
18. Wu, Z., Pan, S., Chen, F., Long, G., Zhang, C., Philip, S.Y.: A comprehensive survey on graph neural networks. IEEE Trans. Neural Netw. Learn. Syst. **32**, 4–24 (2020)
19. Yang, Z., et al.: HotpotQA: a dataset for diverse, explainable multi-hop question answering. In: Proceedings of the 2018 Conference on Empirical Methods in Natural Language Processing, Brussels, Belgium, 31 October–4 November 2018, pp. 2369–2380. Association for Computational Linguistics (2018). https://doi.org/10.18653/v1/d18-1259
20. Zaheer, M., et al.: Big bird: transformers for longer sequences. In: Advances in Neural Information Processing Systems, vol. 33 (2020). https://proceedings.neurips.cc/paper/2020/hash/c8512d142a2d849725f31a9a7a361ab9-Abstract.html

Towards a Methodology for Addressing Missingness in Datasets, with an Application to Demographic Health Datasets

Gift Khangamwa[1]([✉])([iD]), Terence van Zyl[2]([iD]), and Clint J. van Alten[1]([iD])

[1] School of Computer Science and Applied Mathematics,
University of the Witwatersrand, Johannesburg, Johanneburg, South Africa
{gift.khangamwa,clint.vanalten}@wits.ac.za
[2] Institute for Intelligent Systems, University of Johannesburg,
Johannesburg, South Africa
http://www.wits.ac.za/csam,
http://www.uj.ac.za/institute-for-intelligent-systems

Abstract. Missing data is a common concern in health datasets, and its impact on good decision-making processes is well documented. Our study's contribution is a methodology for tackling missing data problems using a combination of synthetic dataset generation, missing data imputation and deep learning methods to resolve missing data challenges. Specifically, we conducted a series of experiments with these objectives; *a*) generating a realistic synthetic dataset, *b*) simulating data missingness, *c*) recovering the missing data, and *d*) analyzing imputation performance. Our methodology used a gaussian mixture model whose parameters were learned from a cleaned subset of a real demographic and health dataset to generate the synthetic data. We simulated various missingness degrees ranging from 10%, 20%, 30%, and 40% under the missing completely at random scheme MCAR. We used an integrated performance analysis framework involving clustering, classification and direct imputation analysis. Our results show that models trained on synthetic and imputed datasets could make predictions with an accuracy of 83% and 80% on *a*) an unseen real dataset and *b*) an unseen reserved synthetic test dataset, respectively. Moreover, the models that used the DAE method for imputed yielded the lowest log loss an indication of good performance, even though the accuracy measures were slightly lower. In conclusion, our work demonstrates that using our methodology, one can reverse engineer a solution to resolve missingness on an unseen dataset with missingness. Moreover, though we used a health dataset, our methodology can be utilized in other contexts.

Keywords: Deep learning · Missing data · Machine learning · Imputation

A. Pillay et al. (Eds.): SACAIR 2022, CCIS 1734, pp. 169–186, 2022.
https://doi.org/10.1007/978-3-031-22321-1_12

1   Introduction

Missing data is a common problem in health datasets and its impact on good decision making processes is well documented. This phenomenon of missing data plagues all scientific research endeavours generally because it is not possible to collect all data in any study, because flaws do exist in both data collection gadgets which can fail and in researchers who by their nature of being human can err. Therefore, the efforts to recover missing data, so as to have data that speaks meaningfully about any specific problem under study are a part of the scientific process.

In this paper we tackled the problem of learning from incomplete data, a problem that exists in population health datasets, which we did in the context of machine learning. We used a demographic and health dataset which has various missing data challenges which we articulate in Subsect. 4. It is a fact that when data are missing even if one imputes all missing cases, the actual data that was missing remains unknown and hence what we create are estimates. Imputation approaches such as multiple imputation attempt to improves on the quality of the imputation value estimates by improving the process of generating the imputation value estimates [14].

Generally, missing data occurs in different ways; a) when there is very little data due to general data unavailability or scarcity, b) when there are missing values in some of the attributes in the data instances because the data was not observed, recorded or was corrupted [5,14], c) or in a case when the observed dataset cannot be made available for research purposes due to privacy, security or confidentiality concerns [21].

However, in this study, we were interested in the first two scenarios of how missing data occurs, which can be simulated by inducing artificial missingness in a synthetic dataset. Consequently, we embarked on a synthetic dataset generation task first, in order to have a complete dataset for use as a ground truth dataset in studying missingness and imputation methods. We, generated a clean synthetic dataset using the Gaussian Mixture Model method based on the parameters that were learned from the statistical distributions of the real observed datasets discussed in Sect. 4. As a result of this, the synthetic dataset provided us with ground truth values to use in evaluating the results of our experiments using various imputation methods.

Moreover, part of our goal was to create a realistic dataset, so that our results on this synthetic dataset can be extended to the observed dataset. We accomplished the task of making the synthetic dataset realistic in two steps, the first being using parameters learned from the real observed dataset and the second step being ensuring that the values generated in the various features matched the original observed dataset in terms of their data type, range of values and the probability distribution.

Therefore, in this paper we propose a methodology for tackling missing data that makes use of synthetic dataset and imputation applied to a demographic and health dataset, which we believe can be generally applicable to other datasets as well.

1.1   Problem Statement

We tackled the problem of missing data using a demographic and health datasets as a study case. These demographic and health dataset are a very important public health surveillance dataset and are invaluable in the measure and assessment of the various international development goals such as sustainable development goals. However, these datasets have various missing data problems such as; 1) missing values, which are indicated using codes for missing values, inconsistent values and unknown values, 2) varied missingness degrees in features indicated as blanks, 3) skewed data due to missing labels in target features such as anemia, which is an indicative feature for presence of malaria, and 4) finally, missing data that is missing under various missingness schemes see Sect. 2.2.

Overall, these missing data problems affect the usage of such important datasets especially in building supervised predictive machine learning models that might be useful in the measurement or forecasting progress in attainment of sustainable development goals or indeed national health targets.

1.2   Objectives

The main objective of the study was to propose a methodology for tackling missing data using a demographic and health dataset as a case study. We compared different imputation methods in-order to determine the best imputation method for our dataset. Specifically, our experiments focused on the following; a) generating a realistic synthetic dataset using the gaussian mixture model trained using parameters extracted from a cleaned subset of the observed dataset, b) simulating data missingness on the synthetic data based on prevalent missingness in the observed dataset, c) recovering the missing data using state of the art imputation methods and deep learning methods, and d) analysing imputation performance using accuracy of classification and imputation using direct performance metrics.

Our hypothesis in the study was as follows; models and methods that perform well on the realistic synthetic dataset with artificially induced missingness should work equally well on our observed real dataset of interest which suffers from missingness challenges.

1.3   Contribution

Our contribution in this paper is the proposition of a methodology for resolving missing data that makes use of synthetic datasets, simulated missing data degrees and missingness schemes, missing data imputation and the usage of deep neural networks as part of the missing data recovery methods. The methodology comes with a step by step approach that is described in the sub Sect. 4.1. We believe that this methodology can be followed in order to address missingness on any novel dataset by using the building blocks that we propose herein.

2   Background

We begin by providing some background material related to the phenomenon of missing data, its causes and character.

2.1   Causes of Missing Data

In the following subsections we highlight how missing data occurs, as earlier discussed this problem can occur under the various data missingness schemes which we get into in sub Sect. 2.2.

Missingness - A General Case: this is the most common case of missing data, where the dataset has missing components in any of its features. This type of missingness is very common in machine learning datasets such that all well known data mining methodologies such as the cross industry standard process for data mining CRISP-DM, knowledge discovery in databases KDD and the sample explore measure model assess SEMMA, all have specific phases with stages and activities dedicated to the task of cleaning the data prior to model building. Some of the activities carried out in these phases include; making imputations, formatting data, removing redundancies and transforming the data as is relevant [9,11].

Skewed Data: class imbalance or skewness occurs when there are more data for one particular target class than the other classes. This leads to under representation of the minority class and might affect the performance of the resultant predictive model built using such data in classifying this minority class.

Skewness or class imbalance data problem may be addressed using sampling techniques. The sampling techniques that are mainly used to address the problem are; a) over-sampling of the under-represented class by reusing the minority class data instances in the model building process to ensure that there is a balance in the number of samples with the majority class, b) under-sampling the majority class so that fewer data instances from this class are used, in order to balance numbers with the minority class [3], c) the other methods combine the two strategies.

Missing Labels: missing labels problem is another problem that is part of the data incompleteness or missingness problem. The challenge that this causes is that though one may have a large number of data instances, the number of instances that have labels are not adequate to build effective supervised machine learning models. This problem differs from the problem discussed in the previous sub section, in that the only component missing here is the target class label from the data instance.

Data Privacy, Confidentiality and Safety Concerns: As indicated earlier, data may be missing even though it is available, when it is withheld due to concerns of security, privacy and confidentiality by those who own such data; in such cases synthetic data generation becomes the only avenue to study such cases see [1,6].

2.2   Categories of Missing Data

According to Ghahramani and Jordan [13], Rubin [5,14] the phenomenon of missing data is categorized as occurring under three different sets of circumstances, leading to categorisation of three unique missingness schemes as follows; missing completely at random MCAR, missing at random MAR and missing not at random MNAR.

Missing completely at random Eq. 1 occurs when there is no systematic difference between instances having missing values and those having the data; in other words the missing data is neither dependent on the observed data nor on any other unobserved data [16,17]. This can be represented as a probability function as follows;

$$p(M|X^{obs}, X^{mis}) = p(M) \tag{1}$$

Missing at random Eq. 2 occurs when there are systematic differences between instances with missing data, and those with data that can be explained by the available data, in other words the missing data is dependent on the observed data only [5,16].

$$p(M|X^{obs}, X^{mis}) = p(M|X^{obs}) \tag{2}$$

Missing not at random Eq. 3 occurs when there are systematic differences between instances with missing data, and those without missing data, which cannot be explained by the available data [2,16]. This implies that, data that is missing is dependent on other missing data and not on data that has been observed [5,14,17], this third category of missingness is harder to resolve and is seldom tackled in most research work on the subject.

$$p(M|X^{obs}, X^{mis}) = p(M|X^{mis}) \tag{3}$$

2.3   Tracking Missing Data

In order to assess the performance of missing data recovery efforts, it is necessary to track the locations of missingness in a dataset whether it is induced or otherwise. Once missing data has been imputed this missingness indicator mechanism helps to compare imputed values against actual ground truth values if known. The same approach has been proposed and utilized by [14] and employed by numerous other researchers [2,5,19,20]. The scheme makes use of a missingness indicator matrix to track the location of missing data in a data matrix $\mathbf{X} = \{\mathbf{x}\}_1^n$. As per [4], any data matrix has two components; a missing component $\mathbf{X}_{miss}$ and an observed component $\mathbf{X}_{obs}$. A missingness indicator matrix:

$$\mathbf{M_{ij}} = \begin{cases} 1, & x_{ij} \text{ missing} \\ 0, & x_{ij} \text{ observed} \end{cases} \tag{4}$$

is used to keep a record of which data is observed and which data is not. Where for each value in the data matrix the missingness indicator matrix keeps track of which value is available and which data value is missing using the row and column indices i and j for each data point $\mathbf{x}$.

The missingness indicator matrix has various uses in studies that employ deep neural network methods; for instance Beaulieu-Jones and Moore [2] who used denoising autoencoders included the missingness indicator matrix in the computation of a cost function for their model. On the other hand, [12,19] included the missingness indicator value for each feature $\mathbf{m}_j$ as part of the training dataset in building models as shown below in Eq. 5.

$$\dot{\mathbf{x}} = \tilde{\mathbf{x}} \odot \bar{\mathbf{m}} + \bar{\mathbf{x}} \odot \mathbf{m} \tag{5}$$

where $\mathbf{m}$ is a missingness indicator and $\bar{\mathbf{m}}$ is the complement of the missingness indicator, $\bar{\mathbf{x}}$ is generated by a neural network, while $\tilde{\mathbf{x}}$ is the observed data point with missingness, $\odot$ is the element-wise multiplication operator these inputs generated the recovered matrix $\dot{\mathbf{x}}$ [12].

3   Related Work

In this section we briefly outline some of the works that are similar to our work. Specifically, we focus on methods that are used in cases where there is missing data; we consider various missing data imputation methods and other techniques for resolving missing data cases i.e. class imbalance, missing labels and others.

We split our discussion into state of the art missing data imputations in Subsect. 3.1, deep learning imputation methods in Sect. 3.1, and finally we look at a few synthetic dataset generation methods that are used to resolve various missing data scenarios. Our approach to the discussion in this section is informed by the two main components of our study which are missing data imputation and synthetic dataset generation.

3.1   Missing Data Imputation Methods

There are several approaches for handling the problem of missing data. The first approach is complete case analysis whereby all data points with missing data are ignored in the analysis. This process involves the removal of all rows or columns that have missing data, thereafter, the clean dataset is used as a complete dataset. The other approaches implement missing data imputation which allows for filling in of missing data, the two main methods in this regard are a) single value imputation approaches that impute a value once, and b) multiple imputation approaches which generate multiple imputed copies of a dataset i.e. 5 copies. Moreover, multiple imputation allows for variability in the imputed values by varying data imputation function parameters, this is done to address the uncertainties regarding what the unknown missing values might actually be [13]. Misra and Yadav [11] provides a good review of the state of the art methods used to address missingness in health datasets.

State of the Art Methods. Multiple imputation remains the major approach that all state of the art machine learning methods employ for addressing missing data. The methods include random forest which is the basis for the Missing Forest imputation algorithm which used multiple trees as estimators to generate multiple copies for missing values [11].

Deep Learning Methods. deep neural networks have been used extensively in the health domain [10]. One major usage of deep neural networks is in the form of deep generative models that make use of a probability distribution function $P(X)$ and generate samples X from some high dimensional space χ [18]. Specifically, the generative capacity of deep neural networks makes these methods to be very suitable for missing data imputation or data recovery. Some of the works that are relevant to our work include; the work by Richardson et al. [12] who devised a method called MCFlow that addressed the missing data imputation problem using a framework that combines deep neural network and Monte Carlo Markov Chains. Beaulieu-Jones and Moore [2] made use of Denoising Autoencoders inorder to address missing data in electronic health records in a clinical trial time series dataset for ALS a progressive neurodegenerative disorder.

Moreover, there are several works that have used generative adversarial networks, these include; Shang et al. [15] who developed VIGAN a view imputation generative adversarial network for imputation of missing views in a multi view environment, where patients health data was collected from a sensor network over multiple time periods. Lin and Tsai [7] developed MisGAN, a missing data imputation method that uses an incomplete dataset to train a GAN network, which is subsequently then used for imputation purposes. A missingness indicator matrix is used to track missing data and recover imputed values from the MisGAN output dataset. Yoon et al. [19] developed GAIN a generative adversarial imputation network that made use of missingness indicator matrix to provide a hint to the discriminator as it tried to determine which samples were imputed and which were observed in their GAN imputation framework.

Our work differs from [7,12,15,19,20] on the basis of the deep neural network that we used which is a denoising autoencoder. Moreover, our methodology incorporates the use of synthetic data as a medium for training models which we intend to use in resolving missingness on an original dataset. We differ with Beaulieu-Jones and Moore [2] on both the type of dataset used and the methodological components in our approach i.e. usage of synthetic data, and the sequence of activities in our approach.

4 Methods

In this section we discuss the methods that were used to accomplish the objectives of the study. Overall, in this study, the use of a realistic synthetic dataset and the simulation of missingness were key to our methodology. This meant that the generation of the synthetic dataset needed to be based on parameters generated from the real dataset with missingness, likewise the artificially induced missingness needed to be similar to the actual missingness of the observed dataset in both degree and scheme of missingness. Moreover, in our approach in Subsect. 4.1, we ensured that artificial targets were generated using a DNN model that was exposed to the original targets. This was done so that these generated artificial targets in the synthetic dataset should be similar and connected to the original targets.

Our discussion is structured as follows; we discuss our datasets in Subsect. 4, thereafter we discuss the approach that was taken in a step by step manner in Subsect. 4.1, and finally we discuss the metrics of interest that we use to present our work in Subsect. 4.2.

Dataset Description. We used the latest DHS survey datasets for the Southern African countries of Malawi, Namibia, Zambia and Zimbabwe. We used these datasets in-order to learn the parameters for generating our synthetic datasets. These countries were selected because they are within the Southern Africa region of Sub-Saharan Africa in the malaria endemic region and have data on malaria, anemia, diabetes and hypertension, which are among the biggest health challenges for the region and the continent [8].

The observed datasets have several data incompleteness problems such as: missing values, which are indicated using codes for missing values, inconsistent values and unknown values; blanks; skewness and missing labels for select target features. Therefore, we selected a file from these observed datasets and cleaned it to have instances that have features in a select set of features only. We thereafter used this cleaned dataset to generate a synthetic dataset as described in the approach in Sect. 4.1. Overall, the dataset had a missingness rate of about 23.8%

4.1 Approach

In this subsection we outline the steps followed in our approach to the study and the specific methods used in the following sequential steps.

1. Our experiment begun by loading a copy of the demographic and health dataset a survey dataset that we selected as our study dataset. Thereafter, we dropped all missing values to create a clean sub-dataset having most of the relevant anemia features. The size of the clean dataset was 2,058 instances and 56 features.
2. Afterwards, we scaled the clean sub-dataset using a minimax scaler to avoid some features from having a dominating effect on the synthetic dataset generation models. Our objective was to ensure that our data be generated by different components of a Gaussian Mixture Model data generator.

3. Next, we used a Gaussian Mixture Model (GMM) to learn hyper parameters for our synthetic dataset generator models. First of all, we performed a parameter search to identify the ideal number of components and an appropriate covariance matrix shape. In this regard, during the search we compared the models using Akaike Information Criterion (AIC) and Bayesian Information Criterion (BIC) inorder to pick the best models. Both AIC and BIC are good tools for model selection, they are defined as given in equations Eq. 8 and Eq. 9. Subsequently, we used the parameters of the best models from the previous steps in the final GMM synthetic dataset generator model that we used to generate a dataset having 20,000 data points. We also created a testing dataset with 5,000 data point which we reserved to test our models later in step 7.

4. We thereafter created targets for the synthetic dataset using partially trained Deep Neural Network (DNN) models. The DNN models were trained using the scaled clean datasets from step 2. We chose a DNN design with two deep layers having 20 nodes each and one dropout layer at 20% dropout rate. Our strategy to avoid over-fitting included using both early stopping and the 20% dropout layer in the training of the target generator model. Moreover we made sure to use all the data points in the training of the generator. We also used model check-pointing to identify and save the best DNN model from the training step. Thereafter, the best DNN model was used to predict labels for the synthetic dataset that was generated using GMM and had 20,000 data points see step 3. Targets were also created for our reserved testing dataset with 5,000 data points using the same approach.

5. The next step was to induce missing completely at random (MCAR) missingness at different degrees in the main synthetic dataset i.e. the one having 20,000 data points, we induced missingness degrees ranging from 10%, 20%, 30%, and, 40% missing values. Each combination of missingness scheme and missingness degree yielded a dataset with missing values. For each missingness dataset we kept track of where exactly in the dataset was the missingness induced using a missingness masking matrix i.e. missingness indicator matrix.

6. In the next step, we proceeded to impute missing values using four methods namely; denoising autoencoder (DAE), random forest based MissForest, k nearest neighbour (KNN) impute and MICE (iterative imputer under Multiple Imputation). The imputation process was carried out five times for each of the datasets having missingness from the previous step 6 above.

7. In the final step, we assessed the outputs of the imputation step using classification and direct analysis of imputation.
   In the classification analysis we used a DNN custom classifier model having a similar structure to the target generator model. Moreover, we repeated our experiments atleast 10 times to get mean performance metrics. Specifically, to train our models we used imputed datasets that were split into a training and validation set. On the other hand, in order to test our models we used datasets that were not used in the training steps at all, we used these datasets a) the clean original dataset from step 1, b) the full synthetic dataset without missingness from step 3, and c) the reserved synthetic test dataset having 5000 data instances.

In the direct assessment of imputation we used the missingness indicator (masking matrix) to locate all missing data locations in the dataset and compared imputed versus actual values. We used various accuracy and error measures such as accuracy, root mean square error, mean absolute percent error and other metrics in our assessment of the imputation results. Finally, in Fig. 1 below we provide a graphical summary of the main steps in our approach.

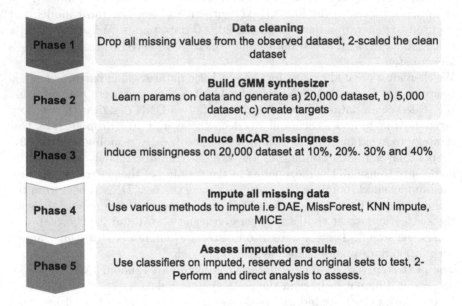

Fig. 1. A phased summary of the steps in the approach taken.

4.2 Metrics of Interest

Our first two metrics of interest were accuracy and the binary cross entropy also referred to as log loss. Accuracy is computed based on confusion matrix output that yields true positives, true negatives, false positives and false negatives. Accuracy as a metric has its weaknesses especially if a dataset is skewed, however, it is a well accepted and intuitive metric. Log loss computation hinges on true value y_i and predicted value $p(y_i)$ and the number of samples N. Using these two metrics we can assess any classification model.

The next set of metrics are mean absolute percent error, and root mean squared error which were used to directly assess the outcome of imputation. In these metrics we consider features in the synthetic dataset as the true values y_i, while the imputed values are the predicted values $\hat{y}$.

The next metrics are Silhouette score Eq. 6 and Rand score, which offer the opportunity to review imputation using KMeans clustering. The silhouette score is based on two distance measures between each sample in relation to other samples in its own cluster and the next cluster and is computed as follows;

$$\text{Silhouette score} = \frac{b - a}{max(a, b)} \tag{6}$$

where a is mean distance to other samples in the same cluster, and b is the mean distance to samples in the next cluster. In this experiment we used euclidean distance to measure these distances.

On the other hand the Rand score compares how effectively the clustering allocates the samples based on their class membership, where ideally members of one class must also be clustered together. Computation of this score is similar to that of accuracy. Where in this case a and b are predicted correctly as belonging to their groups, similar to the true positives and true negatives in a binary classification confusion matrix, while c and d are the incorrect predictions.

$$Rand = \frac{a + b}{a + b + c + d} = \frac{TP + TN}{TP + TN + FP + FN} \tag{7}$$

The method requires true and predicted clusters, where the true clusters are given by the class memberships that exist in the dataset and the predicted are what the cluster algorithm generates.

Finally, metrics Akaike Information Criterion Eq. 8, and Bayesian Information Criterion Eq. 9 were used to determine the best model based on number components for the GMM in the synthetic dataset generation experiment.

$$AIC = \frac{-2}{N} * LL + 2 * \frac{k}{N} \tag{8}$$

$$BIC = -2 * LL + log(N) * k \tag{9}$$

where N is the number of training instances, k is the number of parameters of the model i.e. number of components in our case, and LL is the log likelihood of the model. Moreover, BIC as can be seen above penalizes model complexity.

5   Results and Discussion

5.1   Synthetic Data

The Fig. 2a shows the results of the synthetic dataset generation, the figure visualizes the distribution of the generated synthetic samples by component. Moreover, once the synthetic data generation was complete, the next step was the target generation which was done using a DNN. Figure 2b shows the accuracy obtained in the partial training of target generator as per the methodology described in Subsect. 4.1.

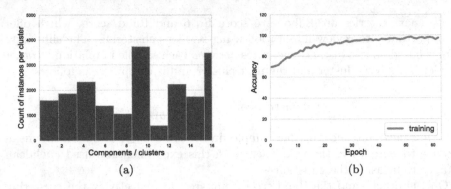

Fig. 2. (a) Gaussian mixture model data generation showing data points generated per component (b) Target generator training accuracy after partial training. Training done using the clean original sub dataset

In the images below we present a visual comparison of the synthetic dataset generated data and the clean original sub-dataset using a K-Means 2 cluster analysis. Specifically in this analysis we plot the Silhouette score for the original clean dataset in Fig. 3a and the synthetic dataset in Fig. 3b in order to have a visual analysis of the goodness of the synthetic dataset in comparison to the original. We note that in both plots there is a skew or imbalance favouring one class.

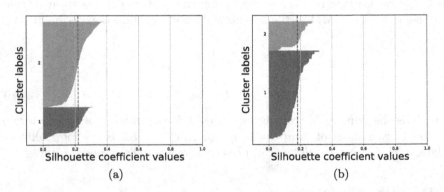

Fig. 3. (a) KMeans silhouette score analysis for the clean original dataset (b) KMeans silhouette score analysis for the synthetic dataset

5.2   Classification

First of all, in Table 1, we present the classification accuracy of the imputed datasets, we also include the imputation accuracy of the full synthetic dataset as our baseline scores. The table presents the training scores given in columns 3 and

4 labelled training and validation respectively. On the other hand testing of the models was done on novel datasets i.e. novel to the training process of the model. Specifically, we tested the models using the full synthetic dataset, the original clean sub-dataset, and the original clean sub-dataset that was balanced using the SMOTE Edited Nearest Neighbour in columns 5, 6, 7 and 8 respectively.

Table 1. Classification accuracy for models built using imputed datasets and tested on the three testing datasets a) full synthetic, b) original clean sub-dataset and balanced original using the SMOTE Edited Nearest Neighbour

Method	Missing	Training	Validation	Synthetic	Testing	Original	Edited NN
DAE	10	87.13	73.75	75.57	71.06	75.51	53.11
DAE	20	87.14	73.76	74.91	71.08	75.83	41.25
DAE	30	82.46	65.82	65.10	63.24	72.74	62.26
DAE	40	87.08	73.92	74.93	70.68	77.55	55.25
MICE	10	93.88	83.47	83.88	79.11	80.08	52.72
MICE	20	93.71	84.02	82.53	78.51	80.91	63.42
II	30	90.42	86.50	72.10	70.33	83.89	67.12
MICE	40	88.00	80.78	74.22	68.95	78.94	59.14
KNN	10	91.05	77.25	80.11	75.67	83.83	51.17
KNN	20	89.12	75.14	78.34	73.51	79.31	48.93
KNN	30	88.09	73.94	76.60	73.25	78.18	55.45
KNN	40	86.89	72.12	75.14	73.09	78.64	64.98
Miss F	10	91.92	80.01	80.21	74.48	81.01	49.81
Miss F	20	91.13	79.69	79.86	75.15	81.64	46.79
Miss F	30	89.69	79.39	78.51	75.13	80.37	56.42
Miss F	40	88.38	78.73	77.42	75.07	81.47	54.96
Synthetic	None	96.28	91.12	NA	88.57	82.66	45.91

Secondly, Table 2 presents the training loss which was computed using the log loss as defined in Subsect. 4.2. The table columns are organised as in the accuracy Table 1 as described above, however, in this table we focus on the classification loss for all the imputation methods and the baseline synthetic dataset.

Overall, our classification results presented in Table 1 and Table 2 show that the models from the imputed datasets yielded classifiers that were able to give a higher performance on the yet unseen test clean sub-dataset as well as the original cleaned observed sub-dataset. Moreover, the DAE imputed methods yielded the lowest log loss in training and also on the other test dataset, an indication of good performance. This shows that the imputation from this process successfully managed to learn from the synthetic dataset a set of parameters i.e. weights which though they represent the synthetic dataset with missingness,

Table 2. Classification loss for models built using imputed datasets and tested on the three testing datasets a) full synthetic, b) original clean sub-dataset and balanced original using the SMOTE Edited Nearest Neighbour

Method	Missing	Training	Validation	Synthetic	Testing	Original	Edited NN
DAE	10	0.088	0.318	0.142	0.644	0.654	6.418
DAE	20	0.082	0.323	0.159	0.713	0.676	10.073
DAE	30	0.240	0.972	1.844	2.697	1.154	11.920
DAE	40	0.073	0.339	0.197	0.859	1.154	11.333
MICE	10	0.146	0.661	1.143	1.467	0.774	8.813
MICE	20	0.151	0.645	1.110	1.467	0.751	7.239
MICE	30	0.224	0.374	3.026	3.227	1.617	7.599
MICE	40	0.250	0.601	8.483	8.296	0.804	13.043
KNN	10	0.193	0.940	1.642	2.243	1.570	7.686
KNN	20	0.236	0.958	1.577	2.220	0.783	11.811
KNN	30	0.254	1.018	2.080	2.776	0.904	13.562
KNN	40	0.283	1.027	2.282	2.917	0.866	10.888
Miss F	10	0.166	0.792	1.395	2.192	0.754	9.514
Miss F	20	0.185	0.818	1.644	2.549	0.793	9.197
Miss F	30	0.221	0.714	1.531	2.179	0.820	9.365
Miss F	40	0.248	0.743	1.973	2.787	0.833	13.938
Synthetic	None	0.097	0.350	NA	0.535	10.050	8.155

some of that knowledge can be transferable to the original dataset with missingness. The other two state of the art missing data imputation methods the KNN imputer and the Miss Forest imputer came second and equally performed well especially on the test dataset and the full synthetic dataset. It should be noted that the though the performance on the original dataset can be improved accross all these methods, we have demonstrated the potential of our approach of using synthetic dataset as an indirect route to resolve missingness in a real observed dataset.

5.3   Direct Analysis of Imputation

In this subsection we present results of the direct analysis of the imputation. We begin by providing results of KMeans clustering of the imputed datasets across all missingness degrees in Table 3. The table contains Rand and Silhouette scores for cluster scenarios i.e. 2, 3 and 4. In both metrics a score of 1 is the best while 0 is the worst.

Our results in Table 3 show that the DAE imputed datasets yielded better clusters on the Silhouette score, while on the Rand score performance varied across the imputed methods. This means that the DAE yielded better clusters and yielded a better approximation of the groups in the dataset for all cluster scenarios. While the same DAE did not emerge a winner on the RAND score that is dependent on actual class label information. However, the scores between the methods on these metrics were not significantly different especially for the 2 cluster scenario, showing that all the methods managed to recover the binary class nature of the dataset.

Table 3. Rand and Silhoutte scores for various cluster sizes on imputed datasets with 30 pct induced missingness, compared with the full synthetic and original clean sub dataset.

	Dataset	Rand scores clusters			Silhoutte score clusters		
		2	3	4	2	3	4
Method	DAE	0.5673	0.3195	0.3195	0.5312	0.5249	0.4958
	Miss forest	0.6184	0.5106	0.5106	0.5295	0.5212	0.5058
	MICE	0.6166	0.5135	0.5135	0.5296	0.5210	0.5068
	KNN	0.5721	0.3071	0.3071	0.5319	0.5238	0.4921

Table 4. Direct analysis of imputation comparing the DAE imputed datasets across missingness degrees.

Dataset	Missing	RMSE	R2	MAPE
DAE	10	547.158	−1.547	1.075×10^{18}
DAE	20	601.237	−0.756	9.531×10^{17}
DAE	30	239.679	0.537	5.429×10^{0}
DAE	40	694.770	−0.511	7.125×10^{17}

Consequently, we decided to investigate the performance of the DAE imputation a bit more closely. Therefore, in Table 4 we focus on the DAE imputed datasets across all missingness degrees. We make a direct comparison between the actual synthetic values versus the values predicted by DAE by showing values for RMSE, R2 and MAPE as defined in Subsect. 4.2. The results in this case were not very conclusive in showing any clear pattern on the performance of the method across missingness degrees and is worth investigating a bit further.

6   Conclusion

Our results on synthetic dataset generation showed that we managed to create a realistic dataset that is similar and comparable to the original dataset based on KMeans analysis of the clustering metrics such as Silhouette score for the two datasets. This was an important step for us because we needed a way to use this dataset with its ground truth values to assess imputation accuracy, because in real observed datasets, when values are really missing there is no ground truth to determine conclusively how effective a data recovery mechanism is. However, if missingness is simulated, ground truth values exist for all missing values and direct assessment of imputation becomes possible. Moreover, through classification experiments we were able to have an objective way to compare performance of imputation for various methods, this was done to allow us the ability to make a recommendation on which method works best for our dataset. We observed that the DAE imputed datasets yielded classifiers that were average performers during training and validation at 87 and 74% accuracy with minimum log loss of 0.073, however, when these models were tested on the original observed cleaned sub dataset the models performed way better than the other methods on lower missingness levels of up to 20% when one considers the loss metric, while on accuracy the performance was slightly lower in comparison to the other methods. On the other hand, some of the state of the art imputation methods such as MICE yielded high accuracy on both the training and testing datasets, however when tested on the unseen clean original sub-dataset the performance was lower.

Therefore since our interest lies not in general classification accuracy but on how a method performed on a yet unseen clean observed sub-dataset the winning method is the DAE imputer on the loss metric especially on misssingness degrees that are closer to the actual missingness in the observed dataset of about 24%. On the other hand the MICE method is more accurate on all the missingness degrees, even though it is not too significantly different from the other state of the art methods i.e. Miss Forest the Random Forest based imputer method and the K Nearest Neighbour based imputer.

Finally, these results indicate a possibility that combining methods and improving on model design especially for DAE method may lead to improved overall performance on the original unseen dataset. In conclusion, through, our experiments on synthetic dataset generation, simulation of missing data based on actual data missingness and usage of various missing data imputation methods, we can propose a methodological approach that has the potential to deal with missingness on unseen real datasets. This method allows one to explore various imputation techniques on a realistic synthetic dataset providing an opportunity to assess and determine what method can be most suitable for a given missingness degree or missingness scheme in some observed dataset.

References

1. Anderson, J.W., Kennedy, K.E., Ngo, L.B., Luckow, A., Apon, A.W.: Synthetic data generation for the internet of things. In: Proceedings - 2014 IEEE International Conference on Big Data, IEEE Big Data 2014, pp. 171–176 (2014)
2. Beaulieu-Jones, B.K., Moore, J.H.: Missing data imputation in the electronic health record using deeply learned autoencoders. In: Pacific Symposium on Biocomputing 2017, pp. 207–218 (2017)

3. Chawla, N.V., Bowyer, K.W., Hall, L.O., Kegelmeyer, W.P.: SMOTE: synthetic minority over-sampling technique. J. Artif. Intell. Res. **16**, 321–357 (2002)
4. García, S., Luengo, J., Herrera, F.: Tutorial on practical tips of the most influential data preprocessing algorithms in data mining. Knowl.-Based Syst. **98**, 1–29 (2016)
5. Ghahramani, Z., Jordan, M.I.: Learning from incomplete data. Technical report. A I Memo No. 1509; C.B.C.L. Paper No. 108, MIT (1994). Dspace.mit.edu publications.ai.mit.edu
6. Lin, P.J., et al.: Development of a synthetic data set generator for building and testing information discovery systems. In: Proceedings of the Third International Conference on Information Technology: New Generations (ITNG 2006), Las Vegas, pp. 1–5 (2006). ISBN 0769524974
7. Lin, W.-C., Tsai, C.-F.: Missing value imputation: a review and analysis of the literature (2006–2017). Artif. Intell. Rev. **53**(2), 1487–1509 (2019). https://doi.org/10.1007/s10462-019-09709-4
8. Manaka, T., Van Zyl, T., Wade, A.N., Kar, D.: Using machine learning to fuse verbal autopsy narratives and binary features in the analysis of deaths from hyperglycaemia. In: Proceedings of SACAIR2021, vol. 1, pp. 90–106 (2022). https://2021.sacair.org.za/wp-content/uploads/2022/02/SACAIR21-Proceedings
9. Marbán, Ó., Mariscal, G., Segovia, J.: A Data mining & knowledge discovery process model. Data Min. Knowl. Discov. Real Life Appl. (February), 1–17 (2009). www.intechopen.com, www.intechweb.org
10. Mathonsi, T., van Zyl, T.L.: A statistics and deep learning hybrid method for multivariate time series forecasting and mortality modeling. Forecasting **4**(1), 1–25 (2022)
11. Misra, P., Yadav, A.S.: Impact of preprocessing methods on healthcare predictions. In: 2nd International Conference on Advanced Computing and Software Engineering (ICACSE-2019), Ml (2019)
12. Richardson, T.W., Wu, W., Lin, L., Xu, B., Bernal, E.A.: McFlow: Monte Carlo flow models for data imputation. In: 2020 Computer Vision and Pattern Recognition (CVPR) (2020). http://arxiv.org/abs/2003.12628
13. Rubin, D.B.: An overview of multiple imputation. In: Proceedings of the Survey Research Methods Section of the American Statistical Association, pp. 79–84 (1988)
14. Rubin, D.R.: Inference and missing data. Biometrika **63**(3), 581–592 (1976)
15. Shang, C., Palmer, A., Sun, J., Chen, K.S., Lu, J., Bi, J.: VIGAN: missing view imputation with generative adversarial networks. In: 2017 IEEE International Conference on Big Data (Big Data), pp. 766–775. IEEE (2017). ISBN 9781538627150
16. Sterne, J.A.C., et al.: Multiple imputation for missing data in epidemiological and clinical research: potential and pitfalls. Res. Methods Rep. 1–11 (2009). https://www.bmj.com/content/338/bmj.b2393
17. Vazifehdan, M., Moattar, M.H., Jalali, M.: A hybrid Bayesian network and tensor factorization approach for missing value imputation to improve breast cancer recurrence prediction. J. King Saud Univ. - Comput. Inf. Sci. **31**(2), 175–184 (2019). ISSN 1319-1578. https://doi.org/10.1016/j.jksuci.2018.01.002
18. Wan, Z., Zhang, Y., He, H.: Variational autoencoder based synthetic data generation for imbalanced learning. In: 2017 - IEEE Symposium Series on Computational Intelligence (SSCI) (2017). ISBN 9781538627266. https://ieeexplore.ieee.org/xpl/conhome/8267146/proceeding
19. Yoon, J., Jordon, J., Schaar, M.V.D.: GAIN: missing data imputation using generative adversarial nets. In: Proceedings of the 35th International Conference on Machine Learning, Stockholm, Sweden (2018)

20. Yoon, S.: GAMIN: generative adversarial multiple imputation network for highly missing data. In: IEEE/CVF Conference on Computer Vision and Pattern Recognition, pp. 8456–8464 (2020)
21. Zheng, X., Wang, B., Xie, L.: Synthetic dynamic PMU data generation: a generative adversarial network approach. In: 2019 International Conference on Smart Grid Synchronized Measurements and Analytics, SGSMA 2019 (2019)

Defeasible Justification Using the KLM Framework

Steve Wang$^{(\boxtimes)}$ (ID), Thomas Meyer (ID), and Deshendran Moodley (ID)

University of Cape Town, Cape Town, South Africa
wngshu003@myuct.ac.za, tmeyer@cs.uct.ac.za, deshen.moodley@uct.ac.za

Abstract. The Kraus, Lehmann and Magidor (KLM) framework is an extension of Propositional Logic (PL) that can perform defeasible reasoning. The results of defeasible reasoning using the KLM framework are often challenging to understand. Therefore, one needs a framework within which it is possible to provide justifications for conclusions drawn from defeasible reasoning. This paper proposes a theoretical framework for defeasible justification in PL and a software tool that implements the framework. The theoretical framework is based on an existing theoretical framework for Description Logic (DL). The defeasible justification algorithm uses the statement ranking required by the KLM-style form of defeasible entailment known as Rational Closure. Classical justifications are computed based on materialised formulas (classical counterparts of defeasible formulas). The resulting classical justifications are converted to defeasible justifications, based on the input knowledge base. We provide an initial evaluation of the framework and the software tool by testing it with a representative example.

Keywords: Knowledge representation · Propositional logic · The KLM framework · Defeasible justification · Rational closure · Defeasible justification tool

1 Introduction

The conclusions produced by reasoning tools are often difficult to understand. Justifications for such conclusions provide users with the exact statements in the knowledge base that is used to deduce the conclusion. Currently, there exist tools that provide explanations for classical entailments [9,10,12]. Such a tool aids users in understanding their knowledge base and reasoning systems [4] as well as providing assistance in debugging their systems [16].

Although there are well-established tools and algorithms to compute explanations for classical reasoning, there are no such tools for defeasible reasoning. While there are many approaches to defeasible reasoning, in this paper we focus only on an approach to defeasible reasoning known as the Kraus, Lehmann and Magidor (KLM) framework [14], which is an extension of Propositional Logic (PL). Chama proposed an algorithm that computes justification for defeasible entailments in Description Logic (DL) [6]. This paper's contribution is extended Chama's work by converting her proposed justification algorithm for DL to PL according to the well-established notions and concepts in the KLM framework. Furthermore, we implement our defeasible justification algorithm which has not been done for the KLM framework. The implementation was tested with a representative example and the result is accurate compared to manual deductions.

2   Background

In classical reasoning with PL, the notion of interpretation and entailment is well-defined [5]. Similarly, for defeasible reasoning with the KLM framework (an extension of PL), the notion of a ranked interpretation was defined [14] and a form of defeasible entailment referred to as Rational Closure was described. Based on these notions, we define an algorithm that computes justifications for defeasible entailment.

Statements in PL are built up with a finite set $\mathcal{P} = \{p, q, ...\}$ of *propositional atoms*. The binary connectives $\wedge, \vee, \rightarrow, \leftrightarrow$ and the negation operator $\neg$ can be applied recursively to form propositional formulas. An *interpretation* $\mathcal{I}$ is a function that maps a propositional atom to either true or false. An interpretation satisfies a formula if the formula evaluates to true under the rules of *satisfiability*. The $\top$ constant in PL denotes a *tautology* that is always interpreted to *true* and the $\perp$ constant is always interpreted to *false*.

A *knowledge base* is a finite set of propositional formulas. An interpretation $\mathcal{I}$ satisfies a knowledge base $\mathcal{K}$ if $\mathcal{I}$ satisfies every formula in $\mathcal{K}$. A knowledge base $\mathcal{K}$ *entails* a formula α, denoted $\mathcal{K} \vDash \alpha$, if and only if every interpretation that satisfies $\mathcal{K}$ also satisfies α. Two propositional formulas α and β are logically equivalent, denoted $\alpha \equiv \beta$, if all interpretations that satisfies α also satisfies β and vice versa.

Horridge defines the notion of *justification* to provide explanations for classical entailments [8]. A subset of formulas $\mathcal{J}$ is a *justification* for the entailment $\mathcal{K} \vDash \alpha$ if $\mathcal{J} \subseteq \mathcal{K}$ such that $\mathcal{J} \vDash \alpha$ and there are no proper subset $\mathcal{J}' \subset \mathcal{J}$ such that $\mathcal{J}' \vDash \alpha$. Horridge defined algorithms to compute justifications for classical entailment [8].

Given a knowledge base $\mathcal{K}$ and an entailed formula η, Horridge identifies all justifications for the entailment by first identifying a single justification (Algorithm 3). A justification for the entailment can be identified by first expanding a subset $\mathcal{S} \subseteq \mathcal{K}$ until $\mathcal{S} \vDash \eta$ (Algorithm 1) then contract $\mathcal{S}$ without breaking the entailment (Algorithm 2). The justification can be used as the root node in the Hitting Set Tree [17] to identify all justification for the given entailment (Algorithm 4). In this paper, the implementation of the Hitting Set Tree is based on Reiter's algorithm.

Note that Algorithms 1, 2, 3 and 4 are slight adjusted and renamed to suit the context of this paper because Horridge's work is done in the context of DL and

Ontologies. These algorithms are used as sub-routines to construct a defeasible justification algorithm.

Algorithm 1. ExpandFormulas

    Input: Knowledge base $\mathcal{K}$ and query η
    Output: Set S

1: **if** $\mathcal{K} \not\models \eta$ **then**
2:     return $\varnothing$
3: **else**
4:     $S := \varnothing$
5:     $S' := \varnothing$
6:     $\Sigma := signature(\eta)$
7:     **while** $S' \neq S$ **do**
8:       $S' = S$
9:       $S = S \cup FindRelatedFormulas(\Sigma, \mathcal{K})$
10:      **if** $S \models \eta$ **then**
11:        return S
12:      **end if**
13:      $\Sigma = signature(S)$
14:     **end while**
15: **end if**
16: return S

There are many approaches to defeasible reasoning. One of the most explored approaches in the literature is the KLM framework suggested by Kraus, Lehmann and Magidor [14]. The KLM framework extends PL by an additional binary connective known as *defeasible implication*, denoted by $\vdash$. Defeasible implications are expressed in the form $\alpha \vdash \beta$ where α and β are propositional formulas. One reads $\alpha \vdash \beta$ as "α typically implies β". Note that $\vdash$ cannot be nested.

A defeasible knowledge base is a finite set of defeasible implications. It is easily shown that a classical formula α is logically equivalent to the defeasible implication $\neg\alpha \vdash \bot$ [11]. Therefore, any knowledge base that contains classical formulas can be converted into a defeasible knowledge base. From here on we assume knowledge bases are defeasible unless stated otherwise explicitly. The material counter-part of a defeasible implication $\alpha \vdash \beta$ is the classical implication $\alpha \rightarrow \beta$. The material counter-part of the knowledge base $\mathcal{K}$, denoted $\overrightarrow{\mathcal{K}}$, is the knowledge base with each of the defeasible implications in $\mathcal{K}$ replaced with its material count-part.

Algorithm 2. ContractFormulas

    Input: Knowledge base $\mathcal{K}$ and entailment η
    Output: Set S

1: return $ContractFormulasRecursive(\varnothing, \mathcal{K}, \eta)$
2: $ContractAxiomsRecusive(S_{support}, S_{whole}, \eta)$
3: **if** $|S_{whole}| == 1$ **then**
4:     return S_{whole}
5: **end if**
6: $S_L, S_R := Splite(S_{whole})$
7: **if** $S_{support} \cup S_L \vDash \eta$ **then**
8:     return $ContractFormulasRecursive(S_{support}, S_L, \eta)$
9: **end if**
10: **if** $S_{support} \cup S_R \vDash \eta$ **then**
11:     return $ContractFormulasRecursive(S_{support}, S_R, \eta)$
12: **end if**
13: $S'_L := ContractFormulasRecursive(S_{support} \cup S_R, S_L, \eta)$
14: $S'_R := ContractFormulasRecursive(S_{support} \cup S'_L, S_R, \eta)$
15: return $S'_L \cup S'_R$

Algorithm 3. ComputeSingleJustification

    Input: Knowledge base $\mathcal{K}$ and entailment η
    Output: Justification $\mathcal{J}$

1: **if** $\eta \in \mathcal{K}$ **then**
2:     return η
3: **end if**
4: $S := ExpandFormulas(\mathcal{K}, \eta)$
5: **if** $S == \varnothing$ **then**
6:     return $\varnothing$
7: **end if**
8: $\mathcal{J} := ContractFormulas(S, \eta)$
9: return $\mathcal{J}$

Similar to classical entailment, defeasible entailment, denoted $\mathrel{|\!\approx}$, is a binary relation over a defeasible knowledge base and a defeasible implication. One reads $\mathcal{K} \mathrel{|\!\approx} \alpha \mathrel{\vdash} \beta$ as "$\mathcal{K}$ *defeasibly entails that α typically implies β*".

Algorithm 4. ComputeAllJustifications

Input: Knowledge base $\mathcal{K}$ and entailment η
Output: Justification $\mathcal{J}$

1: $S_{working} := \mathcal{K}$
2: $X_{explored} := \varnothing$
3: $X_{result} := \varnothing$
4: $\mathcal{J}_{root} := ComputeSingleJustification(S_{working}, \eta)$
5: $X_{result} = X_{result} \cup \{\mathcal{J}_{root}\}$
6: $v_{root} := GetFreshNode(\mathcal{J}_{root})$
7: $Enqueue(v_{root}, Q)$
8: $SetRoot(T_{hst}, v_{root})$
9: **while** $Q \neq \varnothing$ **do**
10:　　$v_{head} = Dequeue(Q)$
11:　　$j_{head} = GetLabel(v_{head})$
12:　　**for** $\alpha \in j_{head}$ **do**
13:　　　　$S_{path} = GetPathToRootLabelSet(v_{head}, T_{hst}) \cup \{\alpha\}$
14:　　　　**if** $S_{path} \notin X_{explored}$ **then**
15:　　　　　　$X_{explored} = X_{explored} \cup \{S_{path}\}$
16:　　　　　　$J' = GetNonIntersectingJustification(S_{path}, X_{result})$
17:　　　　　　**if** $J' == \varnothing$ **then**
18:　　　　　　　　$S_{working} = S_{working} \setminus \{S_{path}\}$
19:　　　　　　　　$J' = ComputeSingJustification(S_{working}, \eta)$
20:　　　　　　　　$S_{working} = S_{working} \cup \{S_{path}\}$
21:　　　　　　**end if**
22:　　　　　　$v_{fresh} = GetFreshNode(J')$
23:　　　　　　$e = GetFreshEdge(\langle v_{fresh}, v_{head}\rangle, \alpha)$
24:　　　　　　$T_{hst} = T_{hst} \cup \{e\}$
25:　　　　　　**if** $J' \neq \varnothing$ **then**
26:　　　　　　　　$X_{result} = X_{result} \cup \{J'\}$
27:　　　　　　　　$Enqueue(v_{fresh}, Q)$
28:　　　　　　**end if**
29:　　　　**end if**
30:　　**end for**
31: **end while**
32: return X_{result}

Lehmann and Magidor suggest the notion of Rational Closure as a form of defeasible entailment and presented an algorithm for Rational Closure [15]. We use the Rational Closure algorithm as a sub-routine to construct a defeasible justification algorithm. The first procedure in the rational closure algorithm is to assign rankings to formulas in the knowledge base. Low ranks are assigned to statements that are less exceptional. The infinite rank is assigned to classical statements. The ranking algorithm is shown in Algorithm 5 [11].

Algorithm 5. Base Rank

    Input: A knowledge base $\mathcal{K}$
    Output: An ordered tuple $(\mathcal{R}_0, ..., \mathcal{R}_{n-1}, \mathcal{R}_\infty, n)$

1:   $i := 0$
2:   $E_0 := \vec{\mathcal{K}}$
3: **repeat**
4:      $E_{i+1} := \{\alpha \to \beta \in E_i | E_i \vDash \neg\alpha\}$
5:      $R_i := E_i \smallsetminus E_{i+1}$;
6:      $i := i + 1$;
7: **until** $E_{i-1} = E_i$
8:   $R_\infty := E_{i-1}$;
9: **if** $E_{i-1} == \varnothing$ **then**
10:      $n := i - 1$;
11: **else**
12:      $n := i$;
13: **end if**
14: **return** $(R_0, ..., R_{n-1}, R_\infty, n)$

Based on the ranking produced by Algorithm 5, the Rational Closure algorithm removed the ranking in ascending order until the remaining formulas entail the negation of the query's antecedent. The knowledge base defeasibly entails the query (the query is in the Rational Closure of the knowledge base) if the (materialised versions of the) remaining formulas classically entail the materialised query. Algorithm 6 is an algorithm for Rational Closure suggested by Kaliski [11].

Algorithm 6. RationalClosure

    Input: A knowledge base $\mathcal{K}$, and a defeasible implication $\eta = \alpha \mathrel{\vrule height 1.2ex depth 0pt width 0pt}\!\!\sim \beta$
    Output:true if $\mathcal{K} \mathrel{\not\vrule height 1.2ex depth 0pt width 0pt}\!\!\sim \alpha \mathrel{\vrule height 1.2ex depth 0pt width 0pt}\!\!\sim \beta$, and **false** otherwise

1:   $(R_0, ..., R_{n-1}, R_\infty, n) := BaseRank(\mathcal{K})$;
2:   $i := 0$
3:   $R := \bigcup_{i=0}^{j<n} R_j$;
4: **while** $R_\infty \cup R \vDash \neg\alpha$ and $R \neq \varnothing$ **do**
5:      $R := R \smallsetminus R_i$;
6:      $i := i + 1$;
7: **end while**
8: **return** $R_\infty \cup R \vDash \alpha \to \beta$;

---

3   Defeasible Justification Algorithm

Chama presented an algorithm that computes defeasible justification in DL [6]. Her algorithm is composed of two sub-algorithms, namely *RationalClosureFor-Justifications* and *ComputeAllJustifications*. We construct a defeasible algorithm for the KLM framework in a similar manner. The algorithm is composed of three sub-algorithms: *RationalClosureForJustification, ComputeAllJustifications* (for classical entailment) and *Dematerialsiation*. Each sub-algorithm is discussed in detail in this section.

We present Algorithm 7 as an algorithm that computes defeasible justification given a knowledge base $\mathcal{K}$ and an entailment $\eta = \alpha \mathrel{\vdash\mkern-7mu\sim} \beta$. The algorithm first ranks the formulas as required by Rational Closure. A slightly adjusted Rational Closure algorithm compared to Algorithm 6, which returns additional parameters, is used. Such parameters include an ordered tuple indicating the ranked formulas and an integer indicating the ranks of formulas discarded to compute the Rational Closure.

In the case where no formulas were discarded in the Rational Closure computation, justifications are computed on the materialised knowledge base $\overrightarrow{\mathcal{K}}$ and query $\overrightarrow{\eta}$ using algorithms mentioned in Sect. 2. Alternatively, in the case where ranks of formulas were discarded in the Rational Closure computation, the discarded formulas are removed from $\mathcal{K}$ before identifying all justifications for the entailment.

Lastly, dematerialisation of justifications is required because results from algorithms mentioned in Sect. 2 are classical formulas. Any formulas in the set of justifications that is not in $\mathcal{K}$ needs to be dematerialised.

Algorithm 7. DefeasibleJustification

Input: Defeasible knowledge base $\mathcal{K}$ and query $\eta = \alpha \mathrel{\vdash\mkern-7mu\sim} \beta$
Output: Justification $\mathcal{J}$

1: $i := 0$
2: $\mathcal{J} := \varnothing$
3: $(\mathcal{R}_0, \mathcal{R}_1, ..., \mathcal{R}_\infty), rank := RationalClosureForJustification(\mathcal{K}, \eta)$
4: **if** $rank == 0$ **then**
5:     $\mathcal{J} = ComputeAllJustification(\overrightarrow{\mathcal{K}}, \overrightarrow{\eta})$
6:     $\mathcal{J} = dematerialise(\mathcal{J}, \mathcal{R}_\infty)$
7:     return $\mathcal{J}$
8: **end if**
9: **while** $i < rank$ **do**
10:     $\mathcal{K} = \mathcal{K} \backslash \mathcal{R}_i$
11:     $i = i + 1$
12: **end while**
13: $\mathcal{J} = computeAllJustifications(\overrightarrow{\mathcal{K}}, \overrightarrow{\eta})$
14: $\mathcal{J} = dematerialise(\mathcal{J}, \mathcal{R}_\infty)$
15: return $\mathcal{J}$

Algorithm 7 can be further enhanced to accept an input knowledge base that contains both classical and defeasible implications. From here on, we refer to a knowledge base with both classical and defeasible formulas as a "mixed knowledge base". As mentioned in Sect. 2, any classical formula α is logically equivalent to the defeasible implication $\neg\alpha \mathrel{\vdash\mkern-7mu\sim} \bot$ and as a result, we can pre-process a mixed knowledge base into a defeasible knowledge base.

Notice that when there are formulas of the form $\neg\alpha \to \bot$ in a given knowledge base, line 5 in Algorithm 5 is $E_{i+1} := \{\neg\alpha \to \bot | E_i \vDash \neg(\neg\alpha)\} = \{\neg\alpha \to \bot | E_i \vDash \alpha\}$. As a result, $\neg\alpha \to \bot$ is always going to be in E_{i+1}. Eventually, when $E_{i-1} = E_i$ then while loop terminates with any formulas of the form $\neg\alpha \to \bot$ in E_{i-1} and line 8 assigns such formulas with the infinity rank. This phenomenon allows us to conclude that all classical formulas in a knowledge base are ranked infinity by Algorithm 5.

Furthermore, consider the knowledge base $\mathcal{K} = \{\alpha \mathbin{\vdash} \beta, \alpha \mathbin{\vdash} \neg\beta\}$ as input to Algorithm 5, the variables are assigned the following values:

- $E_0 = \{\alpha \to \beta, \alpha \to \neg\beta\}$
- $E_1 := \{\alpha \to \beta, \alpha \to \neg\beta\}$ because $E_0 \vDash \neg\alpha$.
- As a result, both $\alpha \to \beta$ and $\alpha \to \neg\beta$ is ranked infinity.

Therefore, we can conclude that when there is a formula of the format $\alpha \mathbin{\vdash} \beta$ and another formula of the form $\alpha \mathbin{\vdash} \neg\beta$ in the input knowledge base then both formulas are ranked infinity by Algorithm 5.

Therefore, Algorithm 5 and consequently Algorithm 7 can be enhanced to accept a mixed knowledge base as input. Algorithm 8 is a *BaseRank* algorithm that accepts a mixed knowledge base.

Algorithm 8. BaseRankForJustification

Input: A mixed knowledge base $\mathcal{K}$
Output: An ordered tuple $(\mathcal{R}_0, ..., \mathcal{R}_{n-1}, \mathcal{R}_\infty, n)$
1: $\mathcal{C} := \{\alpha \to \beta \in \mathcal{K}\} \cup \{\alpha \to \beta, \alpha \to \neg\beta | \alpha \mathbin{\vdash} \beta \text{ and } \alpha \mathbin{\vdash} \neg\beta \in \mathcal{K}\}$
2: $E_0 := \mathcal{K} \setminus \mathcal{C}$
3: $i = 0$
4: **while** $E_{i-1} \neq E_i$ **do**
5:     $E_{i+1} := \{\alpha \mathbin{\vdash} \beta \in E_i | \overrightarrow{E_i} \cup C \vDash \neg\alpha\}$
6:     $R_i := E_i \setminus E_{i+1}$
7:     $i = i + 1$
8: **end while**
9: $R_\infty := \mathcal{C} \cup E_{i-1}$
10: **if** $E_{i-1} = \varnothing$ **then**
11:     $n = i - 1$
12: **else**
13:     $n = i$
14: **end if**
15: **return** $(R_0, R_1, ..., R_\infty, n)$

The Rational Closure algorithm for justification has the same procedures as Algorithm 6 with additional return values. Algorithm 9 is the adjusted algorithm that returns an ordered tuple of ranked formulas from Algorithm 8 and an integer that indicated the ranks of formulas discarded to compute Rational Closure.

Algorithm 9. RationalClosureForJustification

Input: A mixed knowledge base $\mathcal{K}$ and a defeasible implication $\eta = \alpha \mathbin{\vert\!\sim} \beta$

Output: true if $\mathcal{K} \mathbin{\vert\!\approx} \alpha \mathbin{\vert\!\sim} \beta$ and **false** otherwise, rank **i** and an ordered tuple $(\mathcal{R}_0, ..., \mathcal{R}_\infty)$

1: $(R_0, ..., R_{n-1}, R_\infty, n) := BaseRankForJustification(\mathcal{K})$;
2: $i := 0$
3: $R := \bigcup_{i=0}^{j<n} R_j$
4: **while** $R_\infty \cup \vec{R} \vDash \neg\alpha$ and $R \neq \varnothing$ **do**
5:     $R := R \setminus R_i$
6:     $i := i + 1$
7: **end while**
8: **return** $R_\infty \cup \vec{R} \vDash \alpha \rightarrow \beta$, i, $(R_0, ..., R_{n-1}, R_\infty)$

We present an example to illustrate the intuition behind Rational Closure. Consider the following scenario where birds typically fly, penguins are birds, penguins typically cannot fly, robins are birds and birds typically have wings. From the scenario, we can construct the following knowledge base $\mathcal{K} = \{birds \mathbin{\vert\!\sim} fly, penguins \rightarrow birds, penguins \mathbin{\vert\!\sim} \neg fly, robins \rightarrow birds, birds \mathbin{\vert\!\sim} wings\}$.

Firstly, we use Algorithm 8 to assign a ranking to each formula in $\mathcal{K}$.

1. Line 1 of the algorithm extracts all classical formulas from $\mathcal{K}$ and we get the following: $\mathcal{C} = \{penguins \rightarrow birds, robins \rightarrow birds\}$
2. The remaining formulas in $\mathcal{K}$ are collected into E_0. Therefore, $E_0 = \{birds \mathbin{\vert\!\sim} fly, penguins \mathbin{\vert\!\sim} \neg fly, birds \mathbin{\vert\!\sim} wings\}$.
3. The algorithm iterates over the while loop from lines 4 to 8 until $E_{i-1} = E_i$. Over the iterations, we get the following values for each variable.
   (a) $E_1 = \{\alpha \mathbin{\vert\!\sim} \beta | E_0 \vDash \neg\alpha\} = \{penguins \mathbin{\vert\!\sim} \neg fly\}$
   (b) $R_0 = \{birds \mathbin{\vert\!\sim} fly, birds \mathbin{\vert\!\sim} wings\}$
   (c) $E_2 = \varnothing$
   (d) $R_1 = \{penguins \mathbin{\vert\!\sim} \neg fly\}$
   (e) $E_3 = \varnothing$
4. Line 9 of the algorithm assigns $R_\infty = \mathcal{C} \cup E_2 = \{penguins \rightarrow birds, robins \rightarrow birds\}$.

The final rankings of the formulas are shown in Table 1.

For querying whether robins have wings, we construct the defeasible implication $robins \mathbin{\vert\!\sim} wings$. Using the above ranking and Algorithm 9, we can compute if $\mathcal{K} \mathbin{\vert\!\approx} robins \mathbin{\vert\!\sim} wings$. The procedures of Algorithm 9 are as follows:

1. Line 3 of the algorithm assigns $R = \{penguins \mathbin{\vert\!\sim} \neg fly, birds \mathbin{\vert\!\sim} fly, birds \mathbin{\vert\!\sim} wings\}$

Table 1. Ranking of formulas in $\mathcal{K}$

Rank	Formulas
∞	$penguins \rightarrow birds, robins \rightarrow birds$
R_1	$penguins \mathrel{\vdash\!\!\!\sim} \neg fly$
R_0	$birds \mathrel{\vdash\!\!\!\sim} fly, birds \mathrel{\vdash\!\!\!\sim} wings$

2. Since $R_\infty \cup \vec{R} \models \neg robins$ does not hold, the algorithm skips over the while loop from lines 4 to 7.
3. Since $R_\infty \cup \vec{R} \models robins \rightarrow wings$, we can conclude that $\mathcal{K} \mathrel{|\!\approx} robins \mathrel{\vdash\!\!\!\sim} wings$.

For querying whether penguins have wings, we construct the defeasible implication $penguins \mathrel{\vdash\!\!\!\sim} wings$. The procedures of Algorithm 9 are as follows:

1. On line 3, we have $R = \{penguins \mathrel{\vdash\!\!\!\sim} \neg fly, birds \mathrel{\vdash\!\!\!\sim} fly, birds \mathrel{\vdash\!\!\!\sim} wings\}$.
2. $R_\infty \cup \vec{R} \models \neg penguins$ and $R \neq \varnothing$, then R_0 is removed from R and we get $R = \{penguins \mathrel{\vdash\!\!\!\sim} \neg fly\}$
3. Now, $R_\infty \cup \vec{R} \models \neg penguins$ does not hold. Therefore, the while loop terminates.
4. But $R_\infty \cup \vec{R} \not\models penguins \rightarrow wings$, and as a result $\mathcal{K} \mathrel{|\!\not\approx} penguins \mathrel{\vdash\!\!\!\sim} wings$.

The defeasible justification algorithm utilises the classical justification algorithms mentioned in Sect. 2. As expected, the resulting justifications contain only classical formulas and cannot be used directly as the justifications for the defeasible entailment. Algorithm 10 takes a set of ∞-ranked formulas and a set of justifications as inputs. Any formulas in the justifications that are not assigned the infinity rank need to be replaced with their material counterpart.

Algorithm 10. Dematerialise

Input: Set of justifications $\mathcal{J}$ and infinitely ranked formulas R_∞
Output: Set of dematerialised justifications $\mathcal{J}$

1: **for** j in $\mathcal{J}$ **do**
2:     **for** $\eta = (\alpha \rightarrow \beta)$ in j **do**
3:         **if** $\eta \notin R_\infty$ **then**
4:             $\eta = \alpha \mathrel{\vdash\!\!\!\sim} \beta$
5:         **end if**
6:     **end for**
7: **end for**
8: Return $\mathcal{J}$

Again, we illustrate the intuition behind defeasible justification via a representative example. Consider the following knowledge base $\mathcal{K} = \{penguins \rightarrow birds, robins \rightarrow birds, specialpenguins \rightarrow penguins, birds \mathrel{\vdash\!\!\!\sim} fly, birds \mathrel{\vdash\!\!\!\sim} wings, penguins \mathrel{\vdash\!\!\!\sim} \neg fly, specialpenguins \mathrel{\vdash\!\!\!\sim} fly\}$. We use the following 3 queries to demonstrate various cases the algorithm considers:

1. $robins \mathrel{|\kern-0.4em\sim} wings$
2. $penguins \mathrel{|\kern-0.4em\sim} wings$
3. $specialpenguins \mathrel{|\kern-0.4em\sim} fly$.

Firstly, Table 2 shows the formula ranking which Algorithm 8 computes.

Table 2. Ranking of formulas in $\mathcal{K}$

Rank	Formulas
∞	$penguins \rightarrow birds, robins \rightarrow birds, specialpenguins \rightarrow penguins$
R_2	$specialpenguins \mathrel{|\kern-0.4em\sim} fly$
R_1	$penguins \mathrel{|\kern-0.4em\sim} \neg fly$
R_0	$birds \mathrel{|\kern-0.4em\sim} fly, birds \mathrel{|\kern-0.4em\sim} wings$

For query $\eta = robins \mathrel{|\kern-0.4em\sim} wings$ the defeasible justification algorithm executes as follows:

1. Rational Closure:
   (a) $\overrightarrow{R} = \{specialpenguins \rightarrow fly, penguins \rightarrow \neg fly, birds \rightarrow fly, birds \rightarrow wings\}$ and $R_\infty \cup \overrightarrow{R} \models \neg robins$ does not hold. Therefore, no formulas are discarded.
   (b) Since, $R_\infty \cup \overrightarrow{R} \models robins \rightarrow wings$, we have $\mathcal{K} \mathrel{\not\mid\kern-0.4em\approx} robins \mathrel{|\kern-0.4em\sim} wings$.
2. Since no formulas were discarded in the previous step, the entire knowledge base is used to compute the justification.
3. The classical justification for the entailment $\mathcal{K} \models robins \rightarrow wings$ is computed using Algorithms 4 and the following Hitting Set Tree is returned:

   $\{robin \rightarrow bird, bird \rightarrow wings\}$

   $robin \rightarrow bird \qquad bird \rightarrow wings$

   $\varnothing \qquad\qquad\qquad \varnothing$

4. The only justification for the classical entailment $\overrightarrow{\mathcal{K}} \models robins \rightarrow wings$ is $\{robins \rightarrow birds, birds \rightarrow wings\}$.
5. Since the classical formula $birds \rightarrow wings$ is not in R_∞, Algorithm 10 replaces it with the defeasible implication $birds \mathrel{|\kern-0.4em\sim} wings$.
6. The final justification for the defeasible entailment $\mathcal{K} \mathrel{\not\mid\kern-0.4em\approx} robins \mathrel{|\kern-0.4em\sim} wings$ is $\{robins \rightarrow birds, birds \mathrel{|\kern-0.4em\sim} wings\}$.

For query $\eta = penguins \mathrel{|\kern-0.4em\sim} wings$ the defeasible justification algorithm executes as follows:

1. Rational Closure:
   (a) $\overrightarrow{R} = \{specialpenguins \rightarrow fly, penguins \rightarrow \neg fly, birds \rightarrow fly, birds \rightarrow wings\}$ and $R_\infty \cup \overrightarrow{R} \models \neg penguins$. Therefore, R_0 is removed from R and $R = \{penguins \mathrel{|\kern-0.4em\sim} \neg fly, specialpenguins \mathrel{|\kern-0.4em\sim} fly\}$.

(b) Now, $R_\infty \cup \vec{R} \vDash \neg penguins$ does not hold.

(c) $R_\infty \cup \vec{R} \nvDash penguins \rightarrow wings$ and therefore $\mathcal{K} \nvdash penguins \mathrel{\vdash} wings$.

2. The algorithm terminates with no justifications.

The final query $specialpenguins \mathrel{\vdash} fly$ executes as follows:

1. Rational Closure:
   (a) Iterating over lines 4 to 7 of Algorithm 6, both R_0 and R_1 is removed from R such that the condition $R_\infty \cup \vec{R} \vDash \neg\alpha$ does not hold.
   (b) $R_\infty \cup \vec{R} = \{penguins \rightarrow birds, robins \rightarrow birds, specialpenguins \rightarrow penguins, specialpenguins \rightarrow fly\} \vDash specialpenguins \rightarrow fly$.
   (c) Hence, the defeasible entailment $\mathcal{K} \nvDash specialpenguins \mathrel{\vdash} fly$ holds.
2. Classical Justification is calculated based on the remaining materialised formulas $\{penguins \rightarrow birds, robins \rightarrow birds, specialpenguins \rightarrow penguins, specialpenguins \rightarrow fly\}$.
3. The following Hitting Set Tree is constructed:

$\{specialpenguins \rightarrow fly\}$

$|$

$specialpenguins \rightarrow fly$

$|$

$\emptyset$

4. The only justification for the classical entailment $\{penguins \rightarrow birds, robins \rightarrow birds, specialpenguins \rightarrow penguins, specialpenguins \rightarrow fly\} \vDash specialpenguins \rightarrow fly$ is $\{specialpenguins \rightarrow fly\}$
5. Since $specialpenguins \rightarrow fly \notin R_\infty$, Algorithm 10 replaces it with $specialpenguins \mathrel{\vdash} fly$.
6. Final defeasible justification for the defeasible entailment $\mathcal{K} \nvDash specialpenguins \mathrel{\vdash} fly$ is $\{\mathcal{K} \nvDash specialpenguins \mathrel{\vdash} fly\}$.

Notice for the final query, there are two possible justifications:

1. $J_1 = \{specialpenguins \mathrel{\vdash} fly\}$
2. $J_2 = \{specialpenguins \rightarrow penguins, penguins \rightarrow birds, birds \mathrel{\vdash} fly\}$.

However, the only valid justification that supports the query is J_1 because formulas required to make justification J_2 valid are discarded by the defeasible justification algorithm.

4   Defeasible Justification Implementation

We implemented a software tool that uses the algorithms presented in Sect. 3 to compute the justification for a defeasible entailment given a mixed knowledge base and a defeasible implication as a query. The tool is implemented in Java and follows the Model View Controller (MVC) software architecture pattern [13]. The source code for the implementation can be found on GitHub [20]. The tool uses two external packages: the Tweety Project and the SAT4J SAT solver.

Our implementation extends the Tweety Project's Propositional Logic models to construct models and operations required by the KLM Framework. The Tweety Project [2] is a software framework that provides models and operations for most First Order Logic (FOL) including PL [18,19]. Due to restrictions and constants defined by the Tweety Project, we cannot denote the KLM framework operations with the conventional symbols used in the literature. Instead, the negation symbol, $\neg$, is replaced with ! and the binary operations $\wedge, \vee, \rightarrow, \leftrightarrow$ and $\mathrel{|\!\sim}$ are replaced with &&, ||, =>, < = > and ~>, respectively. Furthermore, we constructed a parser that reads Strings such as "$\alpha \sim > \beta$" and produces an instance of *DefeasibleImplication* which allows the software to perform operations such as materialisation.

The classical justification algorithm mentioned in Sect. 2 and Algorithm 9 (*RationalClosureForJustification*) require a tool to compute classical entailment. We used the SAT4J SAT solver [1] to perform classical entailment computations.

4.1   Algorithm Implementation

The implementation of the defeasible justification, Base Rank, Rational Closure and dematerialisation follows from Algorithms 7, 5, 9 and 10, respectively. The algorithm that identifies a single justification for classical entailment follows Horridge's definition mentioned in Sect. 2. Java's Object-Oriented programming style is leveraged to implement Reiter's Hitting Set algorithm [17] to identify all justifications for a classical entailment. In the Hitting Set computation, a tree structure is constructed where each node keeps track of a knowledge base and a justification. Each edge represents a formula in the parent node's justification. The child node's knowledge base is its parent node's knowledge base without the formula represented by the edge between them.

4.2   Testing and Evaluation

The tool is tested with the defeasible justification example mentioned in Sect. 3. Figure 1 shows the output of running the example with the query *Robin* $\mathrel{|\!\sim}$ *Wings*. The tool concludes with the correct defeasible justification {*Bird* $\mathrel{|\!\sim}$ *Wings*, *Robin* $\rightarrow$ *Bird*}.

```
--- exec-maven-plugin:1.5.0:exec (default-cli) @ DefeasibleJustificationForPropositionalLogic ---
Knowledge Base:
{ (Bird->Wings), (Penguin=>Bird), (Robin=>Bird), (Penguin~>!Fly), (SpecialPenguin->Fly), (SpecialPenguin=>Penguin), (Bird->Fly) }
Query:
(Robin~>Wings)
<< Rational Closure Result - BEGIN>>
Entailment holds :       true
Ranked removed :         0
Minimal rank:
R_{INFINITY}    |(Penguin=>Bird),(Robin=>Bird),(SpecialPenguin=>Penguin)
R_{2}           |(SpecialPenguin->Fly)
R_{1}           |(Penguin~>!Fly)
R_{0}           |(Bird->Wings),(Bird~>Fly)
<< Rational Closure Result - END >>
<<ALL possible classical justifications>>
(Bird=>Wings), (Robin=>Bird)
<<Final Justification>>
(Bird->Wings), (Robin=>Bird)
--------------------------------------------------------------------
BUILD SUCCESS
--------------------------------------------------------------------
```

Fig. 1. Program output of the implementation of defeasible justification algorithm

5 Conclusion and Future Work

We present an algorithm that computes defeasible justification for the KLM framework, previously only explored for classical justification. A representative example was used to test the algorithm and the result is correct. Furthermore, we present a software tool that implements the defeasible justification algorithm. The same representative example is used to test the implementation, corresponding with manual calculations. It is the first implementation of the form of defeasible justification described by Chama [6]. Chama's work was a theoretical exercise with DL as the underlying logic. We convert Chama's work to the propositional case and implemented it.

Future work may extend to multiple dimensions in both theoretical and practical aspects of this paper. Several pieces of literature present a problem with Reiter's algorithm for Hitting Set Tree [7,21]. Further investigation is required to determine whether the algorithms and implementations presented in this paper need to be adjusted to account for the issue of Reiter's Hitting Set Tree algorithm.

The defeasible justification algorithm can be extended and adjusted to more complex logics, such as Description Logics [3]. Complex test cases and scenarios can be constructed to test the algorithm's coverage for edge cases. One can conduct a complexity analysis on the defeasible justification algorithm to analyse and improve the efficiency of the algorithm. The algorithm may also incorporate the additional feature of providing the conditionals that contribute to the level of the exceptionality of the query. Enhancements can be made to the software tool to improve its efficiency and accuracy. By applying improved programming techniques and resources, one can scale up the tool's computation by magnitudes. Lastly, a user-friendly interface such as a Graphical User Interface (GUI) can be added to the software tool which allows non-technical users to interact with the tool.

References

1. SAT4J SAT solver. https://www.sat4j.org/index.php. Accessed 29 Aug 2022
2. The tweety project. https://tweetyproject.org/. Accessed 29 Aug 2022
3. Baader, F., Calvanese, D., McGuinness, D., Patel-Schneider, P., Nardi, D., et al.: The Description Logic Handbook: Theory, Implementation and Applications. Cambridge University Press, Cambridge (2003)
4. Biran, O., Cotton, C.: Explanation and justification in machine learning: a survey. In: IJCAI-17 Workshop on Explainable AI (XAI), vol. 8, pp. 8–13 (2017)
5. Büning, H.K., Lettmann, T.: Propositional Logic: Deduction and Algorithms, vol. 48. Cambridge University Press, Cambridge (1999)
6. Chama, V.: Explanation for defeasible entailment. Master's thesis, Faculty of Science (2020)
7. Greiner, R., Smith, B.A., Wilkerson, R.W.: A correction to the algorithm in reiter's theory of diagnosis. Artif. Intell. **41**(1), 79–88 (1989)
8. Horridge, M.: Justification Based Explanation in Ontologies. The University of Manchester (United Kingdom), Manchester (2011)

9. Horridge, M., Parsia, B., Sattler, U.: Explanation of OWL entailments in protege 4. In: ISWC (Posters & Demos) (2008)
10. Horridge, M., Parsia, B., Sattler, U.: Laconic and precise justifications in OWL. In: Sheth, A., et al. (eds.) ISWC 2008. LNCS, vol. 5318, pp. 323–338. Springer, Heidelberg (2008). https://doi.org/10.1007/978-3-540-88564-1_21
11. Kaliski, A.: An overview of KLM-style defeasible entailment (2020)
12. Kalyanpur, A., Parsia, B., Horridge, M., Sirin, E.: Finding all justifications of OWL DL entailments. In: Aberer, K., et al. (eds.) ASWC/ISWC -2007. LNCS, vol. 4825, pp. 267–280. Springer, Heidelberg (2007). https://doi.org/10.1007/978-3-540-76298-0_20
13. Krasner, G.E.: A cookbook for using model-view-controller user interface paradigmin smalltalk-80. J. Object Oriented Program. 1(3), 26–49 (1988)
14. Kraus, S., Lehmann, D., Magidor, M.: Nonmonotonic reasoning, preferential models and cumulative logics. Artif. Intell. 44(1–2), 167–207 (1990)
15. Lehmann, D., Magidor, M.: What does a conditional knowledge base entail? Artif. Intell. 55(1), 1–60 (1992)
16. Moodley, K.: Debugging and repair of description logic ontologies. Ph.D. thesis (2010)
17. Reiter, R.: A theory of diagnosis from first principles. Artif. Intell. 32(1), 57–95 (1987)
18. Thimm, M.: Tweety - a comprehensive collection of java libraries for logical aspects of artificial intelligence and knowledge representation. In: Proceedings of the 14th International Conference on Principles of Knowledge Representation and Reasoning (KR 2014) (2014)
19. Thimm, M.: The Tweety library collection for logical aspects of artificial intelligence and knowledge representation. Künstliche Intelligenz 31(1), 93–97 (2017)
20. Wang, S.: A tool that computes justifications for a defeasible entailment given a knowledge base and a query (2022). https://github.com/SteveWang7596/DefeasibleJustificationForPropositionalLogic. Accessed 29 Aug 2022
21. Wotawa, F.: A variant of Reiter's hitting-set algorithm. Inf. Process. lett. 79(1), 45–51 (2001)

Relevance in the Computation
of Non-monotonic Inferences

Jesse Heyninck[1]([✉]) and Thomas Meyer[2]

[1] Open Universiteit, Heerlen, The Netherlands
jesse.heyninck@ou.nl
[2] University of Cape Town and CAIR, Cape Town, South Africa
tmeyer@cair.org.za

Abstract. Inductive inference operators generate non-monotonic inference relations on the basis of a set of conditionals. Examples include rational closure, system P and lexicographic inference. For most of these systems, inference has a high worst-case computational complexity. Recently, the notion of syntax splitting has been formulated, which allows restricting attention to subsets of conditionals relevant for a given query. In this paper, we define algorithms for inductive inference that take advantage of syntax splitting in order to obtain more efficient decision procedures. In particular, we show that relevance allows to use the modularity of knowledge base is a parameter that leads to tractable cases of inference for inductive inference operators such as lexicographic inference.

1 Introduction

Inductive inference operators generate non-monotonic inference relations $\mathrel{|\!\sim}_\Delta$ on the basis of a set of conditionals Δ of the form $(\psi|\phi)$, read as "if ϕ holds, then typically ψ holds". Examples of inductive inference operators include rational closure (also known as system Z) [8], lexicographic inference [20], system W [15] and c-representations [12]. For these systems, known complexity results point to a high computational complexity [6]. Recently, the property of relevance, as a sub-property of syntax splitting, has been formulated and studied for inductive inference operators [10, 11, 13]. By requiring that inferences in a sublanguage can be made on the basis of the conditionals in a conditional belief base formulated on the basis of that sublanguage, relevance restricts the scope of inferences. In this paper we show how the property of relevance can be used to reduce the computational effort needed in determining inferences on the basis of a conditional belief base relative to an inductive inference operator. In more detail, we show that for a certain class of conditional belief bases called *Horn-conditional* belief bases, the size of the *finest splitting* is a parameter that makes entailment fixed parameter tractable [5,9], which intuitively means that the running time of conditional entailment remains tractable if the modules or partitions of a conditional belief base remain small.

Outline of this Paper: We first state all the necessary preliminaries in Sect. 2 on propositional logic (Sect. 2.1), reasoning with non-monotonic conditionals (Sect. 2.2), inductive inference (Sect. 2.3), system Z (Sect. 2.4), lexicographic inference (Sect. 2.5) and computational complexity (Sect. 2.6). In Sect. 3, we set the stage for this paper by

© The Author(s), under exclusive license to Springer Nature Switzerland AG 2022
A. Pillay et al. (Eds.): SACAIR 2022, CCIS 1734, pp. 202–214, 2022.
https://doi.org/10.1007/978-3-031-22321-1_14

describing existing complexity results for inductive inference. Then, in the first main Sect. 4 of this paper, we explain the computational complexity of lexicographic inference by looking closer at algorithms for computing lexicographic inference. In Sect. 5, we introduce the notion of *splitting size*, which will be used as parameter in the complexity results, and show that lexicographic inference for Horn-bases is fixed parameter tractable. In Sect. 6, we generalize this complexity result to any inductive inference operator based on total preorders that satisfies some minimal conditions. In Sect. 7, we discuss related work, and in Sect. 8 we conclude.

2 Preliminaries

In the following, we briefly recall some general preliminaries on propositional logic, and technical details on inductive inference.

2.1 Propositional Logic

For a set At of atoms let $\mathcal{L}(\text{At})$ be the corresponding propositional language constructed using the usual connectives $\wedge$ (*and*), $\vee$ (*or*), $\neg$ (*negation*), $\rightarrow$ (*material implication*) and $\leftrightarrow$ (*material equivalence*). We sometimes abbreviate a conjunction $\phi \wedge \psi$ by $\phi\psi$. A (classical) *interpretation* (also called *possible world*) ω for a propositional language $\mathcal{L}(\text{At})$ is a function $\omega : \text{At} \rightarrow \{\top, \bot\}$. Let $\Omega(\text{At})$ denote the set of all interpretations for At. We simply write Ω if the set of atoms is implicitly given. An interpretation ω *satisfies* (or is a *model* of) an atom $a \in \text{At}$, denoted by $\omega \models a$, if and only if $\omega(a) = \top$. The satisfaction relation $\models$ is extended to formulas as usual. As an abbreviation we sometimes identify an interpretation ω with its *complete conjunction*, i.e., if $a_1, \dots, a_n \in \text{At}$ are those atoms that are assigned $\top$ by ω and $a_{n+1}, \dots, a_m \in \text{At}$ are those propositions that are assigned $\bot$ by ω we identify ω by $a_1 \dots a_n \overline{a_{n+1}} \dots \overline{a_m}$ (or any permutation of this). For $X \subseteq \mathcal{L}(\text{At})$ we also define $\omega \models X$ if and only if $\omega \models A$ for every $A \in X$. Define the set of models $\text{Mod}(X) = \{\omega \in \Omega(\text{At}) \mid \omega \models X\}$ for every formula or set of formulas X. A formula or set of formulas X_1 *entails* another formula or set of formulas X_2, denoted by $X_1 \models X_2$, if $\text{Mod}(X_1) \subseteq \text{Mod}(X_2)$. Where $\theta \subseteq \Sigma$, and $\omega \in \Omega(\Sigma)$, we denote by ω^θ the restriction of ω to θ, i.e. ω^θ is the interpretation over Σ^θ that agrees with ω on all atoms in θ. Where $\Sigma_i, \Sigma_j \subseteq \Sigma$, $\Omega(\Sigma_i)$ will also be denoted by Ω_i for any $i \in \mathbb{N}$, and likewise $\Omega_{i,j}$ we denote $\Omega(\Sigma_i \cup \Sigma_j)$ (for $i, j \in \mathbb{N}$). Likewise, for some $X \subseteq \mathcal{L}(\Sigma_i)$, we define $\text{Mod}_i(X) = \{\omega \in \Omega_i \mid \omega \models X\}$.

2.2 Reasoning with Nonmonotonic Conditionals

Given a language $\mathcal{L}$, conditionals are objects of the form $(B|A)$ where $A, B \in \mathcal{L}$. The set of all conditionals based on a language $\mathcal{L}$ is defined as: $(\mathcal{L}|\mathcal{L}) = \{(B|A) \mid A, B \in \mathcal{L}\}$. We follow the approach of [7] who considered conditionals as *generalized indicator functions* for possible worlds resp. propositional interpretations ω:

$$((B|A))(\omega) = \begin{cases} 1 & : \; \omega \models A \wedge B \\ 0 & : \; \omega \models A \wedge \neg B \\ u & : \; \omega \models \neg A \end{cases} \tag{1}$$

where u stands for *unknown* or *indeterminate*. In other words, a possible world ω *verifies* a conditional $(B|A)$ iff it satisfies both antecedent and conclusion $((B|A)(\omega) = 1)$; it *falsifies, or violates* it iff it satisfies the antecedence but not the conclusion $((B|A)(\omega) = 0)$; otherwise the conditional is *not applicable*, i. e., the interpretation does not satisfy the antecedent $((B|A)(\omega) = u)$. We say that ω *satisfies* a conditional $(B|A)$ iff it does not falsify it, i.e., iff ω satisfies its *material counterpart* $A \to B$. Given a total preorder (in short, TPO) $\preceq$ on possible worlds, representing relative plausibility, $A \preceq B$ iff $\omega \preceq \omega'$ for some $\omega \in \min_{\preceq}(\mathrm{Mod}(A))$ and some $\omega' \in \min_{\preceq}(\mathrm{Mod}(B))$. This allows for expressing the validity of defeasible inferences via stating that $A \mathrel{|\!\sim}_{\preceq} B$ iff $(A \wedge B) \prec (A \wedge \neg B)$ [22]. As is usual, we denote $\omega \preceq \omega'$ and $\omega' \preceq \omega$ by $\omega \approx \omega'$ and $\omega \preceq \omega'$ and $\omega' \not\preceq \omega$ by $\omega \prec \omega'$ (and similarly for formulas). We can *marginalize* total preorders and even inference relations, i.e., restricting them to sublanguages, in a natural way: If $\Theta \subseteq \Sigma$ then any TPO $\preceq$ on $\Omega(\Sigma)$ induces uniquely a *marginalized TPO* $\preceq_{|\Theta}$ on $\Omega(\Theta)$ by setting

$$\omega_1^{\Theta} \preceq_{|\Theta} \omega_2^{\Theta} \text{ iff } \omega_1^{\Theta} \preceq \omega_2^{\Theta}. \tag{2}$$

Note that on the right hand side of the *iff* condition above $\omega_1^{\Theta}, \omega_2^{\Theta}$ are considered as propositions in the superlanguage $\mathcal{L}(\Omega)$, hence $\omega_1^{\Theta} \preceq \omega_2^{\Theta}$ is well defined [14].

Similarly, any inference relation $\mathrel{|\!\sim}$ on $\mathcal{L}(\Sigma)$ induces a *marginalized inference relation* $\mathrel{|\!\sim}_{|\Theta}$ on $\mathcal{L}(\Theta)$ by setting

$$A \mathrel{|\!\sim}_{|\Theta} B \text{ iff } A \mathrel{|\!\sim} B \tag{3}$$

for any $A, B \in \mathcal{L}(\Theta)$.

An obvious implementation of total preorders are *ordinal conditional functions (OCFs)*, (also called *ranking functions*) $\kappa : \Omega \to \mathbb{N} \cup \{\infty\}$ with $\kappa^{-1}(0) \neq \emptyset$ [26]. They express degrees of (im)plausibility of possible worlds and propositional formulas A by setting $\kappa(A) := \min\{\kappa(\omega) \mid \omega \models A\}$. A conditional $(B|A)$ is accepted by κ iff $A \mathrel{|\!\sim}_{\kappa} B$ iff $\kappa(A \wedge B) < \kappa(A \wedge \neg B)$. Notice that for any TPO $\preceq$, there is an arbitrary number of OCFs $\kappa_{\preceq}$ that *respects* this TPO in the sense that $\kappa_{\preceq}(\omega_1) < \kappa_{\preceq}(\omega_2)$ iff $\omega_1 \prec \omega_2$. An OCF is *convex* if, for every $i, j \in \mathbb{N}$ s.t. $\kappa^{-1}(i) \neq \emptyset$ and $\kappa^{-1}(j) \neq \emptyset$, for every $i \leqslant k \leqslant j$, $\kappa^{-1}(k) \neq \emptyset$. In this paper, when we talk about the *OCF associated with an TPO* $\preceq$ we will assume it is the minimal convex OCF κ that satisfies this condition. This choice is, however, inconsequential. We denote the OCF associated with $\preceq$ by $\kappa_{\preceq}$.

2.3 Inductive Inference

In this paper, we will be interested in inference relations $\mathrel{|\!\sim}_{\Delta}$ parametrized by a conditional belief base Δ. In more detail, such inference relations are *induced by* Δ, in the sense that Δ serves as a starting point for the inferences in $\mathrel{|\!\sim}_{\Delta}$. We call such operators *inductive inference operators*:

Definition 1 ([13]). *An inductive inference operator (from conditional belief bases) is a mapping* $\mathbf{C}$ *that assigns to each conditional belief base* $\Delta \subseteq (\mathcal{L}|\mathcal{L})$ *an inference relation* $\mathrel{|\!\sim}_{\Delta}$ *on* $\mathcal{L}$ *that satisfies the following basic requirement of* direct inference:

DI *If Δ is a conditional belief base and $\mathrel{\vdash\!\!\!\sim}_\Delta$ is an inference relation that is induced by Δ, then $(B|A) \in \Delta$ implies $A \mathrel{\vdash\!\!\!\sim}_\Delta B$.*

Examples of inductive inference operators include system P [18], system Z ([8], see Sect. 2.4), lexicographic inference ([20], see Sect. 2.5) and c-representations ([12]).

As already indicated in the previous subsection, inference relations can be obtained on the basis of TPOs respectively OCFs:

Definition 2. *A model-based inductive inference operator for total preorders (on Ω) is a mapping $\mathbf{C}^{tpo}$ that assigns to each conditional belief base Δ a total preorder $\preceq_\Delta$ on Ω s.t. $A \mathrel{\vdash\!\!\!\sim}_{\preceq_\Delta} B$ for every $(B|A) \in \Delta$ (i.e. s.t. **DI** is ensured). A model-based inductive inference operator for OCFs (on Ω) is a mapping $\mathbf{C}^{ocf}$ that assigns to each conditional belief base Δ an OCF κ_Δ on Ω s.t. Δ is accepted by κ_Δ (i.e. s.t. **DI** is ensured).*

Examples of inductive inference operators for OCFs include system Z ([8], see Sect. 2.4) and c-representations ([12]), whereas lexicographic inference ([20], see Sect. 2.5) is an example of an inductive inference operator for TPOs.

To define the property of *syntax splitting* [13], we assume a conditional belief base Δ that can be split into subbases Δ^1, Δ^2 s.t. $\Delta^i \subseteq (\mathcal{L}_i|\mathcal{L}_i)$ with $\mathcal{L}_i = \mathcal{L}(\Sigma_i)$ for $i = 1, 2$ s.t. $\Sigma_1 \cap \Sigma_2 = \emptyset$ and $\Sigma_1 \cup \Sigma_2 = \Sigma$, writing:

$$\Delta = \Delta^1 \bigcup_{\Sigma_1, \Sigma_2} \Delta^2$$

whenever this is the case.

Definition 3 (Independence (Ind), [13]). *An inductive inference operator $\mathbf{C}$ satisfies (Ind) if for any $\Delta = \Delta^1 \bigcup_{\Sigma_1, \Sigma_2} \Delta^2$ and for any $A, B \in \mathcal{L}_i$, $C \in \mathcal{L}_j$ ($i, j \in \{1, 2\}$, $j \neq i$),*

$$A \mathrel{\vdash\!\!\!\sim}_\Delta B \text{ iff } AC \mathrel{\vdash\!\!\!\sim}_\Delta B$$

Definition 4 (Relevance (Rel), [13]). *An inductive inference operator $\mathbf{C}$ satisfies (Rel) if for any $\Delta = \Delta^1 \bigcup_{\Sigma_1, \Sigma_2} \Delta^2$ and for any $A, B \in \mathcal{L}_i$ ($i \in \{1, 2\}$),*

$$A \mathrel{\vdash\!\!\!\sim}_\Delta B \text{ iff } A \mathrel{\vdash\!\!\!\sim}_{\Delta^i} B.$$

Definition 5 (Syntax splitting (SynSplit), [13]). *An inductive inference operator $\mathbf{C}$ satisfies (SynSplit) if it satisfies (Ind) and (Rel).*

Example 1. Consider the conditional belief base $\{(a|\top), (b|c)\}$. Then we see that:

$$\{(a|\top), (b\top)\} = \{(a|\top)\} \bigcup_{\{a\}, \{b,c\}} \{(b|c)\}$$

For any inductive inference operator, we obtain that $\top \mathrel{\vdash\!\!\!\sim}_\Delta a$ (with **DI**). If the inductive inference operator satisfies (**Ind**), we can also derive that $c \mathrel{\vdash\!\!\!\sim}_\Delta a$. If it satisfies (**Rel**), we can derive that $\top \mathrel{\vdash\!\!\!\sim}_{(a|\top)} a$.

Thus, **Ind** requires that inferences from one sub-language are independent from formulas over the other sublanguage, if the belief base splits over the respective sublanguages. In other words, information on the basis of one sublanguage does not influences inferences made in the other sublanguage. **Rel**, on the other hand, restricts the scope of inferences, by requiring that inferences in a sublanguage can be made on the basis of the conditionals in a conditional belief base formulated on the basis of that sublanguage. **SynSplit** combines these two properties. In this paper, we will focus on the property of relevance. In [13] it was shown that system P and system Z satisfy relevance, but not independence, whereas c-representations satisfy full syntax splitting. In [11] it was shown that lexicographic inference and a new inductive inference operator called system Z^{ind} satisfy syntax splitting, and in [10] it was shown that system W satisfies syntax splitting.

2.4 System Z

We present system Z defined in [8] as follows. A conditional $(B|A)$ is tolerated by a finite set of conditionals Δ if there is a possible world ω with $(B|A)(\omega) = 1$ and $(B'|A')(\omega) \neq 0$ for all $(B'|A') \in \Delta$, i.e. ω verifies $(B|A)$ and does not falsify any (other) conditional in Δ. The Z-partitioning $(\Delta_0, \ldots, \Delta_n)$ of Δ is defined as:

- $\Delta_0 = \{\delta \in \Delta \mid \Delta$ tolerates $\delta\}$;
- $\Delta_1, \ldots, \Delta_n$ is the Z-partitioning of $\Delta \setminus \Delta_0$.

For $\delta \in \Delta$ we define: $Z_\Delta(\delta) = i$ iff $\delta \in \Delta_i$ and $(\Delta_0, \ldots, \Delta_n)$ is the Z-partioning of Δ. Finally, the ranking function κ_Δ^Z is defined via: $\kappa_\Delta^Z(\omega) = \max\{Z(\delta) \mid \delta(\omega) = 0, \delta \in \Delta\} + 1$, with $\max \emptyset = -1$. The resulting inductive inference operator $C_{\kappa_\Delta^Z}^{ocf}$ is denoted by C^Z.

In the literature, system Z has also been called *rational closure* [21].

We now illustrate OCFs in general and system Z in particular with the well-known "Tweety the penguin"-example.

Example 2. Consider $\Delta = \{(f|b), (b|p), (\neg f|p)\}$. This conditional belief base has the following Z-partitioning: $\Delta_0 = \{(f|b)\}$ and $\Delta_1 = \{(b|p), (\neg f|p)\}$. This gives rise to the following κ_Δ^Z-ordering over the worlds based on the signature $\{b, f, p\}$:

ω	κ_Δ^Z	ω	κ_Δ^Z	ω	κ_Δ^Z	ω	κ_Δ^Z
pbf	2	$pb\overline{f}$	1	$\overline{p}bf$	2	$\overline{p}b\,\overline{f}$	2
$\overline{p}bf$	0	$\overline{p}b\overline{f}$	1	$\overline{p}\overline{b}f$	0	$\overline{p}\,\overline{b}\,\overline{f}$	0

As an example of a (non-)inference, observe that e.g. $\top \mathrel{\mid\!\sim}_\Delta^Z \neg p$ and $p \land f \mathrel{\mid\!\not\sim}_\Delta^Z b$.

2.5 Lexicographic Entailment

We recall lexicographic inference as introduced by [20]. For some conditional belief base Δ, the order $\preceq_\Delta^{lex}$ is defined as follows: Given $\omega \in \Omega$ and $\Delta' \subseteq \Delta$, $V(\omega, \Delta') =$

$|(\{(B|A) \in \Delta' \mid (B|A)(\omega) = 0\}|$. Given a set of conditionals Δ Z-partitioned in $(\Delta_0, \ldots, \Delta_n)$, the *lexicographic vector* for a world $\omega \in \Omega$ is the vector $\mathrm{lex}(\omega) = (V(\omega, \Delta_0), \ldots, V(\omega, \Delta_n))$. Given two vectors $(x_1, \ldots, x_n)$ and $(y_1, \ldots, y_n)$, we define $(x_1, \ldots, x_n) \preceq^{\mathrm{lex}} (y_1, \ldots, y_n)$ iff there is some $j \leqslant n$ s.t. $x_k = y_k$ for every $k > j$ and $x_j \leqslant y_j$. We define $\omega \preceq^{\mathrm{lex}}_\Delta \omega'$ iff $\mathrm{lex}(\omega) \preceq^{\mathrm{lex}} \mathrm{lex}(\omega')$. The resulting inductive inference operator $C^{tpo}_{\preceq_{\mathrm{lex}}}$ will be denoted by C^{lex} to avoid clutter.

Example 3 (Example 2 ctd.). For the Tweety belief base Δ as in Example 2 we obtain the following $\mathrm{lex}(\omega)$-vectors:

ω	$\mathrm{lex}(\omega)$	ω	$\mathrm{lex}(\omega)$	ω	$\mathrm{lex}(\omega)$	ω	$\mathrm{lex}(\omega)$
pbf	(0,1)	$pb\overline{f}$	(1,0)	$p\overline{b}f$	(0,2)	$p\overline{b}\,\overline{f}$	(0,1)
$\overline{p}bf$	(0,0)	$\overline{p}b\overline{f}$	(1,0)	$\overline{p}\,\overline{b}f$	(0,0)	$\overline{p}\,\overline{b}\,\overline{f}$	(0,0)

The lex-vectors are ordered as follows:

$$(0,0) \prec^{\mathrm{lex}} (1,0) \prec^{\mathrm{lex}} (0,1) \prec^{\mathrm{lex}} (0,2).$$

Observe that e.g. $\top \mathbin{\vert\!\sim}^{\mathrm{lex}}_\Delta \neg p$ (since $\mathrm{lex}(\top \wedge \neg p) = (0,0) \prec^{\mathrm{lex}} \mathrm{lex}(\top \wedge p) = (1,0)$) and $p \wedge f \mathbin{\vert\!\sim}^{\mathrm{lex}}_\Delta b$.

2.6 Computational Complexity

In this section, we recall the necessary background on parametrized complexity. For more detailed references, we refer to [5]. An instance of a *parametrized problem L* is a pair $(I, k) \in \Sigma^* \times \mathbb{N}$ for some finite alphabet Σ. For an instance $(I, k) \in \Sigma^* \times \mathbb{N}$, we call I the *main part* and k the *parameter*. Where $|I|$ denotes the cardinality of I, L is *fixed-parameter tractable* if there exists a computable function f and a constant c such that $(I, k) \in L$ is decidable in time $O(f(k)|I|^c)$. Such an algorithm is called a *fixed-parameter tractable algorithm*. Thus, the intuition is that when $f(k)$ remains sufficiently small, the problem remains tractable no matter the size of I.

3   Computational Complexity for Inductive Inference

The main complexity problem that will be considered in this paper is that of *entailment under an inductive inference operator*: given an inductive inference operator **C**, decide whether $A \mathbin{\vert\!\sim}_\Delta B \in \mathbf{C}(\Delta)$. We assume for this that there is a constructive method associated with **C**, and are interested in the computational complexity of that constructive method. Of course, one can simply assume $\mathbf{C}(\Delta)$ as a given, but in that case, the decision problem becomes trivial.

To give the appropriate context for this paper, we first recall some complexity results on inductive inference operators as shown by Eiter and Lukasiewicz [6]. It is shown there that in general, both system Z and lexicographic entailment have a reasonably similar complexity: system Z-entailment is solvable by a single call to parallel NP-oracle and lexicographic entailment is P^{NP}-complete. This high complexity is not surprising, as any algorithm for calculating entailment under system Z or lexicographic entailment will need to check for propositional satisfiability (which itself is NP-complete) a number of times. When assuming the conditional knowledge base is of a certain logical form, namely if it consists so-called Horn-conditionals, better computational complexity results for system Z, but not for lexicographic entailment, can be obtained. We first recall the concept of a Horn-conditional knowledge base:

Definition 6 ([6]). *A conditional $(\psi|\phi)$ is* Horn *if ϕ is either $\top$ or a conjunction of atoms and ψ is a conjunction of Horn clauses. Recall that a Horn clause is a disjunction of literals where at most one literal occurs positively. A conditional knowledge base is* Horn-conditional *if every conditional is Horn.*

Notice that the Tweety-knowledge base (Example 2) is a Horn-conditional knowledge base. A more complex example of a Horn-conditional would be $((p \vee \neg q \vee \neg r) \wedge (s \vee \neg q)|p \wedge v \wedge r)$.

For the Horn-case, the complexity of the approaches differ: for system Z, one gets P-completeness whereas for lexicographic entailment, the entailment problem remains P^{NP}-complete. In the next section, we will have a more detailed look as to why the entailment problem for lexicographic entailment remains P^{NP}-hard for conditional knowledge bases that are Horn.

4   Algorithms for Lexicographic Closure

In this section, we look deeper into the reason why, even for Horn-cases, the decision problem for lexicographic entailment remains P^{NP}-complete.

We first recall the algorithm from [24] for determining entailment under lexicographic closure. The auxiliary routine SubsetRank detailed in Algorithm 1 is needed first. It uses a function Subsets(Δ_i, j) which returns all subsets $\Theta \subseteq \Delta_i$ of size j.

Algorithm 1. SubsetRank

Input: A defeasible knowledge base Δ with Z-partitioning $(\Delta_1, \ldots, \Delta_n)$.
Output: An ordered tuple $(R_0, \ldots, R_k, k + 1)$.
 1: $i := 0; k := 0$.
 2: **repeat**
 3:      **for** $j := |\Delta_i|$ to 1 **do**
 4:          $S_{i,j} := \text{Subsets}(\Delta_i, j)$;
 5:          $D_{i,j} := \bigvee_{X \in S_{i,j}} \bigwedge_{(B|A) \in X} A \to B$;
 6:          $R_k := D_{i,j}$
 7:          $k := k + 1$;
 8:      **end for**
 9:      $i := i + 1$;
10: **until** $i := m$
11: **return** $(R_0, \ldots, R_k, k + 1)$

Intuitively, SubsetRank generates, on the basis of a defeasible knowledge base Δ Z-partitioned in $(\Delta_1, \ldots, \Delta_n)$, further levels between each Δ_i and Δ_{i+1}, where every level consists of all subsets of Δ_i of a size between $|\Delta_i| - 1$ and 1. These levels are represented as a disjunction of a conjunction of the material versions of the conditionals in these subsets.

We now proceed to the LexicographicClosure algorithm, that determines whether $A \mathrel{|\!\sim}_{\Delta}^{\text{lex}} B$ holds in Algorithm 2.[1] Intuitively, LexicographicClosure looks

---

Algorithm 2. LexicographicClosure

Input: A defeasible knowledge base Δ and a query $A \mathrel{|\!\sim} B$.
Output: **true** if $A \mathrel{|\!\sim}_{\Delta}^{\text{lex}} B$, and **false** otherwise.

1: $(R_0, \ldots, R_k, k + 1) = $ SubsetRank(Δ)
2: $i := 0$;
3: $\Delta := \bigcup_{j=0}^{k} R_j$.
4: **while** $\Delta \models \neg A$ and $\Delta \neq \emptyset$ **do**
5:     $\Delta := \Delta \setminus R_i$;
6:     $i := i + 1$;
7: **end while**
8: **return** $\Delta \models A \to B$

---

for the lowest level that does not falsify the antecedent of the query, and then checks whether $A \to B$ is implied at this level. We illustrate this first with an example:

Example 4. Recall $\Delta = \{(f|b), (b|p), (\neg f|p)\}$ from Example 2. We first see that:

$$\text{SubsetRank}(\Delta) = (\{b \to f\}, \{p \to b \land p \to \neg f\}, \{p \to b \lor p \to \neg f\})$$

As a first example of a query, we run through LexicographicClosure$(\Delta, p \land b \mathrel{|\!\sim} \neg f)$. We first set $\Delta^1 = \{b \to f, p \to b \land p \to \neg f, p \to b \lor p \to \neg f\}$ and see that $\Delta^1 \models \neg(p \land b)$, and thus proceed with $\Delta^2 = \{p \to b \land p \to \neg f, p \to b \lor p \to \neg f\}$. As $\Delta^2 \not\models \neg(p \land b)$, we check whether $\Delta^2 \models p \land b \to \neg f$. We see that this is the case and thus LexicographicClosure$(\Delta, p \land b \mathrel{|\!\sim} \neg f) = $ **true**.

As a second example, we run through LexicographicClosure$(\Delta, p \land \neg b \mathrel{|\!\sim} f)$. We again see that $\Delta^1 \models \neg(p \land \neg b)$, but now see that $\Delta^2 \models \neg(p \land \neg b)$ as well. We therefore proceed with $\Delta^3 = \{p \to b \lor p \to \neg f\}$ and see that $\Delta^3 \not\models \neg(p \land \neg b)$. We now check whether $\Delta^3 \models (p \land \neg b) \to f$ and see that this is *not* the case, and thus LexicographicClosure$(\Delta, p \land \neg b \mathrel{|\!\sim} f) = $ **false**.

A short inspection of this algorithm also makes clear where the exponential blow-up for lexicographic inference comes from: in the worst case, for every subset of $D_{i,j} \subseteq \Delta$, we have to check whether $\bigcup_{l=k}^{n} R_l \models \neg A$ (where $D_{i,j} = R_k$), thus leading to $2^{|\Delta|}$ checks. This means that, even if we can reduce the complexity of the $\models$-check

---

[1] In [24], conditionals can be assigned an additional rank ∞ to allow for the modelling of strict conditionals. For simplicity, we do not consider strict conditionals, but the results here can be easily adapted to allow them.

(for example by restricting attention to Horn-conditionals), it cannot be ensured that the computational cost of LexicographicClosure reduce. It is perhaps interesting to contrast this with the algorithm for rational closure as given in [24]. There, different levels of conditionals are considered as well, but instead of considering every subset of Δ_i before moving to the next level $i + 1$, we simply look at whether $\bigcup_{j=i}^n \Delta_j \models \neg A$, and then either check $\bigcup_{j=i+1}^n \Delta_j \models A \rightarrow B$ (if $\Delta_i \not\models \neg A$), or move to Δ_{i+1}. Thus, the number of $\models$-checks is linear in n (where Δ is partitioned in $(\Delta_1, \ldots, \Delta_n)$), which is itself bounded by the number conditionals.

5   Splitting a Conditional Knowledge Base

In this section, we recall the notion of syntax splitting, and study the complexity of calculating it. The size of the finest splittings will then be used as a parameter to obtain the fixed parameter tractability in the following sections.

We first recall the concept of a finest syntax splitting [25], adapted to the setting of conditional belief bases in [11].

Definition 7. *Given a conditional belief base Δ, and $\Sigma_1, \ldots, \Sigma_n$ s.t. $\bigcup_{i=1}^n \Sigma_i$ and $\Sigma_i \cap \Sigma_j \neq \emptyset$ for every $i, j = 1, \ldots, n$ s.t. $i \neq j$, $\Sigma_1, \ldots, \Sigma_n$ splits Δ iff $\Delta^i \subset (\mathcal{L}(\Sigma_i) | \mathcal{L}(\Sigma_i))$ for every $i = 1, \ldots, n$ and $\Delta = \bigcup_i^n \Delta^i$. We also say $\Sigma_1, \ldots, \Sigma_n$ is a splitting of Δ.*

Given two partitions $\Sigma_1, \ldots, \Sigma_n$ and $\Sigma'_1, \ldots, \Sigma'_m$ of Σ, we say that $\Sigma_1, \ldots, \Sigma_n$ is finer than $\Sigma'_1, \ldots, \Sigma'_m$ iff for every $1 \leqslant i \leqslant n$, there is some $1 \leqslant j \leqslant m$ s.t. $\Sigma_i \subseteq \Sigma'_j$.

It was shown in [11] that every conditional belief base admits a unique finest splitting:

Proposition 1 ([11]). *For any conditional belief base Δ, there exists a unique finest splitting.*

Where $\Sigma_1, \ldots, \Sigma_n$ is the finest splitting of Δ, we denote by $\mathsf{FinestSplit}(\Delta)$ the set of sets $\{\Delta^1, \ldots, \Delta^m\}$. We finally define $\mathsf{splitSize}(\Delta) = \max\{|\Theta| \mid \Theta \in \mathsf{FinestSplit}(\Delta)\}$. Thus, $\mathsf{splitSize}(\Delta)$ represents the maximal size of a partition of the finest splitting of Δ, and intuitively represents a measure of the modularity of a conditional knowledge base.

Given a formula $A \in \mathcal{L}(\Sigma)$, we can obtain $\Sigma_1 \subseteq \Sigma$ s.t. $A \in \mathcal{L}(\Sigma_1)$ and $\Delta^1 \in \mathsf{FinestSplit}(\Delta)$ by calculating the connected components of the following (undirected) dependency graph:

Definition 8. *Given a conditional belief base Δ, the atomic dependency graph $\mathsf{ADC}(\Delta)$ is defined by $\mathsf{ADC}(\Delta) := (\Sigma, V)$ with $V = \{(a_1, a_2) \mid \exists (B|A) \in \Delta \text{ for which } a_1, a_2 \in \mathsf{Atoms}(AB)\}$.*

Clearly, to calculate $\mathsf{ADC}(\Delta)$ we need $|\mathsf{Atoms}(\Delta)||\Delta|$ checks, i.e. this can be done in linear time.

Proposition 2. *$\Delta^1 \in \mathsf{FinestSplit}(\Delta)$ iff Σ^1 is a connected component of $\mathsf{ADC}(\Delta)$.*

Proof. This is immediate in view of the observation that, where $\mathsf{ADC}(\Delta) = (\Sigma, V)$, $(a_1, a_2) \in V$ if and only if a_1 and a_2 occur together in some conditional $(B|A) \in \Delta$.

Notice that this also means that splitSize(Δ) can be calculated in polynomial time (as connected components can be computed in linear time).

We can now show that for Horn conditional knowledge bases, deciding whether $A \mathrel{|\!\sim}^{\text{lex}}_{\Delta} B$ is fixed-parameter tractable when parameterized by the maximal size of splittings:

Theorem 1. *Deciding $A \mathrel{|\!\sim}^{\text{lex}}_{\Delta} B$ is fixed-parameter tractable with parameter* SplitSize *for the class of Horn-conditional knowledge bases.*

Proof. Since C^{lex} satisfies **Rel**, we can check $A \mathrel{|\!\sim}^{\text{lex}}_{\Delta} B$ by checking $A \mathrel{|\!\sim}^{\text{lex}}_{\Delta^1 \cup \Delta^2} B$, where the finest splitting of Δ is $\Sigma_1, \ldots, \Sigma_n$ and $A \in \Sigma_1$ and $B \in \Sigma_2$. As the number of $\models$-checks (themselves in P for Horn-conditional knowledge bases) needed in LexicographicClosure is now bounded by $2^{|\Delta^1 \cup \Delta^2|}$, we see that SplitSize($\Delta$) determines the number of $\models$-checks needed.

Notice that, in a sense, the restriction of this theorem to Horn-conditional knowledge bases is the best we can hope for. Indeed, any algorithm for virtually any inductive inference operator will involve a number of satisfiability checks of the conditionals (e.g. to determine consistency of the conditional knowledge base). Assuming Horn-conditionals ensures that these checks remain tractable. Giving up this assumption will invariable lead to intractability. Of course, one can move to other classes of conditionals, defined e.g. on the basis of dual Horn or Krom formulas [23]. Such adaptions of the results presented here are left for future work.

6 A General Result

We now slightly adapt the algorithm DefeasibleEntailment from [3] in Algorithm 1 for calculating inferences under an RC-extending inductive inference operator C. In more detail, whereas [3] generate an ordered tuple $(R_0, \ldots, R_{n-1}, R_\infty, n)$ over all formules (up to equivalence) of the language $\mathcal{L}(\Sigma)$, we simply use a ranking over the possible worlds $\Omega(\Sigma)$. Clearly, this is equivalent. We assume that there is access to a subroutine Rank(Δ, ω) which returns, for a TPO-based operator C^{tpo}, the rank $\kappa_{\preceq_\Delta}(\omega)$. As the calculation of $\preceq_\Delta$ is highly dependent on the exact inductive inference operator, we cannot further specify this subroutine.

Proposition 3. *For any inductive inference operator based on TPOs C^{tpo}, Algorithm 1 is sound and complete w.r.t. entailment under C^{tpo}.*

Proof. Assume first DefeasibleEntailment($\Delta, r, A \mathrel{|\!\sim} B$) = **true**. This means that there is some $i \geqslant 0$ s.t. $\{\omega \mid \text{Rank}(\Delta, \omega) = i\} \neq \emptyset$, $\{\omega \mid \text{Rank}(\Delta, \omega) = i\} \cap \text{Mod}(A) \neq \emptyset$ and $\{\omega \mid \text{Rank}(\Delta, \omega) = i\} \subseteq \text{Mod}(A \to B)$. Furthermore, it can be easily checked that for every $j < i$, $\Omega = \{\omega \mid \text{Rank}(\Delta, \omega) = j\} \subseteq \text{Mod}(\neg A)$. This means that $\min_{\preceq_\Delta} \text{Mod}(A) \subseteq \{\omega \mid \text{Rank}(\Delta, \omega) = i\}$ and thus (with $\{\omega \mid \text{Rank}(\Delta, \omega) = i\} \subseteq \text{Mod}(A \to B)$), $\min_{\preceq_\Delta} \text{Mod}(A) \subseteq \text{Mod}(B)$, which implies $A \wedge B \preceq_\Delta A \wedge \neg B$, which on its turn implies $A \mathrel{|\!\sim}_{\preceq_\Delta} B$ and thus $A \mathrel{|\!\sim}_\Delta B \in \mathbf{C}(\Delta)$.

Assume now that DefeasibleEntailment($\Delta, r, A \mathrel{|\!\sim} B$) = **false**. We consider two options: (1) we reached a point where $\Omega = \emptyset$ or (2) we reached a point where

Algorithm 3. DefeasibleEntailment

Input: A defeasible knowledge base Δ, a Δ-faithful rank function r and a query $A \mathrel{|\!\sim} B$.

Output: **true** if $A \mathrel{|\!\sim} B \in \mathbf{C}(\Delta)$, and **false** otherwise.

1: $i := 0$;
2: $\Omega = \{\omega \mid \mathtt{Rank}(\Delta, \omega) = i\}$.
3: **while** $\Omega \subseteq \mathsf{Mod}(\neg A)$ and $\Omega \neq \emptyset$ **do**
4:    $\Omega = \{\omega \mid \mathtt{Rank}(\Delta, \omega) = i\}$.
5:    $i := i + 1$;
6: **end while**
7: **return** $\Omega \subseteq \mathsf{Mod}(A \to B)$

$\Omega \cap \mathsf{Mod}(A) \neq \emptyset$ yet $\Omega \cap \mathsf{Mod}(A \wedge \neg B) = \emptyset$. In the first case, it is easy to see that $\models \neg A$ and thus $A \mathrel{|\!\not\sim}_\Delta B$. In the second case, $A \wedge B \equiv_\Delta A \wedge \neg B$, and thus $A \mathrel{|\!\not\sim}_\Delta B$.

Theorem 2. *Let an inductive inference operator based on TPOs $\mathbf{C}^{tpo}$ be given for which $\preceq_\Delta$ can be calculated exponentially in the size of Δ and that satisfies* **Rel***. Then deciding $A \mathrel{|\!\sim} B \in \mathbf{C}(\Delta)$ is fixed-parameter tractable with parameter* SplitSize *for the class of Horn-conditional knowledge bases.*

Proof. With **Rel**, we can check $A \mathrel{|\!\sim} B \in \mathbf{C}(\Delta)$ by checking $A \mathrel{|\!\sim} B \in \mathbf{C}(\Delta_1 \cup \Delta_2)$, where the finest splitting of Δ is $\Sigma_1, \ldots, \Sigma_n$ and $A \in \Sigma_1$ and $B \in \Sigma_2$. By assumption, calculating $\mathtt{Rank}(\Delta, \omega)$ for every $\omega \in \Omega(\mathsf{At}(A) \cup \mathsf{At}(B))$ takes maximally $2^{|\mathsf{At}(A) \cup \mathsf{At}(B)|}$ time. Checking whether $\Omega \subseteq \mathsf{Mod}(\neg A)$ can be done in $2^{|\mathsf{At}(A) \cup \mathsf{At}(B)|}$ time. Finally, checking whether $\Omega \subseteq \mathsf{Mod}(A \to B)$ can be done in $2^{|\mathsf{At}(A) \cup \mathsf{At}(B)|}$ as we assume Horn-conditional knowledge bases. This means $2^{|\mathsf{At}(A) \cup \mathsf{At}(B)|}$ forms an upper bound on the calculation of `DefeasibleEntailment`.

7   Related Work

To the best of our knowledge, there is only one work where the computational complexity of reasoning with non-monotonic conditionals and the effect of the modularity of knowledge bases on the complexity are studied. In [6], related results were shown. In more detail, [6] introduces the class of *k-feedback-free Horn* knowledge bases, which is, when restricted to conditional knowledge bases without strict premises, coincides with conditional knowledge bases that are Horn and that can be split according to signatures $\Sigma_1, \ldots, \Sigma_n$ s.t. the size of Σ_i has a cardinality of at most k. Eiter and Lukasiewicz then go on to show that if k is kept fixed, entailment under various inductive inference operators including lexicographic inference, is polynomial or even linear (in the case of lexicographic inference). This means that Eiter and Lukasiewicz [6] actually give a more precise result for lexicographic inference. However, in this paper, we made several advances with respect to the work of Eiter and Lukasiewicz: firstly, we use the notion of parametrized complexity, secondly, we make use of the postulate of relevance, studied already for a wide range of inductive inference operators [10,11,13] to show parametric complexity results. Finally, we give a general result (Theorem 2) that shows the decision problem for any TPO-based inductive inference operator that satisfies relevance is fixed-parameter tractable for Horn-conditional knowledge bases. This means our results also apply immediately to any inductive inference operator for which relevance has been shown or will be shown in the future.

8   Conclusion and Future Work

In this paper, we have shown that the size of the finest splitting of a conditional knowledge base can be reasonably viewed as a parameter determining the complexity of inductive inference on the basis of such a knowledge base and an inductive inference operator that satisfies relevance. Intuitively, this means that it is not the size of a conditional knowledge base that determines how hard it is to make inferences on the basis of that knowledge base, but the interconnectedness of the conditionals in that knowledge base. Thus, as long as a conditional knowledge base consists of many reasonably small modules, efficient inference might still be possible. We have shown this by showing parametrized complexity results for Horn-conditional knowledge bases. This is, in a sense, the best result one can hope for, as an algorithm for inductive inference involves a number of satisfiability checks of the conditionals, which will result in intractability when moving beyond Horn-conditional knowledge bases.

The work done in this paper is useful for implementations and applications of inductive inference operators. This includes, among others, applications of non-monotonic reasoning to description logics [1,2,4]. In future work, we plan to adapt the notion of relevance to description logics, and compare the notion of finest splitting with that of modules as it found in the literature on description logics [16,17]. Furthermore, we plan to empirically test the effect of relevance and modularity for non-Horn-conditional knowledge bases by means of implementations of inductive inference [19].

References

1. Britz, K., Casini, G., Meyer, T., Moodley, K., Sattler, U., Varzinczak, I.: Rational defeasible reasoning for description logics. University of Cape Town, Technical report (2018)
2. Britz, K., Casini, G., Meyer, T., Varzinczak, I.: A KLM perspective on defeasible reasoning for description logics. In: Lutz, C., Sattler, U., Tinelli, C., Turhan, A.-Y., Wolter, F. (eds.) Description Logic, Theory Combination, and All That. LNCS, vol. 11560, pp. 147–173. Springer, Cham (2019). https://doi.org/10.1007/978-3-030-22102-7_7
3. Casini, G., Meyer, T., Varzinczak, I.: Taking defeasible entailment beyond rational closure. In: Calimeri, F., Leone, N., Manna, M. (eds.) JELIA 2019. LNCS (LNAI), vol. 11468, pp. 182–197. Springer, Cham (2019). https://doi.org/10.1007/978-3-030-19570-0_12
4. Casini, G., Straccia, U.: Lexicographic closure for defeasible description logics. In: Proceedings of Australasian Ontology Workshop, vol. 969, pp. 28–39 (2012)
5. Downey, R.G., Fellows, M.R.: Fundamentals of Parameterized Complexity, vol. 4. Springer, Heidelberg (2013). https://doi.org/10.1007/978-1-4471-5559-1
6. Eiter, T., Lukasiewicz, T.: Default reasoning from conditional knowledge bases: complexity and tractable cases. Artif. Intell. **124**(2), 169–241 (2000)
7. de Finetti, B.: Theory of Probability, 2 vols. (1974)
8. Goldszmidt, M., Pearl, J.: Qualitative probabilities for default reasoning, belief revision, and causal modeling. AI **84**(1–2), 57–112 (1996)

9. Gottlob, G., Szeider, S.: Fixed-parameter algorithms for artificial intelligence, constraint satisfaction and database problems. Comput. J. **51**(3), 303–325 (2008)
10. Haldimann, J., Beierle, C.: Inference with system W satisfies syntax splitting. arXiv preprint arXiv:2202.05511 (2022)
11. Heyninck, J., Kern-Isberner, G., Meyer, T.: Lexicographic entailment, syntax splitting and the drowning problem. In: Raedt, L.D. (ed.) Proceedings of the Thirty-First International Joint Conference on Artificial Intelligence, IJCAI-22, pp. 2662–2668. International Joint Conferences on Artificial Intelligence Organization (2022). https://doi.org/10.24963/ijcai.2022/369
12. Kern-Isberner, G.: Handling conditionals adequately in uncertain reasoning and belief revision. J. Appl. Non-Classical Logics **12**(2), 215–237 (2002)
13. Kern-Isberner, G., Beierle, C., Brewka, G.: Syntax splitting= relevance+ independence: new postulates for nonmonotonic reasoning from conditional belief bases. In: Proceedings of the International Conference on Principles of Knowledge Representation and Reasoning, vol. 17, pp. 560–571 (2020)
14. Kern-Isberner, G., Brewka, G.: Strong syntax splitting for iterated belief revision. In: Sierra, C. (ed.) Proceedings International Joint Conference on Artificial Intelligence, IJCAI 2017, pp. 1131–1137. ijcai.org (2017)
15. Komo, C., Beierle, C.: Nonmonotonic reasoning from conditional knowledge bases with system W. Ann. Math. Artif. Intell. **90**, 1–38 (2021). https://doi.org/10.1007/s10472-021-09777-9
16. Konev, B., Lutz, C., Ponomaryov, D.K., Wolter, F.: Decomposing description logic ontologies. In: KR (2010)
17. Konev, B., Lutz, C., Walther, D., Wolter, F.: Model-theoretic inseparability and modularity of description logic ontologies. Artif. Intell. **203**, 66–103 (2013)
18. Kraus, S., Lehmann, D., Magidor, M.: Nonmonotonic reasoning, preferential models and cumulative logics. Artif. Intell. **44**(1–2), 167–207 (1990)
19. Kutsch, S.: InfOCF-Lib: a java library for OCF-based conditional inference. In: DKB/KIK@ KI, pp. 47–58 (2019)
20. Lehmann, D.: Another perspective on default reasoning. Ann. Math. Artif. Intell. **15**(1), 61–82 (1995). https://doi.org/10.1007/BF01535841
21. Lehmann, D., Magidor, M.: What does a conditional knowledge base entail? Artif. Intell. **55**(1), 1–60 (1992)
22. Makinson, D.: General theory of cumulative inference. In: Reinfrank, M., de Kleer, J., Ginsberg, M.L., Sandewall, E. (eds.) NMR 1988. LNCS, vol. 346, pp. 1–18. Springer, Heidelberg (1989). https://doi.org/10.1007/3-540-50701-9_16
23. Marek, V.W.: Introduction to Mathematics of Satisfiability. Chapman and Hall/CRC, London (2009)
24. Morris, M., Ross, T., Meyer, T.: Algorithmic definitions for KLM-style defeasible disjunctive datalog. South African Comput. J. **32**(2), 141–160 (2020)
25. Parikh, R.: Beliefs, belief revision, and splitting languages. Logic Lang. Comput. **2**(96), 266–268 (1999)
26. Spohn, W.: Ordinal conditional functions: a dynamic theory of epistemic states. In: Harper, W.L., Skyrms, B. (eds.) Causation in Decision, Belief Change, and Statistics. The University of Western Ontario Series in Philosophy of Science, vol. 42, pp. 105–134. Springer, Heidelberg (1988). https://doi.org/10.1007/978-94-009-2865-7_6

Adaptive Reasoning: An Affect Related Feedback Approach for Enhanced E-Learning

Christine Asaju[1]($\boxtimes$) (iD) and Hima Vadapalli[1,2] (iD)

[1] School of Computer Science and Applied Mathematics,
University of the Witwatersrand, Johannesburg, South Africa
1990591@students.wits.ac.za, hima.vadapalli@wits.ac.za
[2] Academy of Computer Science and Software Engineering,
University of Johannesburg, Johannesburg, South Africa
himav@uj.ac.za
http://www.springer.com/gp/computer-science/lncs

Abstract. Recognition of affective states to enhance e-learning platforms has been a topic of machine learning research. Compared to other input modalities, facial expressions have the potential to reveal nonverbal cues about a learner's learning affect. However, most studies were limited in their analysis of learning affects exhibited by a learner with the possibility of providing appropriate feedback to teachers and learners. This work proposes an adaptive reasoning mechanism that considers the estimated affective states and learning affect in generating feedback with reasoning incorporated. This work utilizes a Convolutional Neural Network-Bidirectional Long-Short Term Memory (CNN-BiLSTM) cascade framework for affective states analysis through processing a live/stored observation of a learner in the form of a temporal signal. Using the proposed ensemble, four affective states were estimated, namely boredom, confusion, frustration, and engagement. Dataset for Affective States in E-Environment (DAiSEE) was used to train, validate, and test the baseline model, which reported an accuracy of 86% on 4305 test samples. In the next stage, mappings between estimated affective states and learning affects (i.e. positive, negative and neutral) were established based on an adaptive mapping mechanism, to consolidate the mapping between affective states and learning affects. Live testing and survey feedback were then used to further validate, adapt and amend the feedback process. Incorporating and interpreting the estimated affective states and learning affect is imperative in providing information to both teachers and learners, and hence potentially improve the existing e-learning platforms.

Keywords: Affective states recognition · E-learning · Adaptive reasoning · Learning affect

A. Pillay et al. (Eds.): SACAIR 2022, CCIS 1734, pp. 215–230, 2022.
https://doi.org/10.1007/978-3-031-22321-1_15

1   Introduction

E-learning has recently risen rapidly and gained popularity due to COVID-related implications. Most universities around the world had taken to emergency online teaching and learning during the COVID-19 pandemic, and online teaching and learning has now become a new norm. Gaining insights into learners' learning processes during these online teaching sessions is crucial for learners, teachers and curriculum developers. Peer evaluations, surveys, open-ended questions, drag-and-drop exercises, quizzes, and gamification are some of the approaches used to analyze the (learning) affect experienced by learners. Despite their widespread use, these methods do not offer teachers feedback in real-time during a teaching and learning session. Teachers in traditional classrooms judge the learning affect experienced by learners mainly based on facial expressions, body language, eye gaze, and other physical characteristics. Teachers also use query sessions to obtain feedback from learners to estimate their level of comprehension. In recent years, researchers have applied machine learning technologies to classify cues expressed by students, such as facial expressions and predict their learning affect in both classroom and online settings. Krithika et al. [21] have indicated the use of emotion and facial expressions to predict student learning affect as compared to other input modalities. However, evaluating a learner's comprehension in a real-time online setting using affective sates expressed and providing timely feedback is a complex problem, and the authors of this paper propose a baseline framework to address this issue. The goal of relevant feedback is based on reasoning between the cues exhibited by the learners and those inferred by the teachers(machine learning model in this context).

Reasoning is generally thought of as an intelligent behaviour displayed by living beings. Human beings naturally develop the ability to reason about the relationships between what they can visually witness, and thereby draw logical inferences about the real condition of the relationship that exists between the items that were being observed. The capacity to extract a meaningful pattern from a stream of data is a difficult problem for machines to achieve according to Santoro et al. in [36]. In recent years, much attention has been paid to the idea of reasoning that leads to many interesting studies in the field of image understanding, signal processing, robotics, and general artificial intelligence. Taking into account diverse observations and interest in reasoning tasks by researchers, it would be beneficial if the notion could be included in the education, particularly in e-learning for learning affect estimation and feedback generation.

James et al. and Julie et al. in [32,33] describe learning affect as the attitudes, interests, and values that a learner exhibits and acquires at school that could have a much greater impact on his/her post-secondary life than the academic results. Reasoning about the learning affect experienced by a learner is important for many reasons, such as: (i) it allows teachers to make an inference about a learner, which might be used to design subsequent lessons [22] (ii) aids in the design of affect-responsive learning environments, which can help students control their emotions and have a more engaging learning experience [11] (iii) to recognize and respond to emotional states that are detrimental to learning, as well as produce those that are beneficial to learning, and to aid and improve learning outcomes [6] (iv) help teachers better understand learning processes by offering non-intrusive, real-time tracking of fleeting affect-cognitive interactions that are missed by traditional self-reporting methods [6] (v) expands the scope of learning phenomena by detection of affects [6] (vi) help provide a better targeted instruction and a successful learning experience as teachers are aware of their learners' emotional traits [15].

Given the significance of learning affect, it is critical to recognize that reasoning about the learning affect of learners is an important tool in developing feedback. Feedback is an important part of the teaching and learning process as it allows learners to identify learning gaps and assess their progress [5]. It is thus envisaged that incorporating the concept of reasoning about a learners' learning affect into existing platforms and providing feedback will help aid the overall teaching and learning process.

The remainder of the paper is structured as follows: Related work on affect analysis and reasoning and its application is provided in Sect. 2. Section 3 describes the proposed framework, which includes data and feature sampling, the training and validation experiments of the proposed model, experimental results on DAiSEE, affective states to learning affect mapping (ASLA), affective states to basic emotions, which is to validate the ASLA mapping, the live test of the proposed model. The conclusions and suggestions for future work are provided in Sect. 4.

2 Related Work: Affect Analysis and Reasoning

2.1 Affect Analysis

Ayvaz et al. [4] presented an information system to recognize students' immediate and weighted emotional states in a virtual classroom based on their facial expressions. Happiness, surprise, fear, disgust, anger, sadness, and neutral emotions were recognized. Happiness and surprise were regarded as positive by the authors, while fear, disgust, anger, and sadness were regarded as negative. The instantaneous emotional states of the students were displayed using various emojis, allowing educators to see the general motivational levels of the virtual classroom with a motivational level chart. Accuracy of 96.38% and 98.24% was reported for the classification of emotional state using K-Nearest Neighbor (KNN) and Support Vector Machine (SVM) respectively.

Zakka et al. [42] investigated a standardized mapping method between facial emotions and learning affects. Their research proposed a framework for estimating and classifying facial emotions based on exhibited facial expressions of anger, disgust, fear, happy, sadness, surprise and neutral. A mapping mechanism to map emotional states to learning affect experienced by a student was also presented. A CNN was trained, validated and tested using Facial Emotion Recognition-2013 dataset (FER2013), and an accuracy of 73.12% was reported on the test set. Asaju et al. [3] proposed a framework that uses a machine learning architecture for handling temporal data gathered during an online session towards estimating a student's learning affect. Emotions related to learning, i.e., boredom, frustration, engagement and confusion were estimated. On 6546 test samples, a baseline CNN-BiLSTM model that was trained and tested using the DAiSEE dataset reported an accuracy of 88%. The estimated emotions were further categorized into affective states such as positive or negative learning affect.

Mukhopadhyay et al. [25] examined students' facial expressions to track emotional changes that take place throughout online learning. Using a CNN model, the authors classified basic emotions of anger, disgust, fear, happy, surprise, sadness and neutral. They subsequently determined the students' state of mind using estimated emotions. For emotion classification and state-of-mind identification, their model reported accuracies of 65% and 62%, respectively. To assess the levels of learning affect and engagement of students, Pise et al. [31] a Temporal Relational Network (TRN) framework which identifies the changes in facial emotions of students during an e-learning session. The use of single-scale, multiscale TRN and an Multi-Layer Perceptron (MLP) model were explored. Samples of basic emotions of anger, disgust, fear, happy, surprise, sadness and neutral from DISFA dataset were used and accuracies of 92.7%, 89.4%, and 86.6% were reported using multiscale TRN, single-scale TRN, and MLP, respectively. Dewan et al. [12] examined how facial images could be used to estimate student engagement in an online context. An InceptionNet based framework reported accuracies of 36.5%, 47.1 %, 70.3%, and 78.3% for recognizing boredom, engagement, confusion and frustration states respectively.

Estimation of affective states using exhibited facial cues is a well researched problem and has a potential to further enhance the e-learning space. It is thus proposed to use facial cues as a modality to estimate the affective state of a learner and thus the learning affect.

2.2 Reasoning and Its Applications

Li et al. [23] presented an Adaptive Hierarchical Graph Network (AHGN) based reasoning model with a semantic coherence technique for video and language inference. Their network does in-depth reasoning in three hierarchies over video frames and subtitles, with the graph reasoning structure being adaptively defined by the semantic structures of the statement. They also introduced a Semantic Coherence Learning algorithm (SCL) for the AHGN to support cross-modal and cross-level semantic coherence. Their AHGN + SCL framework reported 71.38% accuracy using VIOLIN (VideO-and-Language Inference) dataset. Zhang et al. [43] proposed a generative visual dialogue system that has an embedded adaptive reasoning module for multimodal input and reported an accuracy of 77.55%. Their module can accommodate various dialogue scenarios and selects relevant information accordingly. Their generative visual dialogue system was trained using weighted likelihood estimation (WLE). Sailer et al. [35] used an autonomous adaptive feedback mechanism based on Natural Language Processing (NLP) approaches in a simulation setting. The method was used to provide feedback on pre-service teachers' written explanations of their diagnostic reasoning for simulated students with learning disabilities. In this context, diagnostic reasoning is stated in diagnostic accuracy (i.e. whether the diagnosis is correct) and quality of diagnostic justification (i.e., how well relevant supporting information for diagnosis is presented). In their work, the authors compared the effects of automatic adaptive NLP-based feedback against static feedback (i.e., expert solutions). The authors opined that learners' understanding of their existing state of knowledge can be aided by automatic adaptive feedback, which can indicate where and how to improve.

Reviewing the existing literature in the areas of affective states estimation, adaptive reasoning, and feedback mechanisms, it is clear that these concepts have not been fully explored in the domain of e-learning. Current work proposes a framework that captures affective states and learning affect experienced by learners based on the exhibited facial expression, and adapting generic feedback templates based on reasoning. The framework was then tested with live participants who provided their feedback of the estimates made by the model with respect to emotional state and learning affect.

3   Proposed Framework

An overview of the proposed framework is provided in Fig. 1. Objectives include (i) data acquisition, (ii) feature extraction using pre-trained ResNet-152, (iii) recognition of affective states using a Bi-LSTM classifier and estimation of learning affect (ie., positive, negative and neutral) based on initial mapping formulated from literature, (iv) further validation of initial mapping based on basic emotions (i.e., anger, disgust, fear, happy, neutral, sadness, and surprise) and learning affects using samples from DISFA+ dataset, (v) evaluation of the final model using live samples and survey.

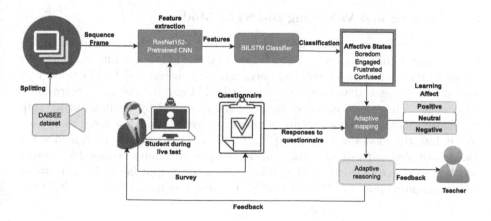

Fig. 1. The proposed framework

3.1 Data and Feature Sampling

Dataset for Affective States in E-learning Environment (DAiSEE) is an open source dataset created to study user engagement and contains 9,068 video sequences from 112 Asian subjects. Crowd annotations were provided as labels representing four affective states related to learning such as boredom, engagement, frustration and confusion. This dataset contains 10 s video samples with a resolution of 640 × 480 pixels/frames and a frame rate of 10 fps. A total of 2,723,882 sequences are available. Benchmark studies by [1, 18] signified the use of DAiSEE in user engagement studies. [18] used DAISEE to test their user engagement prediction model(Long-term Recurrent Neural network) and reported a mean square error (MSE) of 0.0422 and an accuracy of 58.84%. On the other hand, [1] presented an end-to-end spatio-temporal hybrid architecture (ResNet + TCN) to monitor student engagement in an online classroom scenario. They reported state-of-the-art results in classifying student engagement using samples from DAiSEE dataset for model training. Their result reported was 63.9% on 1784 test videos.

Our proposed work uses DAiSEE dataset for baseline training, validation and testing owing to its success in similar studies. A total of 28,634 sequences were used with 24,329 for training and validating the model. The original 480 × 640 × 3 frames were reshaped to 240 × 320 × 3 and were passed through the ResNet-152 [30]. ResNet architecture incorporates a residual learning unit, which helps improve the classification accuracy without adding to the complexity of the model [27]. An output feature vector of 2,048 is sampled at the last max-pool layer for all the 24, 329 samples resulting in a 24,329 × 2,048 size 2D vector for training and validating the model.

3.2 Training and Validating Bi-LSTM Model

The extracted features were then fed to a Bi-LSTM network, a version of the LSTM model that has two LSTMs. The first LSTM takes into account the forward structure of the input sequence, whereas the second LSTM takes into account the reverse structure. This type of LSTM can help a model learn long-term dependencies and improve its accuracy, as stated by Rahman et al. [34]. Evaluation metrics such as accuracy, precision, recall, and F-1 score were used to validate the model as stated in [19]. The Bi-LSTM model was trained using 19,638 samples and validated using 6,546 samples. A binary cross-entropy error function was used to calculate the probability of the output layer. A batch size of 128, maximum epochs of 100 with early stopping was used to avoid overfitting.

3.3 Experimental Results on DAiSEE

The trained model was tested on 4,305 sequence samples from DAiSEE dataset that were not part of train and validation sets. The CNN-BiLSTM model reported an accuracy of 86% and an average F-1 score of 87%, which is a comparable performance to some other state-of-the-art methods [1,16,18,24]. Table 1 shows these comparisons. Figure 2 shows the confusion matrix of the result, while Table 2 provides a summary of the test results. The model reported a misclassification of 1.91% for boredom, 9.49% for confusion, 15.34% for engagement, and 21.38% for frustration affective states. A total of 12.03% misclassifications was reported.

Table 1. Comparison of result with state-of-the art methods

Authors	Dataset	Model	Accuracy reported
[18]	DAiSEE	LRCN	58.84%
[1]	DAiSEE	ResNet + TCN	63.9%
[24]	DAiSEE	SENET +LSTM+GALN	58.84%
[16]	DAiSEE	STGAT	60.32%
Current Study	DAiSEE	CNN-BiLSTM	**86%**

Table 2. Summary of the test reported on DAiSEE dataset

Emotions	Number of samples	Precision	Recall	F1-Score
Boredom	782	0.98	0.91	0.94
Confusion	758	0.89	0.81	0.85
Engagement	1362	0.84	0.79	0.81
Frustrated	1403	0.81	0.94	0.87
Average		**0.88**	**0.86**	**0.87**

3.4 Affective States to Learning Affects (ASLA): Initial Mappings

According to Sheng et al. [7] affective states experienced by learners can be mapped onto positive, negative and neutral learning affects. A number of studies have been carried out in this regard. Daschmann et al. [10] reported that students who experience high levels of boredom are more likely to have negative outcomes such as poor grades, absentism, and eventually dropping out from school, indicating a negative affect on learning. On similar lines, D'Mello et al. [13] opined that students become bored when the lecture does not interest them or when they are unable to handle any of the tasks that are part of the learning process, resulting in a poor learning affect. In their subsequent study, D'Mello et al. [14] opined that confusion is often caused by a lack of understanding of the subject, especially when students face a challenge during their learning activities or are unclear about the knowledge being disseminated, concluding that confusion is a sign of negative learning affect. When students are confused, especially

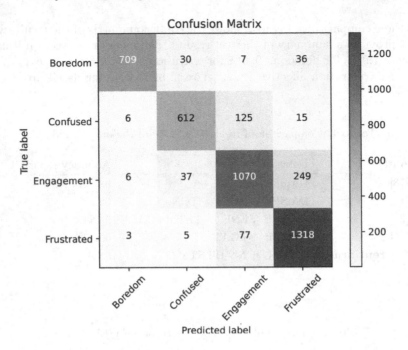

Fig. 2. Confusion matrix showing the test experiment's results

when there is a need to understand or comprehend a lecture, they can become frustrated and make repeated mistakes [13,41]. As a result, frustration was also associated with a negative learning affect. The authors of [9,28] associated an intense feeling of flow-like engagement, especially when the learning outcomes are clear and the lecture content is well understood, with a positive learning affect. Table 3 illustrates mapping between affective states and learning affect based on the works carried out in this space.

3.5   Affective States to Basic Emotion: Validating ASLA Mapping

Experiments to consolidate the mapping between affective states and learning affects were carried out by investigating the mapping between affective states and emotions. Previous work on mapping between basic emotions and learning affects (as shown in Table 4) was used as a baseline for the verification and validation of the mappings between affective states and learning affects.

The objective behind this experiment was to determine whether the estimated affective states of engagement, frustration, boredom, and confusion are mapped to the relevant basic emotions (e.g., anger, disgust, fear, happy, surprise, sadness, and neutral). To achieve this objective, authors tested the DAiSEE trained model on a set of 4755 unlabelled samples from the DISFA + facial expression dataset. The results gathered from using samples labelled with emo-

Table 3. Mapping of affective states to learning affects based on work by Dachmnn et al. [10], Asaju et al. [2,3], D'Mello et al. [14], Stein et al. [41], Csikszentmihalyi et al. [8,9], Shernoff et al. [39], Guo et al. [17] and Nakamura et al. [26]

Affective states	Literature	Mapped to
Boredom	[2,3,10]	Negative
Confusion	[2,3,14]	Negative
Frustration	[2,3,8,9,13,41]	Negative
Engagement	[2,3,9,17,26,39]	Positive

Table 4. Mapping of facial emotion to learning affect by Sathik et al. [38], Kapoor et al. [20], Pan et al. [29], Zakka et al. [42] and Asaju et al. [2,40]

Emotions	[38]	[20]	[29]	[42]	[2,40]
Anger	Negative	Negative	Positive	Negative	Negative
Disgust	–	Negative	Negative	Negative	Negative
Fear	Positive	Positive	–	Positive	Positive
Happy	–	Positive	Positive	Positive	Positive
Sadness	Negative	Negative	–	Negative	Negative
Surprise	Positive	Positive	Positive	Positive	Positive
Neutral	–	–	Positive	Positive	Neutral

tion states to estimate affective states are provided in Table 5 and are discussed as follows:

Samples labelled as depicting "happy" and "surprise" emotions (from DISFA+) were classified as "engagement" affective states by our trained model. Work by [20,29,37,42] maps "happy" and "surprise" emotions to a positive learning affect, consolidating our mapping of "engagement" affective state to a positive learning affect. This is in line with the mappings provided in [8,9,17,26,39]. Generally, it is believed that when students concentrate or are engaged during a learning session, it shows that they have some level of understanding of the lecture. This can result in positive learning affect and can eventually cause a better performance in that subject. This study opines that the expressions of happy, surprise, and /or affective states of engagement shown by students during learning are highly likely to result in a positive learning affect.

Samples labelled as depicting "sadness" emotion (from DISFA+) were estimated as affective state of "confusion" by our trained model. The works by [20,38,42], mapped sadness emotion to a negative learning affect, consolidating our mapping of "confusion" affective state to a negative learning affect. This is inline with the mappings provided by D'Mello et al. [14]. Traditionally, sadness emotions depict a level of uncertainty or unhappiness or dis-satisfaction about a situation, and so exhibit affective states of confusion. This could mean that they may not have a good understanding of the lecture and, therefore, the likelihood of a negative learning affect is very high.

Samples labelled as depicting "fear" emotion (from DISFA+) were estimated as "engagement" affective states by our trained model. Fear was considered positive learning affect by the authors in [20,38,42], consolidating our mapping of "engagement" affective states as a positive learning affect. This corresponds with the mappings of works in [8,9,17,26,39]. Traditionally, fear could be considered as depicting a negative condition, but through a literature survey and results from experiments and the survey that was carried out using students, it was suggested that fear that was detected during learning means a positive learning affect.

Samples labelled as depicting "disgust" emotion (from DISFA+) were estimated as "boredom" affective state by our trained model. Disgust emotions were considered negative emotions by the authors in [3,20,29], consolidating our mapping of "boredom" affective state to a negative learning affect. This corresponds to the mappings provided in [10,13]. This study agrees with the mapping of disgust emotion and the affective state of boredom as a negative learning affect, as provided by the literature and the proposed model. This study infers that when students become bored during a learning session, they can depict disgust emotions, and this could result in a likelihood of a negative learning affect.

Samples labelled as depicting "anger" emotion (DISFA+) were estimated as "frustration" affective state by our trained model. The authors of [20,38,42], mapped "anger" emotion to a negative learning affect, consolidating our mapping of "frustration" affective state to a negative learning affect. This is in line with the mappings provided in [9,26]. Generally, emotions of "anger" and affective states of "frustration" are negative signs in any scenario, and this study opines that the expressions of "anger" and affective states of "frustration" shown by students during learning are highly likely to result in a negative learning affect.

Table 5. Validating mapping between Affective States and Learning Affects using existing mapping between emotions and learning affects

DISFA+ Labels	Sample size	Estimated Affective states by proposed model	Mapped Learning Affect
Anger	822	Frustration	Negative
Disgust	782	Boredom	Negative
Fear	758	Engagement	Positive
Happy	851	Engagement	Positive
Sadness	751	Confusion	Negative
Surprise	791	Engagement	Positive
Total	4755		

3.6   Live Testing of the Proposed Model

It is imperative to validate models through live testing to understand the effectiveness and pitfalls of a proposed framework. The authors propose to evaluate and examine the workings of the model by conducting experiments in semi-controlled laboratory settings, which involve computer science students at the institution where this research is carried out. As this involves human participants, an application for ethical clearance was sent to the University's Human Research Ethics Committee (non-medical) for approval. Survey questionnaires, participation and consent forms used as part of our application are provided in Appendix. Ethics clearance No: H21/05/02.

The proposed system was tested by 22 students from the Computer Science department. After going through the participation and consent forms, participants were required to face a video recorder while watching a 10-min lecture on a given topic. The recorded video sequence depicting the facial cues exhibited by the participant is then fed to the proposed model to estimate an affective state. The affective states that were depicted more in a sample were considered to be the highest depicted affective states in a sample. The second highest depicted affective states were also recorded. This study considered the highest occurring affective states for analysis. After this session, questionnaires were administered to participants to compare the affective states as experienced by the participants themselves (in terms of their basic emotions) with those estimated by the model. This is to investigate the relationship between affective states and basic emotions and to validate the previous mapping of affective states to basic emotions. The obtained results are illustrated in Table 6.

The estimates from our trained model indicated that some affective states were depicted more frequently than others. It was observed that a majority of the participants/learners(59.09%) expressed engagement affective state as the highest occurring affective state, followed by frustration (13.64%) and confusion (27.27%). On the other hand, 40.91% of the respondents estimated happy emotions as their feelings during the lecture, and an additional 9.09% of the participants reported surprise. This totals to a 50% towards the affective state of engagement. 13.64% of the participants reported sadness as their experienced emotion as they did not understand what was being taught. This was inline with the number of participants who were labeled as exhibiting an affective state of frustration, which is a negative learning affect. 36.36% of participants reported neutral expression.

We summarize the survey results to validate our mappings as validated through various experiments and literature in this study in Table 7.

Drawing on several studies and experiments carried out, the current study confirms that when students are experiencing learning, they can experience a variety of affective states that can provide feedback to both teachers and students. Furthermore, research reveals that affects expressed and learning affects are intertwined.

Table 6. Model and participants' estimates during live test.

Participant	Estimates affective states with the highest probability	Estimated affective state with second highest probability	Participant's estimates in terms of basic emotions
1	Engagement	Engagement	Happy
2	Frustration	Confusion	Neutral
3	Confusion	Boredom	Sadness
4	Engagement	Engagement	Happy
5	Confusion	Frustration	Sadness
6	Confusion	Frustration	Sadness
7	Engagement	Engagement	Happy
8	Engagement	Confusion	Neutral
9	Confusion	Engagement	Neutral
10	Engagement	Confusion	Neutral
11	Engagement	Engagement	Happy
12	Frustration	Boredom	Neutral
13	Frustration	Boredom	Neutral
14	Engagement	Engagement	Happy
15	Engagement	Confusion	Happy
16	Confusion	Engagement	Neutral
17	Engagement	Engagement	Surprise
18	Engagement	Confusion	Happy
19	Confusion	Engagement	Neutral
20	Engagement	Frustration	Surprise
21	Engagement	Engagement	Happy
22	Engagement	Engagement	Happy

Table 7. Adaptive reasoning summary

Affective state/Emotions	Learning affect
Sadness, Frustration, Confusion	Negative
Happy, Surprise, Engagement	Positive
Neutral	Neutral

4   Conclusion and Future Work

This paper proposed an adaptive reasoning framework utilizing a machine learning approach. The framework was able to classify four affective states of boredom, frustration, engagement, and confusion. Using DAiSEE dataset, an accuracy of 86% was reported. Various deep learning approaches have used DAiSEE dataset, such as work in [1,16,18,24] etc., the approach used in this work has achieved a comparable performance to other state-of-the-art methods. Estimated affective states were mapped into learning affect of positive, negative and neutral categories, based on a survey of relevant literature. The study further tested the proposed framework with emotion dataset; DISFA+. The purpose was to estimate the labels for samples from DISFA+ dataset that had been labelled for basic emotions, and to verify if the estimated affective states mapped into the relevant basic emotions, and thus to verify the initial mapping of affective states. Live testing experiments and survey feedback were also carried out. The general findings of the study revealed that there is a correlation between facial emotions and learning affect. The work's findings show that using estimation of emotion/affective states and mapping to learning affect for analysis provides a better understanding of the student's learning, particularly in e-learning systems where physical contact is not possible. The proposed system is intended to serve as a first step toward the development of complete e-learning platforms with adaptive feedback incorporated. The authors also would like to explore the use of eye gaze, head pose, physiological signals, body posture, and other input signals along with feedback generation based on reasoning.

Appendix

See Fig. 3.

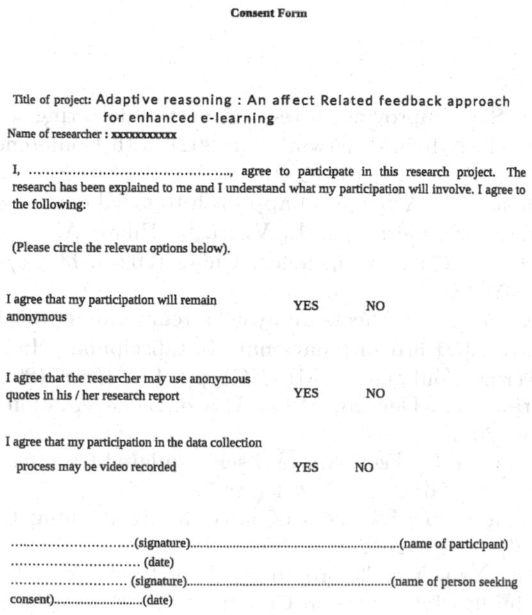

Consent Form

Title of project: Adaptive reasoning : An affect Related feedback approach for enhanced e-learning

Name of researcher : xxxxxxxxxxx

I, ..., agree to participate in this research project. The research has been explained to me and I understand what my participation will involve. I agree to the following:

(Please circle the relevant options below).

I agree that my participation will remain anonymous YES NO

I agree that the researcher may use anonymous quotes in his / her research report YES NO

I agree that my participation in the data collection process may be video recorded YES NO

...........................(signature)...(name of participant)
.... (date)
.... (signature)..(name of person seeking consent)..........................(date)

Fig. 3. Consent form and questionnaire

Adaptive reasoning : An affect Related feedback approach for enhanced e-learning

Questionnaire

This study intends to improve upon the current set-up in the e-learning environment by the analysis of facial emotion recognition, learning affect estimations and reasoning regarding a student's understanding. As part of the studies, I am investigating the emotional changes in students during an E-learning class under the supervision of XXXXXX. In the study after conducting a live test, which includes detection and classification of the emotions, will map the emotions that is being expressed into positive, negative, and neutral learning affects. The intention is to apply this system to the e-learning domain in order to estimate the learning affects of students. As part of the work, the researcher would like you to answer the following questions in attempt to carry out a survey to get a further understanding of the accuracy of the system that was trained, validated, tested and which a live testing is being conducted.

Please answer the following questions which is based on the tutoring video, the emotion and the learning affect estimation. Please note that your answers will be kept anonymous.

Thank you.

Questions	Answers							
1. How do you feel about the lecture? Please respond to this question by picking from the list of emotions given.	Anger	Disgust	Fear	Happy	Neutral	Sadness	Surprise	
2. What category of learning affect will you classify your learning outcomes from the lecture?	Positive	Negative	Neutral					
3. Give a rating of the lecture you have watched. For example, how well it was presented.	30%	40%	50%	60%	70%	80%	90%	100%
4. Is facial emotion expression a good method to evaluate a student in an online learning?	YES	NO	I don't know	Comment				
5. Do you think it is a good measure to improve upon the e-learning environment by estimating the learning affects of students through their facial emotion expression?	YES	NO	I don't know	Comment				
6. Can this method provide an insight to learner's comprehension level.	YES	NO	I don't know	Comment				
7. Can this study help teachers to adequately evaluate the students during online teaching sessions?	YES	NO	I don't know	Comment				
8.Suggest other criteria that could be used to estimate learning affect in online learning platform.								
9. Can this system be improved upon in the future?	YES	NO	I don't Know	Comments				
10. If yes for question(9)Suggest ways by which it can be improved upon.	It can be improved upon by..							

Fig. 3. (*continued*)

References

1. Abedi, A., Khan, S.S.: Improving state-of-the-art in detecting student engagement with ResNet and TCN hybrid network. In: 2021 18th Conference on Robots and Vision (CRV), pp. 151–157. IEEE (2021)
2. Asaju, C., Vadapalli, H.: A temporal approach to facial emotion expression recognition. In: Jembere, E., Gerber, A.J., Viriri, S., Pillay, A. (eds.) SACAIR 2021. CCIS, vol. 1551, pp. 274–286. Springer, Cham (2022). https://doi.org/10.1007/978-3-030-95070-5_18
3. Asaju, C.B., Vadapalli, H.: Affects analysis: a temporal approach to estimate students' learning. In: 2021 3rd International Multidisciplinary Information Technology and Engineering Conference (IMITEC), pp. 1–7. IEEE (2021)
4. Ayvaz, U., Gürüler, H., Devrim, M.O.: Use of facial emotion recognition in e-learning systems (2017)
5. Butler, D.L., Winne, P.H.: Feedback and self-regulated learning: a theoretical synthesis. Rev. Educ. Res. **65**(3), 245–281 (1995)
6. Calvo, R.A., D'Mello, S.: Frontiers of affect-aware learning technologies. IEEE Intell. Syst. **27**(6), 86–89 (2012)
7. Chen, S., Dai, J., Yan, Y.: Classroom teaching feedback system based on emotion detection. In: 9th International Conference on Education and Social Science (ICESS 2019), pp. 940–946 (2019)
8. Csikszentmihalyi, M.: The contribution of flow to positive psychology (2000)
9. Csikszentmihalyi, M., Larson, R.: Flow and the Foundations of Positive Psychology, vol. 10. Springer, Heidelberg (2014). https://doi.org/10.1007/978-94-017-9088-8

10. Daschmann, E.C., Goetz, T., Stupnisky, R.H.: Testing the predictors of boredom at school: development and validation of the precursors to boredom scales. Br. J. Educ. Psychol. **81**(3), 421–440 (2011)

11. DeFalco, J.A., et al.: Detecting and addressing frustration in a serious game for military training. Int. J. Artif. Intell. Educ. **28**(2), 152–193 (2018). https://doi.org/10.1007/s40593-017-0152-1

12. Dewan, M.A.A., Murshed, M., Lin, F.: Engagement detection in online learning: a review. Smart Learn. Environ. **6**(1), 1–20 (2019). https://doi.org/10.1186/s40561-018-0080-z

13. D'Mello, S.: A selective meta-analysis on the relative incidence of discrete affective states during learning with technology. J. Educ. Psychol. **105**(4), 1082 (2013)

14. D'Mello, S.K., Graesser, A.C.: Confusion. In: International Handbook of Emotions in Education, pp. 299–320. Routledge (2014)

15. Ferreira, M., Martinsone, B., Talić, S.: Promoting sustainable social emotional learning at school through relationship-centered learning environment, teaching methods and formative assessment. J. Teach. Educ. Sustain. **22**(1), 21–36 (2020)

16. Fwa, H.L.: Fine-grained detection of academic emotions with spatial temporal graph attention networks using facial landmarks (2022)

17. Guo, Y., Klein, B., Ro, Y., Rossin, D.: The impact of flow on learning outcomes in a graduate-level information management course. J. Glob. Bus. Issues **1**(2), 31–39 (2007)

18. Gupta, A., D'Cunha, A., Awasthi, K., Balasubramanian, V.: DAiSEE: towards user engagement recognition in the wild. arXiv preprint arXiv:1609.01885 (2016)

19. Joshi, R.: Accuracy, precision, recall & f1 score: interpretation of performance measures (2016). Accessed 1 Apr 2018

20. Kapoor, A., Mota, S., Picard, R.W., et al.: Towards a learning companion that recognizes affect. In: AAAI Fall Symposium, vol. 543, pp. 2–4 (2001)

21. Krithika, L., Lakshmi Priya, G.G.: Student emotion recognition system (SERS) for e-learning improvement based on learner concentration metric. Procedia Comput. Sci. **85**, 767–776 (2016)

22. Lederman, N.G.: Teachers' understanding of the nature of science and classroom practice: factors that facilitate or impede the relationship. J. Res. Sci. Teach. Off. J. Natl. Assoc. Res. Sci. Teach. **36**(8), 916–929 (1999)

23. Li, J., et al.: Adaptive hierarchical graph reasoning with semantic coherence for video-and-language inference. In: Proceedings of the IEEE/CVF International Conference on Computer Vision, pp. 1867–1877 (2021)

24. Liao, J., Liang, Y., Pan, J.: Deep facial spatiotemporal network for engagement prediction in online learning. Appl. Intell. **51**(10), 6609–6621 (2021). https://doi.org/10.1007/s10489-020-02139-8

25. Mukhopadhyay, M., Pal, S., Nayyar, A., Pramanik, P.K.D., Dasgupta, N., Choudhury, P.: Facial emotion detection to assess learner's state of mind in an online learning system. In: Proceedings of the 2020 5th International Conference on Intelligent Information Technology, pp. 107–115 (2020)

26. Nakamura, J., Csikszentmihalyi, M.: The concept of flow. In: Nakamura, J., Csikszentmihalyi, M. (eds.) Flow and the Foundations of Positive Psychology, pp. 239–263. Springer, Dordrecht (2014). https://doi.org/10.1007/978-94-017-9088-8_16

27. Nguyen, L.D., Lin, D., Lin, Z., Cao, J.: Deep CNNs for microscopic image classification by exploiting transfer learning and feature concatenation. In: 2018 IEEE International Symposium on Circuits and Systems (ISCAS), pp. 1–5. IEEE (2018)

28. Oliveira, W.: Towards automatic flow experience identification in educational systems: a human-computer interaction approach. In: Extended Abstracts of the Annual Symposium on Computer-human Interaction in Play Companion Extended Abstracts, pp. 41–46 (2019)
29. Pan, M., Wang, J., Luo, Z.: Modelling study on learning affects for classroom teaching/learning auto-evaluation. Science **6**(3), 81–86 (2018)
30. Peng, J., et al.: Residual convolutional neural network for predicting response of transarterial chemoembolization in hepatocellular carcinoma from CT imaging. Eur. Radiol. **30**(1), 413–424 (2020). https://doi.org/10.1007/s00330-019-06318-1
31. Pise, A.A., Vadapalli, H., Sanders, I.: Estimation of learning affects experienced by learners: an approach using relational reasoning and adaptive mapping. Wirel. Commun. Mobile Comput. **2022** (2022)
32. Pooley, J.A., O'Connor, M.: Environmental education and attitudes: emotions and beliefs are what is needed. Environ. Behav. **32**(5), 711–723 (2000)
33. Popham, W.J.: Assessment literacy for teachers: faddish or fundamental? Theory Into Pract. **48**(1), 4–11 (2009)
34. Rahman, M., Watanobe, Y., Nakamura, K., et al.: A bidirectional LSTM language model for code evaluation and repair. Symmetry **13**(2), 247 (2021)
35. Sailer, M., et al.: Adaptive feedback from artificial neural networks facilitates preservice teachers' diagnostic reasoning in simulation-based learning. Learn. Instruct. 101620 (2022)
36. Santoro, A., et al.: A simple neural network module for relational reasoning. In: Advances in Neural Information Processing Systems, p. 30 (2017)
37. Sathik, M.M., Sofia, G.: Identification of student comprehension using forehead wrinkles. In: 2011 International Conference on Computer, Communication and Electrical Technology (ICCCET), pp. 66–70. IEEE (2011)
38. Sathik, M., Jonathan, S.G.: Effect of facial expressions on student's comprehension recognition in virtual educational environments. Springerplus **2**(1), 1–9 (2013). https://doi.org/10.1186/2193-1801-2-455
39. Shernoff, D.J., Csikszentmihalyi, M., Schneider, B., Shernoff, E.S.: Student engagement in high school classrooms from the perspective of flow theory. In: Shernoff, D.J., Csikszentmihalyi, M., Schneider, B., Shernoff, E.S. (eds.) Applications of Flow in Human Development and Education, pp. 475–494. Springer, Dordrecht (2014). https://doi.org/10.1007/978-94-017-9094-9_24
40. Singh, S., Mujinga, M., Lotriet, H., Tait, B.: SAICSIT conference 2021: proceedings of the South African institute of computer scientists and information technologists (2021)
41. Stein, N.L., Levine, L.J.: Making sense out of emotion: the representation and use of goal-structured knowledge. In: Psychological and Biological Approaches to Emotion, pp. 63–92. Psychology Press (2013)
42. Zakka, B.E., Vadapalli, H.: Estimating student learning affect using facial emotions. In: 2020 2nd International Multidisciplinary Information Technology and Engineering Conference (IMITEC), pp. 1–6. IEEE (2020)
43. Zhang, H., et al.: Generative visual dialogue system via adaptive reasoning and weighted likelihood estimation. arXiv preprint arXiv:1902.09818 (2019)

TransFusion: Transcribing Speech with Multinomial Diffusion

Matthew Baas[(✉)] [iD], Kevin Eloff [iD], and Herman Kamper [iD]

MediaLab, Electrical and Electronic Engineering, Stellenbosch University,
Stellenbosch, South Africa
{20786379,20801769,kamperh}@sun.ac.za

Abstract. Diffusion models have shown exceptional scaling properties
in the image synthesis domain, and initial attempts have shown similar
benefits for applying diffusion to unconditional text synthesis. Denoising diffusion models attempt to iteratively refine a sampled noise signal
until it resembles a coherent signal (such as an image or written sentence).
In this work we aim to see whether the benefits of diffusion models can
also be realized for speech recognition. To this end, we propose a new
way to perform speech recognition using a diffusion model conditioned on
pretrained speech features. Specifically, we propose TransFusion: a transcribing diffusion model which iteratively denoises a random character
sequence into coherent text corresponding to the transcript of a conditioning utterance. We demonstrate comparable performance to existing
high-performing contrastive models on the LibriSpeech speech recognition benchmark. To the best of our knowledge, we are the first to apply
denoising diffusion to speech recognition. We also propose new techniques
for effectively sampling and decoding multinomial diffusion models. These
are required because traditional methods of sampling from acoustic models are not possible with our new discrete diffusion approach.

Keywords: Denoising diffusion · Speech recognition · Diffusion
decoding

1 Introduction

Automatic speech recognition (ASR) is the task of transcribing a speech utterance into the words being said. The current paradigm for high-performance ASR
involves the use of supervised training of large neural networks with a connectionist temporal classification (CTC) loss [5]. Intuitively, these models predict
a probability of a character occurring in a particular time window within an
utterance. While this method is the current state-of-the-art for ASR [3,4], the
question remains whether better methods might exist. We aim to approach ASR
from a new perspective and evaluate how closely such an initial attempt can
approach the state-of-the-art CTC-based models.

M. Baas and K. Eloff—Equal contribution.

We have seen in other domains such as image and audio synthesis [12, 20] that denoising diffusion probabilistic models (or 'diffusion models') [21] have exceptional scaling and performance properties. Diffusion models are trained to iteratively denoise a signal sampled from a known noise distribution until it resembles a coherent signal of interest (e.g. images or audio). Efforts like [7, 19] have shown that we can condition this denoising process to correspond to some desired signal (e.g. a text description of an image). In this work we aim to determine whether applying diffusion to speech recognition yields similar properties, and to what extent it can compete with the current best CTC-type models.

Concretely, we attempt to formulate ASR as a conditional diffusion task. Conditioned on speech features from a self-supervised speech representation model, our system attempts to iteratively denoise a random character sequence to ultimately resemble the transcript of the utterance associated with the speech features. Self-supervised speech representation models process speech into a sequence of vectors that represent high-level information about the speech [3]. Our model uses these speech features as conditioning in a multinomial diffusion task [8] – a discrete variant of diffusion – whereby the model predicts a distribution of characters occurring at each position in a transcript. Since our model transcribes speech with a diffusion task, we dub it TransFusion. To the best of our knowledge, we are the first to apply diffusion to the task of ASR. Furthermore, the typical decoding methods used in ASR are not readily applicable to our new type of model. So, we also go on to propose initial new techniques for improving sampling of diffusion-type ASR acoustic models.

We compare our model to existing high-performing CTC-type models such as wav2vec 2.0 [2] on the standard LibriSpeech ASR benchmark [16]. We do not use a language model or external lexicon, as typical methods of combining acoustic and language models have not yet been developed for our new diffusion approach. In this evaluation setting, we demonstrate comparable word error rate (WER) performance to existing high-performing CTC-type models of similar size (`test-other` WER of 8.8%) despite the dearth of decoding and sampling heuristics available for our new diffusion-type ASR approach. In summary, we find that the scaling properties of diffusion models in other domains are also present in ASR. We also recognize the need for future development of larger ASR diffusion models and for methods to combine language models with diffusion acoustic models. Code, models, and demo: https://github.com/RF5/transfusion-asr.

2   Related Work

Modern high-performance ASR systems operate in the time-domain, typically using end-to-end deep neural networks to transcribe an utterance. In particular, current state-of-the-art methods such as [2, 4, 9] first use a large convolutional encoder to downsample a waveform into a vector sequence with each vector typically corresponding to 10 ms to 50 ms of audio. This sequence is then refined in a large transformer variant [22] to yield output features. These models are also typically trained in two phases: a pretraining phase with unlabeled speech, and a fine-tuning phase with labeled speech (i.e. audio where the transcript is known). The pretraining phase is formulated in a variety of ways, but often involves a contrastive or masked language modelling task whereby these output features must accurately predict what information is present in that time window even if that portion of the input audio is masked [2,9]. The fine-tuning process to perform ASR is done with CTC, discussed next.

Meanwhile, diffusion models (outlined later) have been almost exclusively applied to continuous domains such as image synthesis [19,20] and music or audio synthesis [12]. Movellan et al. (1999) [14] was the first to apply diffusion with textual data, where they attempted to classify the word spoken in short videos of people saying one of four possible words. However, the diffusion framework referenced in this work is *not* a denoising diffusion probabilistic model. Rather, they define their own concept of 'diffusion networks' as a continuous stochastic version of recurrent neural networks [14]. The second, more recent, work considering diffusion with textual data defined the key formulation for diffusion on discrete units such as text characters – aptly named multinomial diffusion [8]. While the authors of [8] show considerable performance at unconditional text synthesis, they leave the question open as to how effective such diffusion methods will be when applied to ASR – the goal of this work.

2.1 Connectionist Temporal Classification

The fine-tuning step of most existing end-to-end ASR systems involves the use of connectionist temporal classification (CTC). CTC is a method to model the probability of one sequence given a different, possibly unaligned sequence [5]. For speech recognition, the sequence of output features produced by the pretraining step discussed earlier is used to model the probability of the sequence of characters (target transcript). These sequences are unaligned since each output feature from the model corresponds to a small window of time (e.g. 10 ms), while an English character may correspond to a long period of time (e.g. 300 ms). Essentially, each item in the output sequence produced by the model parameterizes a categorical distribution over the characters in the alphabet and a special ϵ character. CTC then allows for many-to-one alignments of this output sequence to the ground-truth transcript by collapsing repeated characters and removing ϵ characters. A loss is formed as the negative log likelihood of all possible alignments between the model output and target character sequence, where dynamic programming is used to make the computation tractable [6]. So, the fine-tuning process of the current large self-supervised models such as [2,9] involves maximizing the likelihood of the ground-truth transcript, given the model's output features.

2.2 Denoising Diffusion Probabilistic Models

One of the newer techniques that is rising in popularity for speech and image synthesis models is that of denoising diffusion probabilistic models, or simply 'diffusion models'. Concretely, a diffusion model [21] defines a Markov process of T steps from $t \in \{0, ..., T-1\}$. The modelled data (e.g. waveforms or images) is defined as the signal at the first timestep $\mathbf{x}_0$, and the last timestep is defined as a known noise distribution, e.g. $\mathbf{x}_T = \mathcal{N}(\mathbf{0}, \mathbf{I})$ for images. The diffusion process consists of a *forward* and *reverse* function to move the signal from $\mathbf{x}_t$ to $\mathbf{x}_{t+1}$ and from $\mathbf{x}_t$ to $\mathbf{x}_{t-1}$, respectively. The forward diffusion process is defined by a function $q(\mathbf{x}_t|\mathbf{x}_{t-1})$ which iteratively *adds* noise to a signal until – at timestep T – it resembles a pure noise distribution. Similarly, the reverse diffusion process $p(\mathbf{x}_{t-1}|\mathbf{x}_t)$ iteratively denoises the signal until at $t = 0$ it resembles a coherent signal. This reverse process is parameterized with a large neural network [21].

Specifically, the diffusion network (the model associated with the reverse diffusion process p) is trained to predict the noise added through the forward process – i.e. to predict the difference between the desired coherent signal and the signal after adding varying amounts of noise. At each inference step the diffusion network is called to parameterize $p(\mathbf{x}_{t-1}|\mathbf{x}_t)$ and then sample the slightly denoised next step $\mathbf{x}_{t-1}$. Diffusion models have recently been shown to scale very well to large model sizes and datasets [19,20], and we hypothesize that it will yield similar beneficial properties when applied to ASR. One issue with typical diffusion is that it is formulated in the continuous domain, such as denoising a continuous pixel or audio sample value slightly in each step. For discrete signals like text, we must use a recent discrete variant of diffusion – multinomial diffusion.

2.3   Multinomial Diffusion

Diffusion is typically used for continuous signals such as images or waveforms [12,20]. However, in 2021, Hoogeboom et al. introduced a method to perform diffusion on signals with discrete alphabets: multinomial diffusion [8].

Concretely, multinomial diffusion defines the input to a diffusion model as a sequence of discrete units (i.e. letters or words) represented as one-hot encoded vectors $\mathbf{x}$. We index the diffusion timestep with $t \in \{0, ..., T-1\}$ and the position in the sequence (the character index in the transcript) with $i \in \{0, ..., N-1\}$ such that $\mathbf{x}_{t,i} \in \{0, 1\}^K$ is the one-hot encoding of the character represented at diffusion timestep t and sequence position i using a K-sized alphabet. However, all diffusion operations proposed by [8] are independent across sequence length, thus when we omit the index i it indicates that the statement applies to the entire sequence independent of sequence index. Multinomial diffusion defines the forward noising process $q(\mathbf{x}_t|\mathbf{x}_{t-1})$, the posterior $q(\mathbf{x}_{t-1}|\mathbf{x}_t, \mathbf{x}_0)$, and the reverse diffusion process $p(\mathbf{x}_{t-1}|\mathbf{x}_t)$ for a K-sized alphabet as:

$$q(\mathbf{x}_t \mid \mathbf{x}_{t-1}) = \mathcal{C}\left(\mathbf{x}_t \mid (1 - \beta_t)\mathbf{x}_{t-1} + \beta_t/K\right)$$

$$q(\mathbf{x}_{t-1} \mid \mathbf{x}_t, \mathbf{x}_0) = \mathcal{C}\left(\mathbf{x}_{t-1} \mid \frac{1}{A}\left[\alpha_t\mathbf{x}_t + (1 - \alpha_t)/K\right] \odot \left[\bar{\alpha}_{t-1}\mathbf{x}_0 + (1 - \bar{\alpha}_{t-1})/K\right]\right)$$

$$p(\mathbf{x}_{t-1} \mid \mathbf{x}_t) = \mathcal{C}\left(\mathbf{x}_{t-1} \mid \frac{1}{A}\left[\alpha_t\mathbf{x}_t + (1 - \alpha_t)/K\right] \odot \left[\bar{\alpha}_{t-1}\hat{\mathbf{x}}_0 + (1 - \bar{\alpha}_{t-1})/K\right]\right)$$

where $\mathcal{C}$ denotes a categorical distribution with category probabilities specified after $|$. β_t is the diffusion noise schedule defined in the original binomial diffusion work [21], and $\alpha_t = 1 - \beta_t$, $\bar{\alpha}_t = \prod_{\tau=0}^{t} \alpha_\tau$. The fraction $\frac{1}{A}$ is a normalizing constant to ensure the probabilities sum to one [8]. The final-timestep sequence $\mathbf{x}_0$ is the one-hot encoding derived from the ground-truth text for the posterior, and $\hat{\mathbf{x}}_0$ is the predicted probabilities over the vocabulary for each position in the sequence. This is how the diffusion process is parameterized with a neural network: at each diffusion timestep, the model predicts a distribution over the vocabulary $\hat{\mathbf{x}}_0$ of the *fully denoised transcript* at $t = 0$. To get an intuitive idea how the reverse process iteratively denoises a sample transcript using multinomial diffusion, see Fig. 3 (explained later). In this work we phrase the task of speech recognition as a speech-feature-guided multinomial diffusion task.

3   Model

Our model is a denoising probabilistic diffusion model [21] that transcribes utterances from a provided sequence of speech features extracted from a self-supervised speech representation model. So, as our model is a transcribing diffusion model, we dub it **TransFusion**. Concretely, it adapts classifier-free guidance [7] and multinomial diffusion [8] to allow a discrete diffusion model to be conditioned on a sequence of speech features extracted from the self-supervised speech representation model WavLM [3].

3.1   Conditioning Diffusion on Speech Representations

The training and inference setup of TransFusion is shown in Fig. 1. During training, we have an input utterance waveform and its associated ground-truth transcript denoted $\mathbf{x}_0$. During each training step, a noised version of the transcript is calculated for diffusion timestep t using $q(\mathbf{x}_t|\mathbf{x}_0) = \mathcal{C}\left(\mathbf{x}_t \mid \bar{\alpha}_t\mathbf{x}_0 + (1 - \bar{\alpha}_t)/K\right)$ [8], where t is sampled uniformly at random from $\{0, ..., T - 1\}$. Intuitively, the input text fed to the model has its characters randomly flipped, with increasing randomness until at the highest timestep $t = T - 1$, the transcript fed to the model is entirely random. The waveform is converted into a sequence of high-level speech features $\mathbf{c}$ using a fixed pretrained WavLM model [3]. This sequence of features $\mathbf{c}$ is then used to condition the main TransFusion model. TransFusion's architecture is that of a transformer variant (discussed later) and maps the noisy input characters to a predicted distribution of output characters for the desired transcript $\mathbf{x}_0$. More formally,

$$p(\hat{\mathbf{x}}_0|\mathbf{x}_t, \mathbf{c}) = \text{TransFusion}(\mathbf{x}_t, \mathbf{c})$$

During inference (Fig. 1, right), given speech features $\mathbf{c}$ from an utterance with unknown transcript, we sample a random sequence of characters at $\mathbf{x}_T$. We then iteratively denoise the transcript by using TransFusion and the diffusion parameters to compute a distribution for the reverse process $p(\mathbf{x}_{t-1}|\mathbf{x}_t)$. The slightly denoised transcript is then sampled from this distribution as $\mathbf{x}_{t-1}$ and used as the input to the model in the next iteration. This process continues until $t = 0$ and $\mathbf{x}_0$ is a refined prediction of the transcript of the utterance.

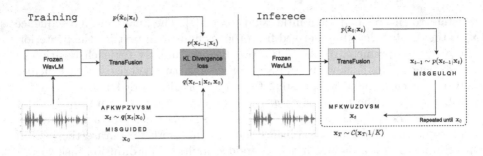

Fig. 1. TransFusion diagram. Speech features from an utterance computed using a frozen WavLM encoder [3] are used to condition TransFusion. During training (left) the model is trained according to multinomial diffusion [8] to minimize the KL divergence between the reverse process $p(\mathbf{x}_{t-1}|\mathbf{x}_t)$ and posterior process derived from the ground truth utterance $q(\mathbf{x}_{t-1}|\mathbf{x}_t, \mathbf{x}_0)$. During inference (right), a uniformly random sampled transcript $\mathbf{x}_T$ is iteratively denoised using TransFusion until $\mathbf{x}_0$ is the predicted output transcript.

3.2 Training Task

The loss follows that of multinomial diffusion (Sect. 2.3). Specifically, using the diffusion parameters β, α, $\bar{\alpha}$ at timestep t and the current noised inputs, the posterior $q(\mathbf{x}_{t-1}|\mathbf{x}_t, \mathbf{x}_0)$ is computed. Likewise, with the model's prediction and the diffusion parameters, the reverse distribution $p(\mathbf{x}_{t-1}|\mathbf{x}_t)$ also provides a distribution over $\mathbf{x}_{t-1}$. For TransFusion to accurately undo the noise added between timestep $t-1$ and t, the predicted reverse distribution $p(\mathbf{x}_{t-1}|\mathbf{x}_t)$ should be close to the posterior $q(\mathbf{x}_{t-1}|\mathbf{x}_t, \mathbf{x}_0)$ (which has access to the ground-truth transcript). So, a training loss is formed as the Kullback-Leibler (KL) divergence:

$$\mathcal{L} = \mathrm{KL}\left(\, q(\mathbf{x}_{t-1}|\mathbf{x}_t, \mathbf{x}_0) \,\|\, p(\mathbf{x}_{t-1}|\mathbf{x}_t)\,\right)$$

Furthermore, the theory of multinomial diffusion also requires an additional term be added when $t = 0$ [8]. Namely the cross-entropy between the one-hot ground-truth distribution $\mathbf{x}_0$ and the predicted probabilities from TransFusion $\hat{\mathbf{x}}_0$ is added to the loss when $t = 0$ [8]. Intuitively, this pushes the distribution predicted by the model to be close to the one-hot ground truth targets when $t = 0$. This loss is readily computed and the model can be trained through backpropagating through the predictions $\hat{\mathbf{x}}_0$ used to compute $p(\mathbf{x}_{t-1}|\mathbf{x}_t)$.

3.3 Architecture

TransFusion's architecture is based on a transformer [22] and is depicted in Fig. 2. It draws on the diffusion conditioning paths proposed for images in [20], and incorporates the relative positional encoding used by wav2vec 2.0 [2]. Concretely, the model consists of 24 transformer layers where the vector sequence used for the self-attention block [22] is carefully crafted to incorporate timestep and conditioning information. For discrete inputs like the character sequence $\mathbf{x}_t$ and the timestep t, we first embed them into a contin-

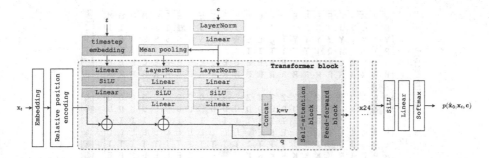

Fig. 2. TransFusion architecture. The sequence of input characters $\mathbf{x}_t$ is passed through an embedding and positional encoding layer and then into a sequence of 24 transformer blocks before being projected to an output distribution $p(\hat{\mathbf{x}}_0|\mathbf{x}_t, \mathbf{c})$ over the vocabulary for each character in the sequence. In each transformer block, the mean-pooled WavLM features $\mathbf{c}$ and processed timestep embedding is added to each vector in the sequence which acts as the query to a self-attention block [22]. The key and value are formed by concatenating the transformed sequence of WavLM vectors with the main sequence derived from the characters. `SiLU`, `LayerNorm`, `Concat` layers refer to Sigmoid Linear Unit [18], layer normalization [1], and concatenation across sequence length, respectively.

uous vector space using regular learnt embedding layers (for characters) or fixed sinusoidal embeddings [15] (for the timestep). The timestep embedding (after a few layers as in Fig. 2) is summed with each character embedding to condition the sequence on the current timestep.

The frozen WavLM model computes the sequence of features $\mathbf{c}$ associated with the utterance, producing a vector for every 20 ms of the utterance. To condition TransFusion on these features $\mathbf{c}$, we adapt the technique proposed for image synthesis in [20] and compute two streams of information as shown in Fig. 2. First, we sum the vector derived from the mean across the entire sequence $\mathbf{c}$ with each character embedding. And second, we concatenate the entire sequence of vectors $\mathbf{c}$ (after passing it through a few layers) with the sequence of character embeddings, and use this longer sequence as the keys and values for the self-attention block. Intuitively, the mean-pooled vector is meant to provide TransFusion with a summary of the entire utterance, while providing the full sequence $\mathbf{c}$ to the attention layer allows for the model to learn more fine-grained spelling as it can attend to specific parts of the utterance for each query vector.

This series of operations is all encompassed in a single transformer block, and the full TransFusion comprises of 24 such blocks and a final softmax projection head to yield the final distribution $p(\hat{\mathbf{x}}_0|\mathbf{x}_t, \mathbf{c})$. Finally, since concatenating the entire sequence of WavLM-derived vectors in each transformer block entails a very large memory and compute time cost, we only apply the `Concat` layer in Fig. 2 in a select few transformer blocks (detailed in Sect. 5).

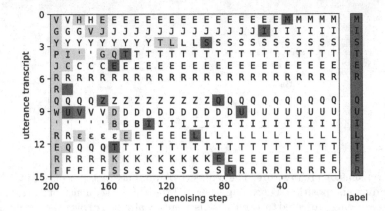

Fig. 3. Example denoising process of the TransFusion model, starting from a random sequence $\mathbf{x}_T$ (leftmost column) and denoising until $\mathbf{x}_0$ (rightmost column, excluding label). Green blocks indicate transitions to the ground truth target transcription. Red and yellow blocks indicate transitions from a right character to a wrong character, and a wrong character to another wrong character respectively. (Color figure online)

4   Diffusion Decoding

To perform ASR with TransFusion, we must define the process of decoding a new utterance input through the diffusion model. The simplest form of the diffusion decoding process, described briefly in Sect. 3.1 and shown in Fig. 1, iteratively computes $\mathbf{x}_{t-1}$ given $\mathbf{x}_t$. An example of this decoding is given in Fig. 3, where the speech features $\mathbf{c}$ contain the linguistic content "MISTER QUILTER". Starting at a random sequence of characters at $t = T = 200$, the model iteratively denoises until the final transcript is reached at $t = 0$. This approach, while effective, often results in errors related to the position of words in the overall sequence. If the model mistakes word placement early on, it is unable to make corrections because it is unable to easily shift characters. To this end, we propose new techniques for effectively decoding multinomial diffusion models.

4.1   Resampling

RePaint [13] introduced the idea of resampling for the reverse diffusion process to improve inpainting performance during image synthesis. With RePaint, instead of linearly denoising from $t = T - 1$ to $t = 0$, they jump back and forth applying both forward and reverse diffusion on a set schedule. They found that repeatedly denoising and adding noise (i.e. diffusing) an image improved image generation quality, allowing the model to improve local coherency. This resampling schedule can be used directly in our decoding. Concretely, we define two constants, jump length L and number of jumps J. During decoding, we alternate between denoising (applying the reverse function $p(\mathbf{x}_{t-1}|\mathbf{x}_t)$) and diffusing (applying the forward noising process $q(\mathbf{x}_t|\mathbf{x}_{t-1})$), each lasting L timesteps and repeated J times. This is repeated until $\frac{T}{L} - 1$ times linearly along t. See [13] for precise details.

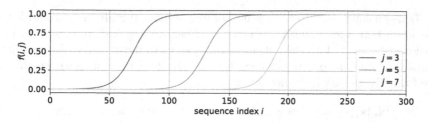

Fig. 4. Example of sequentially progressive diffusion scaling function f for $J = 10$, shown over the character sequence for $j \in \{3, 5, 7\}$. This function modifies the variance scale based on both resampling jump and sequence index. The actual function used is $f(i, j) = \sigma((i - \frac{jN}{J} + 2J)/8)$. The constant offset $2J$ is added to ensure the full sequence is diffused at $j = 0$, and 8 was analytically chosen to ensure a smooth overlap between sequential jumps j.

4.2 Sequentially Progressive Diffusion

Another method introduced in [13] is inpainting or known region conditioning. The idea of inpainting is to predict the missing pixels under a masked region of an image. Influenced by this, we considered using a similar approach by decoding word-by-word from the beginning of the transcript. The predicted words would be then treated as the inpainted region and the model then must predict the rest of the transcript. While effective, this approach is computationally expensive, as the entire denoising process must be repeated for every word in the transcript.

Instead, we develop a new tractable method to implement alongside resampling. We call this sequentially progressive diffusion: instead of applying forward diffusion uniformly along the sequence length, we now scale β along the length of the transcript. This allows us to retain earlier parts of the utterance while diffusing/noising the later parts. To scale diffusion noise based on sequence position, we define the scaled diffusion noise schedule $\beta'_{t,i}$ for each character position i:

$$\beta'_{t,i} = \beta_t \cdot f(i, j)$$

where f defines the scaling function and $j \in \{0, 1, \ldots, J - 1\}$ is the current resample jump. f is chosen such that the diffusion is applied uniformly at $j = 0$, and slides linearly along the sequence length, only diffusing the end of the transcript at $j = J - 1$. We implemented f as a shifted sigmoid as shown in Fig. 4 to make a smooth transition between the retained and diffused regions.

4.3 Classifier-Free Guidance

As with other conditional diffusion efforts in the image synthesis domain [19, 20], we utilize classifier-free guidance [7] to improve the alignment between the output transcript and speech features $\mathbf{c}$. In our context, classifier-free guidance attempts to force TransFusion to learn both a conditional TransFusion($\mathbf{x}_t, \mathbf{c}$) and unconditional TransFusion($\mathbf{x}_t$) reverse diffusion process. This is achieved by randomly dropping out the conditioning information $\mathbf{c}$ with some small probability (in our case 0.1 – the same as those found to work well in [7, 20]). In this way, TransFusion learns to both generate unconditional realistic text and learns to generate text aligned with a transcript. Then during ASR inference, at each diffusion step, we update the output $p(\hat{\mathbf{x}}_{t-1} \mid \mathbf{x}_t, \mathbf{c})$ to move in the direction *from* the unconditional output *to* the conditional output with a 'guidance weight' [7]. More formally, during inference we set:

$$p(\hat{\mathbf{x}}_0 | \mathbf{x}_t, \mathbf{c}) = w\,\text{TransFusion}(\mathbf{x}_t, \mathbf{c}) + (1 - w)\,\text{TransFusion}(\mathbf{x}_t)$$

The reasoning is as follows: if the model output (logits of $\hat{\mathbf{x}}_0$) *conditioned* on information about the utterance is $\text{TransFusion}(\mathbf{x}_t, \mathbf{c})$ and *not conditioned* on the utterance information is $\text{TransFusion}(\mathbf{x}_t)$, then intuitively the linear direction from $\text{TransFusion}(\mathbf{x}_t)$ to $\text{TransFusion}(\mathbf{x}_t, \mathbf{c})$ corresponds to the direction of increasing conditioning information. We can then improve the strength of the conditioning – in our case, to improve alignment of output transcript with utterance – by linearly adjusting the conditional output in this direction [7]. Note that we apply this linear combination before the output softmax of the model in Fig. 2 to ensure the adjusted output still is a valid probability distribution. With $w = 1$ in the above equation there is no guidance while increasing values $w > 1$ strengthens the guidance effect. We found $w = 1.5$ to yield the best results based on decoding ablations on our validation set (following the same ASR setup as described in Sect. 5). We use this setting in all our evaluations.

4.4   Full Inference Process

The full decoding process combines the methods defined above to perform ASR on an utterance from speech features $\mathbf{c}$. The core of the inference is resampling with sequentially progressive noise scaling in the forward diffusion steps. From the same validation decoding ablation experiments, we found resampling worked best with $J = 10$ and $L = 10$, which we use for our final resampling decoding in the next section. The reverse diffusion step utilises classifier-free guidance to improve alignment of the output transcript. We also note that we can use arbitrary sequence lengths at inference due to the model using relative positional encoding. In our final inference we use a sequence length $N = 400$ to ensure we cover all transcripts in the LibriSpeech test and dev datasets described in Sect. 5.1 (>99% of LibriSpeech transcripts are shorter than 400 characters). In all our diffusion training and evaluation experiments we set β_t according to the cosine noise schedule from [15] using the recommended value of $s = 0.008$.

5   Experimental Setup

We compare our model against other common ASR models on a standard speech recognition benchmark dataset: LibriSpeech [16]. Namely, we compare against the high-performing self-supervised speech representations models wav2vec 2.0 [2], Conformer [23], and w2v-BERT models [4].

5.1 Dataset and Metrics

We perform our experiments on the LibriSpeech dataset [16]. It consists of 960 h of spoken audiobooks by multiple speakers with varying amounts of noise and audio quality. We train our model on the full 960 h `train` split of LibriSpeech and evaluate it on the official `dev` and `test`-splits. For the frozen WavLM model, we use the `WavLM-Large` pretrained model from the original authors [3]. Note that this model has only been pretrained with a masked prediction task on raw audio, and has not been fine-tuned to perform ASR (i.e. it has never been exposed to transcripts of utterances). To evaluate our model we use the standard ASR metrics of word error rate (WER) and character error rate (CER), however ASR papers typically focus on the WER metric [3,9] so we focus on WER for comparison. We compute the mean WER for our model sampled using various decoding strategies and compare it against the baseline models (described next) using their best results reported by the original authors.

5.2 Baseline Models

We compare against three state-of-the-art models for ASR: wav2vec 2.0 [2], Conformer [23], and w2v-BERT [4]. The first two are large transformers trained in two phases. First, they are trained in a self-supervised fashion using large amounts of unlabeled audio on a masked token prediction task, and then they are fine-tuned with a CTC objective on the LibriSpeech dataset of labeled (i.e. transcribed) audio. The w2v-BERT model also follows this pretraining-finetuning setup, but also incorporates additional tricks to improve performance such as self-training with noisy student training [17]. These practical techniques to squeeze out additional performance from ASR models have been developed primarily for CTC-type ASR models and are largely undeveloped for diffusion-type ASR models. This is because CTC-type models (including all the baseline models) ultimately produce a probability distribution for a character or phoneme being present at a **certain time** in the utterance. Meanwhile, our diffusion-type model produces a probability distribution for a character being present at a **certain position** in the transcript. It remains as future work to develop and adapt the practical techniques to improve performance of such diffusion-type models.

Furthermore, decoding an acoustic model with a typical language model and lexicon has also been developed with these CTC-type models in mind, making them not very effective when applied directly to our model which predicts characters at fixed positions in the transcript. The primary reason for this is that any insertion or deletion of characters early on in the transcript will cause a substantial change in the predicted likelihood of the rest of the characters in the utterance. For example, if the modal model prediction for the first two words is "SQUEZE IT", and the lexicon is naively used to decode the first word to "SQUEEZE", then the diffusion acoustic model will have a very high likelihood of "SQUEEZE T" and a very low likelihood for the desired correction "SQUEEZE IT". Again, this problem stems from our different way of phrasing the ASR problem as predicting characters at fixed positions in a transcript. So, if we insert a character early on in the decoding process against the acoustic model's recommendation (i.e. not using its modal prediction), the acoustic likelihood will incentivise the dropping of a character elsewhere in the transcript to retain a high likelihood score for the ground-truth transcript. Because decent techniques for language model decoding have not yet been developed for diffusion-type models, all experiments in this paper do not use a lexicon or language model—the models that we compare to are also used without a language model or lexicon.

5.3   TransFusion Implementation

Layers: For the relative positional encoding layer at the input to the transformer, we use the formulation provided in [2]. As mentioned in Sect. 3, we do not apply the `Concat` layer in every transformer block. To save compute resources and to allow for an improved attention weighting (discussed next), we only apply the `Concat` operation in every 4th block. So, with 24 transformer blocks, the `Concat` layer is present in layers $1, 5, 9, 13, 17, 21$. The self-attention and feed-forward blocks follow the same architecture as in the original attention article [22].

Model Hyperparameters: Our model uses a 29-sized character alphabet and contains 24 transformer blocks. The output dimension of all embedding, linear, and attention layers is 768. The feed-forward blocks have a dimension of $4 \times 768 = 3072$ and each self-attention operation uses 8 attention heads. Transformer attention and feed-forward blocks use a dropout of 0.1, and we also completely dropout all conditioning information with probability 0.1 (in line with classifier-free diffusion guidance [7]). For the relative positional encoding [2], we use a 256-size convolution kernel with 32 convolution groups. The conditioning sequence **c** from `WavLM-Large` model is defined as the average of the activations of the last 9 layers from the model pretrained on LibriLight [10], since [3] found these last layers to be most important for representing linguistic information.

Optimization: We train TransFusion on the full LibriSpeech 960 h training subset for 350k updates using a batch size of 720 with Adam optimization [11] with $\beta = (0.9, 0.999)$. Further, we use a constant learning rate of 3×10^{-5} with a linear warmup of 10k updates and clip the global gradient norm at 10. As this is an initial foray into ASR with diffusion, we do not use any data augmentation such as SpecAugment. This differs from the baselines, all of which have been trained with substantial data augmentation to further improve performance [2,4,23]. Even without augmentations, the model's validation performance (WER on internal validation split) was still improving at the end of training – we hypothesize that training a larger model for longer with all the typical data augmentation techniques used for other CTC-type models will yield further improvements to results listed in the next section. To demonstrate the level of improvement gained from longer training, we continue training our model up to 462k updates and show its performance compared to the base 350k update model in the next section. All training is done on three NVIDIA Quadro RTX 6000 devices with mixed FP16/FP32 precision.

Table 1. WER results on the LibriSpeech dev and test splits for ASR models trained on the full 960 h LibriSpeech training set. Decoding for prior models is done with lexicon-free CTC-decoding [5], while several decoding strategies are applied to our diffusion model (Sect. 4). No language models are used in decoding, and all models are fine-tuned on the full LibriSpeech 960 h training data. The unlabeled data used to pretrain each model is specified as the LibriSpeech 960h dataset (LS-960h) [16] or LibriLight 60 000 h dataset (LV-60kh) [10].

Model	Pretraining	Params (M)	dev set		test set	
			clean	other	clean	other
Including pretraining						
wav2vec 2.0 Base [2]	LS-960h	95	3.2	8.9	3.4	8.5
wav2vec 2.0 Large [2]	LS-960h	317	2.6	6.5	2.8	6.3
wav2vec 2.0 Large [2]	LV-60kh	317	2.1	4.5	2.2	4.5
Conformer XXL [23]	LV-60kh	1000	1.6	3.2	1.6	3.3
w2v-BERT XXL [4]	LV-60kh	1000	**1.5**	**2.7**	**1.5**	**2.8**
No pretraining						
wav2vec 2.0 Large [2]	None	317	2.8	7.6	3.0	8.5
Conformer L [23]	None	103	**1.9**	**4.4**	**2.1**	**4.3**
TransFusion (ours, 350k updates)	None	253				
basic decoding (Sect. 3.1)			9.6	12.1	10.5	12.5
+ classifier free guidance			9.4	11.7	10.2	12.2
+ resampling			8.4	10.7	9.0	11.0
+ sequentially progressive diffusion			8.1	10.5	8.9	10.8
+ further trained to 462k updates			**6.1**	**8.3**	**6.7**	**8.8**

6 Results

The ASR results are given in Table 1. First we observe that our new model using our best decoding strategy does not beat the current state-of-the-art CTC-type models such as Conformer L or w2v-BERT. However, the performance is still considerable given that this is a first investigation of an entirely new approach to ASR using discrete diffusion. Concretely, we achieve a `test-clean` and `test-other` WER of 6.1% and 8.8%, respectively. The nature of many of our model's mistakes follows the issue outlined in Sect. 5.2. Specifically, often if the model decodes a character by an erroneous insertion/deletion early on in the transcript, it will attempt to drop or insert another erroneous character later on to ensure that the alignment for the rest of the characters is still correct. While this worsens the WER, the effect on CER is less impactful, where TransFusion achieves a CER of 3.2% and 3.6% on `test-clean` and `test-other`, respectively.

Furthermore, we observe from Table 1 that each of our decoding techniques cumulatively improves the results, with our final addition of sequentially progressive diffusion yielding a more than 1.5% absolute WER improvement over basic decoding. This demonstrates the effectiveness of our initial decoding methods and suggests that even further improved performance may be achievable given better decoding approaches. Further training our model to 462k updates also substantially improves performance (last row of Table 1), bringing the `dev-other` subset results above that of wav2vec 2.0 `Base`. This suggests that – in line with our motivation about diffusion scaling in Sect. 2 – even greater performance is likely achievable with increased compute and model sizes.

It is also interesting to observe that the difference in WER for TransFusion between the less noisy `dev-/test-clean` sets and the more noisy `dev-/test-other` sets is much less than all the CTC-type models. For the CTC models, WER on the clean subsets are often half of the result on the noisy subset, while with the diffusion model the difference is much smaller. While we are not sure of the precise reason for this, we speculate that our decoding method for diffusion-type acoustic models does not draw out the full performance possible from TransFusion, unlike the powerful CTC decoding possible with CTC-type models. In other words, if the performance trend of the CTC-type models is representative of the difficulty difference between `clean` and `other` subsets, then we should be able to achieve a better clean performance once more optimal decoding methods are developed for diffusion-type models. Finally, while TransFusion does use features from a pretrained WavLM model, it does not fine-tune the WavLM encoder which provides these features unlike the other methods considered, hence we denote the trained weights of TransFusion as not including any pretraining in Table 1.

7   Conclusion

In this paper we proposed TransFusion – a model that utilizes multinomial diffusion to phrase the task of speech recognition as a conditional discrete diffusion task. Our model iteratively denoises an arbitrarily noised transcript until it resembles coherent text corresponding to the transcript of a provided utterance. This is done by providing speech features associated with the utterance to condition a large transformer model predicting a categorical distribution over a character alphabet. Since we are the first to phrase ASR in this way, we proposed new methods to decode such diffusion-type ASR models during inference. We showcase TransFusion's performance on the LibriSpeech dataset and compare it to existing state-of-the-art CTC-type models and demonstrate comparable performance. While we do not outperform the best large CTC-type models we compare to, we achieve a 8.8% WER/3.6% CER on the LibriSpeech `test-other` set. This is noteworthy given the completely new method for ASR proposed here. Future work will develop better decoding strategies and methods for combining the diffusion acoustic model with a language model. We will also consider training on standard speech features instead of using WavLM.

Acknowledgements. All experiments were performed on Stellenbosch University's High Performance Computing (HPC) cluster. This work is supported in part by the National Research Foundation of South Africa (grant no. 120409).

References

1. Ba, J.L., Kiros, J.R., Hinton, G.E.: Layer normalization. arXiv preprint arXiv:1607.06450 (2016)
2. Baevski, A., Zhou, Y., Mohamed, A., Auli, M.: wav2vec 2.0: a framework for self-supervised learning of speech representations. In: NeurIPS (2020)
3. Chen, S., et al.: WavLM: large-scale self-supervised pre-training for full stack speech processing. IEEE J. Sel. Topics Signal Process. **16**, 1505–1518 (2022)
4. Chung, Y.A., Zhang, Y., Han, W., Chiu, C.C., Qin, J., Pang, R., Wu, Y.: W2v-BERT: combining contrastive learning and masked language modeling for self-supervised speech pre-training. In: ASRU (2021)
5. Graves, A., Fernández, S., Gomez, F., Schmidhuber, J.: Connectionist temporal classification: labelling unsegmented sequence data with recurrent neural networks. In: ICML (2006)
6. Hannun, A.: Sequence modeling with CTC. Distill (2017). https://doi.org/10.23915/distill.00008. https://distill.pub/2017/ctc
7. Ho, J., Salimans, T.: Classifier-free diffusion guidance. In: NeurIPS Workshop on Deep Generative Models and Downstream Applications (2021)
8. Hoogeboom, E., Nielsen, D., Jaini, P., Forré, P., Welling, M.: Argmax flows and multinomial diffusion: Learning categorical distributions. In: NeurIPS (2021)
9. Hsu, W.N., Bolte, B., Tsai, Y.H.H., Lakhotia, K., Salakhutdinov, R., et al.: HuBERT: self-supervised speech representation learning by masked prediction of hidden units. arXiv preprint arXiv:2106.07447 (2021)
10. Kahn, J., Rivière, M., Zheng, W., Kharitonov, E., Xu, Q., et al.: Libri-light: a benchmark for ASR with limited or no supervision. In: ICASSP (2020)
11. Kingma, D.P., Ba, J.: Adam: a method for stochastic optimization. In: ICLR (2015)
12. Kong, Z., Ping, W., Huang, J., Zhao, K., Catanzaro, B.: DiffWave: a versatile diffusion model for audio synthesis. arXiv preprint arXiv:2009.09761 (2020)
13. Lugmayr, A., Danelljan, M., Romero, A., Yu, F., Timofte, R., Gool, L.V.: RePaint: inpainting using denoising diffusion probabilistic models. CoRR (2022)
14. Movellan, J.R., Mineiro, P.: A diffusion network approach to visual speech recognition. In: AVSP (1999)
15. Nichol, A.Q., Dhariwal, P.: Improved denoising diffusion probabilistic models. In: PMLR (2021)
16. Panayotov, V., Chen, G., Povey, D., Khudanpur, S.: Librispeech: an ASR corpus based on public domain audio books. In: ICASSP (2015)
17. Park, D.S., et al.: Improved noisy student training for automatic speech recognition. In: Interspeech (2020)
18. Ramachandran, P., Zoph, B., Le, Q.V.: Searching for activation functions. arXiv preprint arXiv:1710.05941 (2017)
19. Ramesh, A., Dhariwal, P., Nichol, A., Chu, C., Chen, M.: Hierarchical text-conditional image generation with CLIP latents. arXiv preprint arXiv:2204.06125 (2022)
20. Saharia, C., Chan, W., Saxena, S., Li, L., Whang, J., et al.: Photorealistic text-to-image diffusion models with deep language understanding. arXiv preprint arXiv:2205.11487 (2022)
21. Sohl-Dickstein, J., Weiss, E., Maheswaranathan, N., Ganguli, S.: Deep unsupervised learning using nonequilibrium thermodynamics. In: ICML (2015)
22. Vaswani, A., et al.: Attention is all you need. In: NeurIPS (2017)
23. Zhang, Y., et al.: Pushing the limits of semi-supervised learning for automatic speech recognition. arXiv preprint arXiv:2010.10504 (2020)

Fine-Tuned Self-supervised Speech Representations for Language Diarization in Multilingual Code-Switched Speech

Geoffrey Frost[1](✉)(iD), Emily Morris[2](iD), Joshua Jansen van Vüren[1](iD), and Thomas Niesler[1](iD)

[1] Department of E and E Engineering, Stellenbosch University, Stellenbosch, South Africa
{gfrost,jjvanvueren,trn}@sun.ac.za
[2] Cape Town, South Africa

Abstract. Annotating a multilingual code-switched corpus is a painstaking process requiring specialist linguistic expertise. This is partly due to the large number of language combinations that may appear within and across utterances, which might require several annotators with different linguistic expertise to consider an utterance sequentially. This is time-consuming and costly. It would be useful if the spoken languages in an utterance and the boundaries thereof were known before annotation commences, to allow segments to be assigned to the relevant language experts in parallel. To address this, we investigate the development of a continuous multilingual language diarizer using fine-tuned speech representations extracted from a large pre-trained self-supervised architecture (WavLM). We experiment with a code-switched corpus consisting of five South African languages (isiZulu, isiXhosa, Setswana, Sesotho and English) and show substantial diarization error rate improvements for language families, language groups, and individual languages over baseline systems.

Keywords: Language diarization · Code-switched speech · Low-resource · WavLM

1 Introduction

Prevalent in multilingual societies, code-switched (CS) speech is the phenomenon where two or more languages are used within the same conversation or utterance [25]. The development of automatic speech recognition (ASR) systems to model such speech typically require large amounts of data, which is often challenging to collect. Past attempts at compiling such a multilingual CS corpus have relied on a complex iterative process to transcribe audio recordings [30]. First, a principal transcriber segments a file into utterances and annotates their primary language. All unlabelled segments are passed to the next transcriber, who will in turn annotate their primary language. This iterative process is repeated until there are no more unlabelled segments. Prior knowledge of the specific language boundaries within a given utterance would allow this iterative process to be parallelised, saving time and reducing the human capital needed to orchestrate the previously described complex process. With the aim of extending a current multilingual CS corpus, we investigate the development of a simple end-to-end language diarization (LD) system to aid in this task.

There is limited recent literature that addresses LD for CS speech [16]. Moreover, speech representations extracted from large pre-trained self-supervised acoustic models have yet to be leveraged in this domain despite being shown to perform well on a variety of downstream acoustic tasks [32]. In this work, we investigate the application WavLM [5], a recent architecture of this kind that achieves state-of-the-art performance on a suite of down-stream language tasks, for LD [32]. We experiment with a corpus of low-resource multilingual code-switched soap opera speech, comprising English and four other low-resource South African languages (isiZulu, isiXhosa, Setswana, Sesotho). Although WavLM is pre-trained on monolingual English, we show that it transfers well to low-resource LD, and substantially improves upon previously proposed architectures for the same task.

2   Background

Broadly, both prelexical information (phonetic repertoire, phonotactics, rhythm and intonation) and lexical-semantic knowledge (meaning) is utilised when determining a spoken language [22,33], with inexperienced human listeners effectively able to identify languages relying only on the former, implying language identification (LID) can be performed with minimal content understanding [20,27]. Both phonotactic and acoustic features are effective representations when building systems to perform LID [19,34]. Typically, phonotactic systems utilise a bank of monolingual large vocabulary continuous speech recognition (LVCSR) systems run in parallel, one for each language. The one producing the highest log-likelihood for the recognised word sequence is selected as the language spoken [12,18,23]. This requires corpora with which to train each LVCSR system, which is not feasible for the low-resource languages often found in CS speech.

Unlike phonotactic approaches, acoustic systems attempt to learn a distribution across languages directly from acoustic features, and as such are more relevant to this work. Whilst early work relied on classical algorithms (GMMs, SVMs or LR) [2,15,21,26,31], more recently, attention has shifted to end-to-end architectures using deep neural networks [3,8–10,17,24,29]. However, these systems tend to solve a sequence-to-one problem, whereby a single embedding is extracted from a variable length utterance and hence assume only a single language is being spoken. This is insufficient for our task, for which spoken languages can change throughout a single utterance.

Whilst to the best of our knowledge there has been no work in LD for South African languages, recently one study has successfully demonstrated the use of end-to-end acoustic based systems for code-switched LD [16]. In this approach, popular deep-learning-based speaker diarization approaches were applied to LD, yielding promising results. Long short-term memory networks and transformer-based systems were trained and evaluated on a corpus comprising 52 hours of speech from three bilingual CS subcorpora (Gujarati-English, Tamil-English and Telugu-English). Systems trained for bilingual diarization and multilingual diarization performed well, with accuracies exceeding 80%.

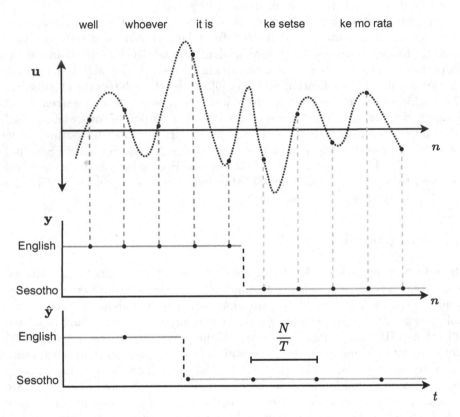

Fig. 1. A simple representation of a sampled code-switched utterance waveform **u** between English and Sesotho and the corresponding ground truth language labels **y**, and those predicted by a LD system **ŷ**.

2.1 Language Diarization

As it is fundamental to our work, we formally describe the task of LD, reinforced by a simple example. Given a sampled utterance waveform $\mathbf{u} = (u_n \in \mathbb{R}|1, ..., N)$ with language labels $\mathbf{y} = (y_n \in [C]|n = 1, ..., N)$, where N is the number of samples and C is the set of languages that are to be identified, let the function $G(\mathbf{u}; \theta)$, where θ are learnable parameters, be the function that estimates language labels $\hat{\mathbf{y}} = (\hat{y}_t \in [C]|1, ..., T)$ for non-overlapping segments of **u** of sample length $\frac{N}{T}$, where T is the number of segments. We illustrate this with simple CS example in Fig. 1. Note how the period of predicted label segments is larger than the ground truth.

3   Corpus

We perform our LD experimentation using a multilingual CS corpus compiled
from South African soap operas [30]. The corpus consists of 14.3 h of annotated
and segmented speech taken from 626 South African soap opera episodes, divided
into four language-balanced subcorpora. Each subcorpus contains monolingual
and CS utterances in English and one of four Bantu languages: isiZulu, isiXhosa,
Setswana and Sesotho. We will refer to these subcorpora as English-isiZulu (E-
Z), English-isiXhosa (E-X), English-Setswana (E-Se) and English-Sesotho (E-
So). The subcorpora are each split into training, development, and test sets.
The four Bantu languages in the corpus represent the two most widely spoken
South African Bantu language groups. Namely, the Nguni languages (IsiZulu and
isiXhosa), and the Sotho-Tswana languages (Setswana and Sesotho). Language
groups are collections of languages with similar linguistic roots and characteristics. It is worth noting the proportional spread of the utterances across the
subcorpora. Although each subcorpus is language-balanced, the E-Z subcorpus
contains roughly twice as much data as the other subcorpora. The exact break-
down in terms of subcorpora is presented in Table 1.

The corpus contains utterances with a mixture of intersentential and intrasentential code switches, with the former occurring between sentences and the latter
within sentences. Intrasentential code-switching can occur at a morpheme level
within words, with a word such as *amasponsors* being indicated as a switch from
isiZulu to English. The rapid nature of the code-switching combined with the
relatively fast pace of the soap opera speech requires high resolution LD which
is hard to achieve [30].

Even for soap opera speech, where language switches are frequent, many
utterances remain monolingual. In total, only 6.3 of the 14.3 h of speech correspond to utterances with code switches. Of this, only ≈4.5 h appear in the
training set. However, while CS utterances are relatively sparse in the training set, the corpus design has ensured that they form a larger portion of the
development set, whilst the test set is entirely comprised of CS speech.

Table 1. The amount of data (in minutes) in the training (train), development (dev)
and test sets of the four subcorpora.

Subcorpus	Train	Dev	Test	Total
E-Z	288.60	8.00	30.40	327.00
E-X	160.54	13.68	14.34	188.58
E-Se	139.74	13.83	17.83	171.60
E-So	141.72	12.77	15.54	169.80

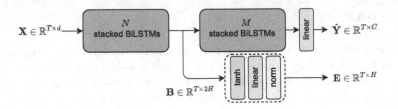

Fig. 2. Network diagram for the two-stage baseline BiLSTM LD system.

4   Models

We consider the application of three architectures for end-to-end LD. The first two approaches, which serve as our baselines, respectively utilise a two-stage bidirectional long short-term memory (BiLSTM) network and an x-vector transformer architecture as proposed in [16]. We also introduce the pre-trained self-supervised acoustic representation model considered in this work, WavLM.

4.1   BiLSTM

Initially proposed for speaker diarization [7], a two-stage BiLSTM architecture has been shown to perform well for LD [16]. A sequence of T acoustic feature vectors $\mathbf{X} = (\mathbf{x}_t \in \mathbb{R}^d | t = 1, ..., T)$ are extracted from an utterance. A set of N BiLSTM layers are used to generate language representations $\mathbf{B} = (\mathbf{b}_t \in \mathbb{R}^{2H} | t = 1, ..., T)$ from $\mathbf{X}$, followed by M BiLSTM layers to estimate the the sequence of language labels $\hat{\mathbf{Y}} = (\mathbf{y}_t \in \mathbb{R}^{|C|} | t = 1, ..., T)$ where C is the set of languages that are to be identified. In addition to computing the frame-wise cross-entropy (CE) loss between the ground truth labels $\mathbf{Y} = (y_t \in [C] | 1, ..., T)$ and $\hat{\mathbf{Y}}$, a deep-clustering (DC) loss [11] is used to encourage $\mathbf{B}$ to be language discriminative as shown in Eq. 1 where $\mathbf{e}_t \in \mathbb{R}^H$ is an emending of $\mathbf{b}_t$ and α is a regularisation parameter. A high-level depiction of this architecture is presented in Fig. 2.

$$\mathcal{L} = \alpha L_{CE}(y_t, \hat{\mathbf{y}}_t) + (1 - \alpha) L_{DC}(y_t, \mathbf{e}_t) \tag{1}$$

We use the same architectural hyper-parameters as previous applications of this architecture to LD of CS speech, with H, N, M, and α set to 256, 2, 3, and 0.5 respectively [16].

4.2   X-vector Self-Attention

X-vector Self-Attention (XSA) is another end-to-end architecture proposed for LD in [16]. First, x-vectors $\mathbf{E} = (\mathbf{e}_t \in \mathbb{R}^H | t = 1, ..., S)$ are extracted for non-overlapping segments of length s of acoustic feature representations $\mathbf{X}' = (\mathbf{x}_t \in \mathbb{R}^{s \times d} | t = 1, ..., S)$ for a given utterance. Note x-vectors simply refer to a fixed-sized neural embedding extracted for an arbitrary length of speech.

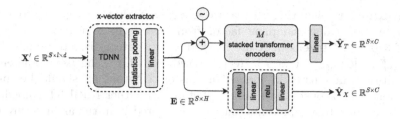

Fig. 3. Network diagram for the XSA LD system.

$\mathbf{E}$ is then sinusoidally positionally encoded, and passed through a series of M stacked self-attention (transformer encoder) modules, as illustrated in Fig. 3. The x-vector extractor's time-delayed neural network (TDNN) is made up of a series of temporal convolutional layers, the outputs of which are pooled over the time dimension (mean and variance) and linearly projected. The cross-entropy between $\mathbf{Y} = (y_t \in [C]|1, ..., S)$ and segment level language predictions made with the output of the transformer network $\hat{\mathbf{Y}}_T = (\hat{\mathbf{y}}_t^T \in \mathbb{R}^{|C|}|t = 1, ..., S)$, and the output of the x-vector extractor $\hat{\mathbf{Y}}_X = (\hat{\mathbf{y}}_t^X \in \mathbb{R}^{|C|}|t = 1, ..., S)$ are used in a multi-objective loss as shown in Eq. 2 for a single segment, where α is a regularisation parameter. Whilst $\hat{\mathbf{Y}}_X$ is not used to make actual predictions come test time, its inclusion encourages the x-vector extractor to learn segment-level language information.

$$\mathcal{L} = \alpha L_{CE}(y_t, \hat{\mathbf{y}}_t^T) + (1 - \alpha)L_{CE}(y_t, \hat{\mathbf{y}}_t^X) \tag{2}$$

We use the same architectural hyper-parameters as in previous applications of this architecture to LD of code-switched speech, with H, M, and α set to 256, 4, and 0.5 respectively [16].

4.3   WavLM

Large transformer-based acoustic language models have proven successful in capturing complex contextualised representations for a multitude of speech tasks [32]. WavLM [5] resembles other self-supervised speech representation frameworks such as wav2vec2.0 [1] and HuBERT [13], whereby a temporal convolutional feature extractor extracts audio representations $\mathbf{X} = (\mathbf{x}_t \in \mathbb{R}^H|t = 1, ..., T)$ directly from a 16 kHz waveform $\mathbf{u}$. These are subsequently fed into a large transformer encoder consisting of L blocks, where the last encoder outputs a sequence of contextual representations $\mathbf{C}^L = (\mathbf{c}_t^L \in \mathbb{R}^H|t = 1, ..., T)$ which are used to solve a masked learning objective. Whilst other frameworks have achieved great success in speech recognition, they disregard information important for other speech tasks such as paralinguistics, speaker identity and semantics. WavLM addresses these shortcomings by introducing a masked speech denoising and prediction HuBERT-like loss term. For a given utterance $\mathbf{u}$, noisy or overlapped speech is manually simulated by sampling noise or other utterances from the batch to produce $\mathbf{u}'$ which is then fed into the convolutional feature extractor generating the feature sequence $\mathbf{X}$ that corresponds to a 20ms framerate (i.e. a downsampling factor from waveform to feature representation of 320×). M tokens in $\mathbf{X}$ are masked, and a set of discrete pseudo-labels $\mathbf{z}^k = (z_t^k \in [C^k] | t = 1, ..., T)$ are predicted from $\mathbf{C}^L$. The pseudo-labels are the acoustic unit cluster in the set of clusters C^k that c_t^L should belong to, discovered in an acoustic unit discovery step with either HuBERT embeddings (first-stage), or latent representations $\mathbf{C}^{L'}$ extracted from the architecture itself (second-stage). Predictions are made for K acoustic unit sets, each with a different granularity (number of clusters) to facilitate the learning of different speech attributes. The distribution over the pseudo-labels p is parameterized in the same way as HuBERT, resulting in the below loss function being used for self-supervised learning.

$$\mathcal{L} = \sum_{k \in K} \sum_{t \in M} \log p^k(z_t^k | \mathbf{c}_t^L) \tag{3}$$

By keeping the original pseudo-labels for induced noisy/overlapped speech in an utterance, the network is forced to denoise the input, resulting in improved robustness for complex acoustic environments. Additionally, each transformer encoder consists of a convolution-based relative position embedding layer, which uses a relative position bias to allow the positional encoding to change based on the content of the input sequence [6].

WavLM is trained on an extremely large English corpus, which includes 94k hours from LibriLight [14], 10k hours from GigaSpeech [4] and 24k hours from VoxPopuli [28]. We use two of the released pre-trained models of varying size in our work, namely *WavLM-base+* ($H = 768$, $L = 12$) and *WavLM-large* ($H = 1024$, $L = 24$). For both, we use the contextual embeddings from the last transformer encoder layer $\mathbf{C}^L$ for language diarization by attaching a simple linear layer that maps each embedding $\mathbf{c}_t^L$ to a distribution over the possible C languages being spoken over time $\hat{\mathbf{Y}}$. As with the BiLSTM baseline, we compute the frame-wise CE loss between $\hat{\mathbf{Y}}$ and $\mathbf{Y}$. A high-level depiction of the network is presented in Fig. 4.

5   Experimental Procedure

To investigate the extent to which accurate diarization of code-switched speech can be achieved in the presence of multiple low-resource Bantu languages, each architecture is applied to three hierarchical diarization tasks:

1. **English/Bantu**: All four Bantu languages are grouped and the network determines whether the language spoken in a segment is English or belongs to the Bantu family.
2. **English/Nguni/Sotho-Tswana**: The Bantu languages are grouped according to their respective language groups and the network determines whether the language spoken in a segment is English, a Nguni language or a Sotho-Tswana language.

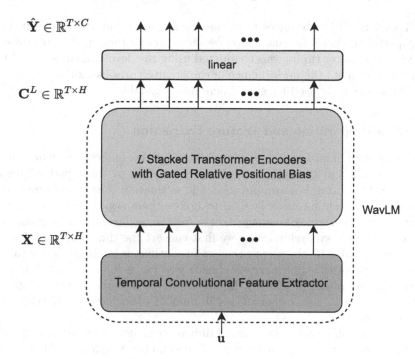

Fig. 4. Network diagram for the WavLM based LD system.

3. **English/isiZulu/isiXhosa/Setswana/Sesotho**: The network determines whether the language spoken in a segment is English, isiZulu, isiXhosa, Setswana or Sesotho.

Given the limited amount of training data and the substantial linguistic similarities between these languages, differentiating between various Bantu languages is a challenging task. In turn, the aforementioned three hierarchical tasks can be seen as increasing in complexity as we increase the number of Bantu language categories to be identified. Furthermore, while the monolingual utterances in the training data will still allow the architectures to learn underlying representations of the languages, the lack of CS utterances could affect their ability to correctly categorise segments in the presence of rapid language changes.

Both baseline systems are trained with the same configurations used in [16], with the exception that we increase the batch size to 64. When training the WavLM models, we use a learning rate of 1×10^{-4} and a weight decay of 1×10^{-4} with the AdamW optimizer. We use a batch size of 4 with 16 gradient accumulation steps, and train for 16 epochs. The learning rate is increased linearly for the first 1000 steps, followed by an exponential decay. We include label smoothing for all cross-entropy loss terms, set to 0.1. When training systems for tasks 2 and 3, we initialise model parameters with the corresponding architectures weights attained from training for the previous task, as it always resulted in improved performance during development. Additionally, for tasks 2 and 3 we remove all

monolingual English utterances from the training set in an attempt to increase
the proportional Bantu language representation across the set. All optimization
and hyper-parameter tuning was conducted using the development set whilst we
present and evaluate the performance of the architectures using the test set. We
make the source code used for experimentation available[1].

5.1   Data Preparation and Feature Extraction

Samples in each mini-batch are zero-padded to the longest sequence in the
respective batch and predictions made corresponding to these padded regions
are disregarded during loss computation and evaluation. For transformer-based
architectures, a padding mask is used to ignore these regions when computing
self-attention. To acquire language labels for the segment-level language predictions made by each architecture, we first convert the time-stamped language
boundaries provided by the corpus to a set of continuous language labels (a label
for each sample of the utterance waveform) $\mathbf{y} = (y_n \in [C]|n = 1, ..., N)$, where
N is the number of samples in the digital waveform. $\mathbf{y}$ is then down-sampled to
$\tilde{\mathbf{y}} = (\tilde{y}_t \in [C]|t = 1, ..., T)$, where T is the number of segment predictions made
by each architecture.

For both baseline systems we use 23-dimensional mel-spectrograms as acoustic feature vectors, with a frame length of 25 ms and hop length of 10 ms as in [16].
Since the corpus we use provides language boundaries for each utterance and not
labels for discrete segments, we do not have to further divide mel-spectrograms
into 200 ms (19 frames) segments as in [16] for the BiLSTM architecture. However, we do have to conduct this division for the XSA architecture as the x-vector
extractor is specifically designed to extract representations for such segments. In
this case, the corresponding language label for each segment is the frame-wise
language label that occurs the most often.

5.2   Evaluation Metrics

We use the error rate as our primary evaluation metric, which quantifies the proportion of incorrectly identified language segments as shown in Eq. 4. Although a
good representation of general system performance, the global error rate (GER),
which computes the proportion of incorrect predictions across the entire evaluation set, can potentially be dominated by longer utterances. To quantify error
rates on a per-utterance level, we also compute the mean error rate (MER) across
utterances.

$$ER = \frac{Incorrect\ Predictions}{Total\ Predictions} \tag{4}$$

---

[1] https://github.com/GeoffreyFrost/code-switched-language-diarization.

Table 2. Test set GER and MER (%) for each respective diarization task. *Task 1* denotes *English/Bantu* diarization, *Task 2* diarization denotes *English/Nguni/Sotho-Tswana*, and *Task 3* denotes diarization of all languages.

	#Params	Task 1		Task 2		Task 3	
		GER	MER	GER	MER	GER	MER
BiLSTM	9M	32.50	33.07	47.22	47.97	57.76	58.76
XSA	12M	36.50	38.17	48.90	51.73	59.70	62.82
WavLM-base+	95M	12.57	14.87	15.96	19.17	33.55	37.18
WavLM-large	317M	**10.05**	**11.94**	**12.93**	**16.12**	**32.80**	**36.76**

6 Results and Discussion

We present the error rates for the three diarization tasks using our chosen architectures in Table 2. The two variations of the WavLM model achieve substantial improvements over the baseline architectures across all three tasks. The WavLM-large architecture provides the best overall performance, with the lowest GER and MER achieved being for task 1 (10.05% and 11.94% respectively). The performance of all the architectures degrades with an increase in the granularity of the language categories which also results in a decrease in the number of training utterances per category. This is particularly prevalent in our two baseline models, with absolute GER increases of 14.72% from task 1 to task 2 and by 10.54% from task 2 to task 3 for the BiLSTM, and absolute increases of 12.4% from task 1 to task 2 and by 10.80% from task 2 to task 3 for the XSA architecture. These baseline architectures use randomly initialised weights and do not benefit from the same pre-training scheme as the WavLM architectures, making them more reliant on a larger amount of training data to achieve good results. However, the WavLM architectures do see a larger relative decrease in performance between task 2 and task 3. We also note that for all networks and tasks the GER is lower than the MER indicating that diarization error is dependent on the length of the utterance. This makes sense, as longer segments of speech have more contextual language information, which makes the language diarization task easier to perform.

To further analyse the performance of WavLM and investigate the potential cause of the large performance degradation from task 2 to 3, we present the confusion matrices for both with predictions generated by the WavLM-large architecture in Fig. 5. By analysing the confusion matrix for task 3 (Fig. 5b), its clear the degradation in performance is a result of incorrect language identification within the Nguni and Sotho-Tswana language groups. Albeit that WavLM's acoustic representations are comparatively effective for LD, these are grounded in English through its monolingual self-supervised pre-training scheme. Thus, the language structure that distinguishes languages within the same group, especially aspects that could not be learnt during English pre-training (e.g. syntax and phonology), are harder to learn.

Comparing the presented confusion matrices provides further insight into the influence of language groups on LD performance. Despite there being more training data available for the Nguni language group than for the Sotho-Tswana language group, by observing the confusion matrix presented in Fig. 5a it is clear that accuracies for the two classes are essentially the same for task 2. The confusion matrix for task 3 presented in Fig. 5b shows how the Nguni languages benefit from this additional data, with improved isiZulu and isiXhosa accuracy compared to Sesotho and Setswana. However, as already discussed, there is a large degree of confusion within both Bantu language groups. Clearly the distinct nuances that distinguish languages within the same group are substantially more difficult to discern compared to those that differentiate groups within the same family. In addition to previously described effects of WavLM's monolingual English pre-training, this behaviour is potentially exacerbated by the use of the weights attained from task 2 to initialise the model for task 3, although the subsequent training should have tuned the model to correctly differentiate between the two classes within the Bantu and Sotho-Tswana language groups.

WavLM-large performs particularly well on identifying isiZulu, potentially due to the language being over-represented during training compared to other Bantu languages. This is further reinforced when considering the amount of isiXhosa misidentified as isiZulu, noting that for this language there is roughly half the amount of training data. In contrast, the equal (and lower) representation of both Sesotho and Setswana during training results in similar (and higher) confusion between the two.

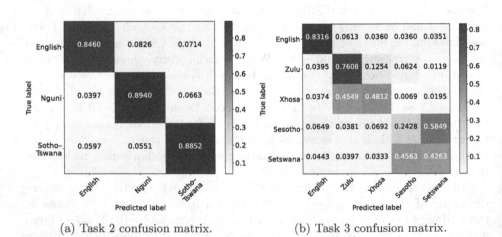

(a) Task 2 confusion matrix.   (b) Task 3 confusion matrix.

Fig. 5. Confusion matrices depicting the accuracy of the WavLM-large architecture on Task 2 (English/Nguni/Sotho-Tswana) and Task 3 (all languages).

7 Conclusion

In this work, we investigated the application of fine-tuned speech representations extracted from a large pre-trained self-supervised architecture (WavLM) for language diarization of code-switched speech. Through experimentation conducted with a code-switched corpus comprising five South African languages, we showed that utilising such an architecture can improve upon previously proposed systems for the same task. Despite being pre-trained on a monolingual corpus (English), WavLM was able to improve upon baseline systems when tasked with diarizing English/Bantu, English/Nguni/Sotho-Tswana and English/isiZulu/isiXhosa/Setswana/Sesotho coded-switched speech, reducing error rates by between 21.13% and 31.85% absolute compared to the best performing baseline system. Whilst individual language accuracies are too low to aid in fully-parallelized corpora annotation, surprisingly good performance was observed when performing language group diarization. Such performance may be sufficient to assign segments of utterances to language group streams (i.e. English, Nguni and Sotho-Tswana) reducing the number of language experts a segment may need to be sequentially reviewed by.

7.1 Limitations and Future Work

Due to computation and time constraints, this study was limited to exploring only one self-supervised architecture for LD. Additionally, disproportionate amounts of training data for each language may have influenced results. In future work, we aim to more rigorously investigate the application of self-surprised models to LD. This includes investigating if monolingual pre-training hinders the ability to reliably learn the discrete differences between languages within the same group by comparing the use of multilingual self-supervised models for the same task (such as wav2vec2-XLSR).

References

1. Baevski, A., Zhou, Y., Mohamed, A., Auli, M.: wav2vec 2.0: a framework for self-supervised learning of speech representations. In: Advances in Neural Information Processing Systems, vol. 33, pp. 12449–12460 (2020)
2. Brummer, N.: Measuring, refining and calibrating speaker and language information extracted from speech. Ph.D. thesis, University of Stellenbosch, Stellenbosch (2010)
3. Cai, W., Cai, Z., Liu, W., Wang, X., Li, M.: Insights in-to-end learning scheme for language identification. In: Proceedings of IEEE International Conference on Acoustics, Speech and Signal Processing (ICASSP), pp. 5209–5213 (2018)
4. Chen, G., et al.: Gigaspeech: an evolving, multi-domain ASR corpus with 10,000 hours of transcribed audio. In: Proceedings of Interspeech (2021)
5. Chen, S., et al.: WavLM: large-scale self-supervised pre-training for full stack speech processing. IEEE J. Sel. Topics Signal Process. **6**, 1505–1518 (2022)
6. Chi, Z., et al.: XLM-E: cross-lingual language model pre-training via electra. arXiv preprint arXiv:2106.16138 (2021)

7. Fujita, Y., Kanda, N., Horiguchi, S., Nagamatsu, K., Watanabe, S.: End-to-end neural speaker diarization with permutation-free objectives. In: Proceedings of Interspeech (2019)
8. Gelly, G., Gauvain, J.L.: Spoken language identification using LSTM-based angular proximity. In: Proceedings of Interspeech, pp. 2566–2570 (2017)
9. Geng, W., et al.: End-to-end language identification using attention-based recurrent neural networks. In: Proceedings of Interspeech, pp. 2944–2948 (2016)
10. Gonzalez-Dominguez, J., Lopez-Moreno, I., Moreno, P.J., Gonzalez-Rodriguez, J.: Frame-by-frame language identification in short utterances using deep neural networks. Neural Netw. **64**, 49–58 (2015)
11. Hershey, J.R., Chen, Z., Le Roux, J., Watanabe, S.: Deep clustering: discriminative embeddings for segmentation and separation. In: Proceedings of IEEE International Conference on Acoustics, Speech and Signal Processing (ICASSP), pp. 31–35. IEEE (2016)
12. Hieronymus, J.L., Kadambe, S.: Spoken language identification using large vocabulary speech recognition. In: Proceedings of Fourth International Conference on Spoken Language Processing (ICSLP), pp. 1780–1783 (1996)
13. Hsu, W.N., Bolte, B., Tsai, Y.H.H., Lakhotia, K., Salakhutdinov, R., Mohamed, A.: Hubert: self-supervised speech representation learning by masked prediction of hidden units. IEEE/ACM Trans. Audio Speech Lang. Process. **29**, 3451–3460 (2021)
14. Kahn, J., et al.: LIBRI-LIGHT: a benchmark for ASR with limited or no supervision. In: Proceedings of IEEE International Conference on Acoustics, Speech and Signal Processing (ICASSP), pp. 7669–7673. IEEE (2020)
15. Li, H., Ma, B., Lee, K.A.: Spoken language recognition: from fundamentals to practice. Proc. IEEE **101**(5), 1136–1159 (2013)
16. Liu, H., et al.: End-to-end language diarization for bilingual code-switching speech. In: Proceedings of Interspeech, pp. 1489–1493 (2021)
17. Lopez-Moreno, I., Gonzalez-Dominguez, J., Martinez, D., Plchot, O., Gonzalez-Rodriguez, J., Moreno, P.J.: On the use of deep feedforward neural networks for automatic language identification. Comput. Speech Lang. **40**, 46–59 (2016)
18. Mendoza, S., Gillick, L., Ito, Y., Lowe, S., Newman, M.: Automatic language identification using large vocabulary continuous speech recognition. In: Proceedings of IEEE International Conference on Acoustics, Speech, and Signal Processing (ICASSP), pp. 785–788 (1996)
19. Muthusamy, Y.K., Barnard, E., Cole, R.A.: Reviewing automatic language identification. IEEE Signal Process. Mag. **11**(4), 33–41 (1994)
20. Muthusamy, Y.K., Jain, N., Cole, R.A.: Perceptual benchmarks for automatic language identification. In: Proceedings of IEEE International Conference on Acoustics, Speech and Signal Processing (ICASSP), pp. I-333 (1994)
21. Nakagawa, S., Ueda, Y., Seino, T.: Speaker-independent, text-independent language identification by HMM. In: Proceedings of Second International Conference on Spoken Language Processing (1992)
22. Ramus, F., Mehler, J.: Language identification with suprasegmental cues: a study based on speech resynthesis. J. Acoust. Soc. Am. **105**(1), 512–521 (1999)
23. Schultz, T., Rogina, I., Waibel, A.: LVCSR-based language identification. In: Proceedings of IEEE International Conference on Acoustics, Speech, and Signal Processing (ICASSP), pp. 781–784 (1996)
24. Trong, T.N., Hautamäki, V., Lee, K.A.: Deep language: a comprehensive deep learning approach to end-to-end language recognition. In: Proceedings of Odyssey: The Speaker and Language Recognition Workshop, vol. 2016, pp. 109–116 (2016)

25. Van Dulm, O.: The grammar of English-Afrikaans code switching: a feature checking account. Ph.D. thesis, External Organizations (2007)
26. Van Leeuwen, D.A., Brummer, N.: Channel-dependent GMM and multi-class logistic regression models for language recognition. In: Proceedings of Odyssey: The Speaker and Language Recognition Workshop, pp. 1–8 (2006)
27. Van Leeuwen, D.A., De Boer, M., Orr, R.: A human benchmark for the NIST language recognition evaluation 2005. In: Proceedings of Odyssey: The Speaker and Language Recognition Workshop, p. 12 (2008)
28. Wang, C., et al.: VoxPopuli: a large-scale multilingual speech corpus for representation learning, semi-supervised learning and interpretation. arXiv preprint arXiv:2101.00390 (2021)
29. Watanabe, S., Hori, T., Hershey, J.R.: Language independent end-to-end architecture for joint language identification and speech recognition. In: Proceedings of IEEE Automatic Speech Recognition and Understanding Workshop (ASRU), pp. 265–271 (2017)
30. van der Westhuizen, E., Niesler, T.: A first South African corpus of multilingual code-switched soap opera speech. In: Proceedings of the Eleventh International Conference on Language Resources and Evaluation (LREC) (2018)
31. Yan, Y.: Development of an approach to language identification based on language-dependent phone recognition. Oregon Graduate Institute of Science and Technology (1995)
32. Yang, S.W., et al.: Superb: speech processing universal performance benchmark. In: Proceedings of Interspeech (2021)
33. Zhao, J., Shu, H., Zhang, L., Wang, X., Gong, Q., Li, P.: Cortical competition during language discrimination. Neuroimage 43(3), 624–633 (2008)
34. Zissman, M.A.: Comparison of four approaches to automatic language identification of telephone speech. IEEE Trans. Speech Audio Process. 4(1), 31 (1996)

Evaluating Automated and Hybrid Neural Disambiguation for African Historical Named Entities

Jarryd Dunn[✉][iD] and Hussein Suleman[iD]

University of Cape Town, Cape Town, South Africa
dnnjar001@myuct.ac.za, hussein@cs.uct.ac.za

Abstract. Documents detailing South African history contain ambiguous names. These may be due to people having the same name or the same person being referred to by different names. Thus, when searching for information about a particular person, the name used may affect the results. This problem may be alleviated by using a Named Entity Disambiguation (NED) system to disambiguate names by linking them to a knowledge base. Hence, a multilingual language model-based NED system was developed to disambiguate people's names within a historical South African context using documents from the 500 Year Archive (FHYA) written in English and isiZulu. The multilingual language model-based system improved on a probability-based baseline and achieved a micro F1-score of 0.726. However, the system performed worse on documents written in isiZulu compared to the English documents. Thus, the system was augmented with handcrafted rules, resulting in a small but significant improvement in F1-score.

Keywords: Natural language processing · Named entity disambiguation · Machine learning · South African languages · Transformers

1 Introduction

Historical documents often contain many references to entities such as people, places and organisations which are of particular interest to researchers. However, African historical documents frequently contain entities with ambiguous names. Names can be ambiguous as multiple entities may have the same name (homonyms), or the same entity may have multiple names (synonyms). Ambiguous people names can be particularly problematic within a historical context as the narrators of historical accounts may use different names for the same person. Thus, using a particular name biases the information associated with a person. This paper explores limiting this problem by disambiguating the names of people within a collection of historical documents from a South African context. For this paper, only people are considered as entities.

Names can be disambiguated using a Named Entity Disambiguation (NED) system that first detects mentions within a document, where a mention is a span of text that refers to an entity. The NED system then disambiguates the mentions by linking them with entities in a knowledge base, this process can be referred to as Named Entity Linking (NEL) [1].

There has been little previous work in performing either NEL or NED within a historical South African context. However, performing other Natural Language Processing (NLP) tasks have highlighted several difficulties associated with the construction of NED systems in a historical South African context. Firstly, it is common for both nouns and adjectives to be used as names [2,3]. Thus, models need to be able to use the context of a word to determine if it is an entity mention. Furthermore, names are also likely to come from various languages and cultures, making it more difficult for the names to be detected. Two systems were developed to explore the ability of language model-based NED systems to deal with the challenges posed by performing NED in a historical South African context. These systems follow a modular architecture with three main components: entity identification, candidate generation and entity selection (see Fig. 1). Entity identification typically uses a Named Entity Recognition (NER) system to identify named entities' mentions within the text. The second and third modules then performed entity linking to link the mentions found by the mention detection module to entries in a knowledge base. First, the candidate generation module generates a candidate list of entities to which a mention may refer. Finally, the entity selection module determines which entity is referred to by the mention. Entity selection is typically made by ranking the entities in the candidate list according to the likelihood that they are the entity mentioned in the text. To explore the use of neural techniques for NED in a historical South African context, the following research questions were investigated:

Fig. 1. Typical modular architecture used for labelling named entities within the text. Here the mention Shaka is linked with the correct entity.

How accurate are transformer-based language models for NED applied to the text from the 500 Hundred Year Archive (FHYA)? The FHYA consists of a collection of historical South African artefacts and their metadata records. Although transformer-based language models have been successfully used for many NLP tasks [4], they had not been evaluated for NED within a historical South African context. Since NED systems tend to be domain-specific [5], the language model-based NED systems were compared with a baseline system, which covered the simplest cases of entity disambiguation.

What are the effects of handcrafted features on the accuracy of the NED model? The more accurate the NED models are, the more valuable they will be to the

FHYA. Thus, handcrafted rules may be added to increase the performance of the NED model as they capture more information that the model can use for disambiguation.

2   Related Works

2.1   Historical NED

Historical documents pose several additional challenges to NER on contemporary documents. Firstly, the text tends to be noisier due to Optical Character Recognition (OCR) errors when digitising the original documents. Previous NED systems have used regular expressions [6], Levenshtein distance [7], or word recognition systems [8] to correct the errors before using the text.

The HIPE (Identifying Historical People, Places and other Entities) campaign run by CLEF resulted in several NER and NEL systems for historical English, French and German newspaper articles. The system developed by Boros et al. [7] (team L3I) produced the top performance for 50 out of the 52 scoreboards in the competition by using a BERT-based model for identifying entities and a bi-directional LSTM-based model to link the entities. While Labusch and Neudecker [9] used two separate BERT models for generating embeddings and evaluating candidate entities for entity mentions within the text. Their system performs similarly to the L3I model in terms of precision but had lower recall scores, which they attributed to an insufficient knowledge base.

Outside the HIPE campaign, Heino et al. [6] developed a NED system for documents from a knowledge graph based on the Second World War in Finland (WarSampo[1]). The knowledge graph contains resources similar to those found in the FHYA. Their system used handcrafted rules including normalising names to have the same surface form and filtering candidates based on their lifespans. Finally, they used a threshold to determine if a name should be linked to the top candidate or a NIL entity. This approach results in a high overall F1 score for the disambiguation of people's names.

2.2   Low-Resource NED

NED systems often require large amounts of data to be trained. For high-resource languages, such as English, this data is readily available. However, there are typically very few purpose-built datasets for NED in low-resource language (LRL)s. Zero-shot linking techniques require less in-domain data for training as they classify previously unseen classes at test time [10]. Logeswaran et al. [11] and Wu et al. [12] both created zero-shot NEL systems that only required a collection of entities and descriptions for the entities. Their models use two stages: candidate generation and entity selection.

The Logeswaran et al. [11] model performed candidate generation using an information retrieval approach where BM25 is used to measure the similarity

---

[1] https://www.ldf.fi/dataset/warsa.

between the mention and the candidate documents. However, Wu et al. [12] subsequently found that measuring the cosine similarity between the encodings generated using a bi-encoder increased the recall for the candidate list by 13%.

Both models took a similar approach to entity selection, by creating a joint representation of the mention and a candidate entity. Logeswaran et al. [11] used a BERT-based model, which took a concatenation of the mention and entity embeddings as input. The output from the [CLS] token produced by the model was used as the embedding for the mention-entity pair. The final score was produced by feeding the embedding into a single-layer feed-forward neural network. Wu et al. [12] take a similar approach using a linear layer on top of a cross encoder to produce the final score for a mention-entity pair.

2.3 South African NLP

Although there does not appear to have been any work done on NED within a South African context, some research has been done on related NLP tasks such as NER. Louis, De Waal, and Venter [2] produced a NER system using a Bayesian network using rules based on linguistic analysis and gazetteers. More recently, a corpus for ten of the eleven official South African languages was produced [13]. This corpus includes labelled data for NER, which was used by Eiselen [3] to train NER systems for each of the languages using a CRF-based model. These NER systems were fairly successful, with F1 scores between 0.64 and 0.77. Subsequently, Hanslo [14] achieved similar results on the same dataset using the multilingual language model XLM-RoBERTa (XLM-R). Their NER system produced f1-scores above 60% for all the languages except Sesotho.

3 Data Collection

A testbed consisting of a collection of annotated documents written in English and isiZulu and a knowledge base (KB) was produced to evaluate the NED systems. The documents for the testbed came from a snapshot of the FHYA project[2] taken on the 1st of April 2021. The FHYA project is a research tool that aims to provide access to artefacts from 500 years of South African history. These items include written texts in several languages, audio recordings and excavated artefacts, among others. There is also a metadata record for each item that contains additional information about the item. These metadata records are always in English. The NED testbed consisted of the text from the metadata and machine-readable versions of the documents. A annotated subset was created by using volunteers to crowd-source the annotations.

The second major component of the testbed is the knowledge base. The knowledge base effectively sets an upper limit for the recall of the NED system [9], since only entities within the knowledge base can be linked. Thus, since existing knowledge bases had insufficient coverage of the entities from the FHYA,

---

[2] https://fhya.org/.

a new knowledge base was constructed as part of the annotation process. The knowledge base took the form of a database with a single entry for each entity within the text made up of a unique ID and a brief description of the entity. The entities and their descriptions were collected from either an authoritative list maintained by the FHYA or by volunteers during the annotation process.

3.1 Document Selection

Based on the recommendations from members of the FHYA, PDF documents were taken from the uMgungundlovu and FHYA Depot collections and converted to plain text using Tesseract[3] for English documents and the CTexT isiZulu OCR engine[4] for isiZulu documents. However, several documents were removed as the extracted text was unreadable. The extracted text was then cleaned using basic rules to correct common OCR errors. Finally, offensive or derogatory terms were censored by replacing all but the first and the last letters with an asterisk.

A subset of the unlabelled documents containing approximately a quarter of a million words – the annotation dataset – was annotated with the entities to which the text referred. Since there was considerably less text in the metadata records than in the documents written in English or isiZulu, all the metadata records were used in the annotation dataset. The rest of the dataset consisted of passages selected using random sampling from the English and isiZulu documents, such that there were approximately the same number of words from English and isiZulu documents.

3.2 Document Annotation

A custom built application (the Annotator[5]) was used to crowdsource annotations for the document collection. The Annotator enabled entity labels to be created using a custom database and ensured that the system would generate all the required data. The Annotator allowed users to identify names from within passages of text from the FHYA documents, and label each name with the identifier of the person to whom the name referred. Each document was annotated by three independent users to improve the quality of the labels. The final labels were then determined based on the majority label assigned to each word.

The Annotator was open for approximately two and a half months. However, some of the documents had not been annotated at the end of this period. Thus, the final labelled dataset[6] exported from the Annotator consisted of 675 document segments out of the original 858 document segments from the annotation dataset. All of the metadata segments were annotated and the rest of the documents were fairly evenly split with 97136 (48.7%) and 92434 (46.4%) words from the English and isiZulu documents respectively. Thus, documents

---

[3] https://opensource.google/projects/tesseract.
[4] https://hlt.nwu.ac.za.
[5] https://gitlab.com/jwdunn/mastersannotator.
[6] https://doi.org/10.25375/uct.19029692.

were not removed to balance the number of words since this would have reduced the amount of training data.

3.3   Fold Creation

The Automatic, Hybrid and Baseline systems were evaluated using 10-fold cross-validation using the same folds for each experiment. The document segments and labels exported from the Annotation system were split into ten folds containing similar number of mentions. Additionally, since it is unrealistic that all the mentions will correspond to an entity in the knowledge base, some entities in each fold were hidden for both training and evaluation. For each fold, entities appearing in the fold were randomly selected and hidden using a uniform distribution until the entities corresponding to 10% of the mentions were hidden.

4   Baseline

It is difficult to compare the performances of NED systems that were evaluated on different datasets as the nature of the mentions in the dataset has a significant effect on the system's performance. For instance, if a dataset has few ambiguous mentions then systems will typically be more accurate. Thus, following the proposal of Ilievski, Vossen, and Schlobach [15], a probability-based baseline was produced to benchmark the NED systems.

4.1   Architecture

The baseline NED system linked the entity to the mention it most frequently referred to in the labelled dataset created by human annotators, as shown in Fig. 2. Thus, a particular mention will always be linked to the same entity regardless of its context, meaning that it can only deal with the simplest form of NED - common and unambiguous mentions. A comparison

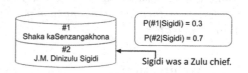

Fig. 2. Overview of the baseline system. In this case, the baseline system assigns the mention Sigidi to the wrong entity.

with the baseline should indicate how well a NED system can disambiguate more complex cases, while the results from the baseline indicate how much ambiguity exists in the dataset. The baseline system consists of two modules: mention detection and entity selection.

4.2   Mention Detection

Mention detection is performed based on string similarity using a list of names. The list of names is produced based on the entities labelled in the training data. Initially, all the mentions from the training data are added to the names list. The list is then expanded by splitting each name based on white space and adding the resulting parts to the list. Thus, if the names list was ["James Stuart"] it would become ["James Stuart", "James", "Stuart"]. Finally, each of the names is converted to lowercase and duplicates are removed. The similarity between each token and name is calculated using the Levenshtein ratio[7].

Mentions are identified by finding the longest span that matches one of the names in the names list. A span is considered to match a name if the similarity is greater than 0.9. Once the mentions have been identified in all documents, they are passed on to the entity selection module.

4.3   Entity Selection

Entity selection is performed using a simple lookup table with entries for each mention, and its corresponding entity. During training, the baseline system is given access to the labels for the training portion of the labelled dataset. These labels are used to determine the conditional probability of an entity being referred to by a particular mention. This is calculated as $\frac{m \cap e}{m}$. Where $m \cap e$ is the number of times the mention m refers to entity e and m is the number of times the mention appears. The final table is then constructed by pairing each mention with the entity ID to which it was most likely to refer.

5   Automatic NED System

Language models have been successfully used in a variety of NLP tasks [4] including NED [7]. However, much of this research is done on high-resource languages. Thus, this experiment seeks to answer the first research question – *"How accurate are transformer-based language models for NED applied to text from the 500 Year Archive?"*.

5.1   Architecture

The Automatic NED system makes use of three modules: mention detection, candidate generation and entity selection. Each of these modules makes use a language-model which is trained independently of the other models used in the other modules.

Mention Detection. A single XLM-R based NER model (CoNLL-Zu) was used for mention detection for all documents. The NER model was first trained for token classification on the CoNLL03 training dataset [16]. The CoNLL03 dataset contains data from Reuters news articles with labels for the entities [16]. The model was then further trained on the government domain isiZulu dataset to increase its performance on South African names. Finally, the model was fine-tuned on the training data from the FHYA collection. The CoNLL-Zu model

---

[7] https://pypi.org/project/python-Levenshtein/.

was fine-tuned, rather than being trained from scratch, so it could benefit from general concepts learnt in the larger isiZulu and English corpora while picking up patterns specific to the documents in the FHYA collection.

During training, the documents had to be split into chunks since XLM-R models accept a maximum of 512 tokens as input. The chunks contained 128 words since a word-piece tokenizer is used and thus one word can be split into multiple tokens. Each chunk had an overlap of 10 words to the left and right to ensure there was a context window of at least ten words for each token. A word-level context window was used rather than a token-level context window as the tokeniser was only trained on English data; thus, it is unclear how closely the isiZulu tokens align with semantically meaningful linguistic tokens for isiZulu texts.

Candidate Generation. Once the mentions had been identified, they were passed to the candidate generation module along with the context within which they appeared. The context consisted of the text 10 words to either side of the mention. Following Wu et al. [12], the Automatic model used a bi-encoder architecture, implemented using SBERT [17] and the multilingual language model XLM-R [18], to compare the similarity between the mention and entities in the knowledge base. SBERT was used as it has been shown to produce more semantically meaningful outputs [17] than using BERT's CLS embedding or an aggregate of the token embeddings in the final layer of the language model. Thus, the similarity between the embeddings generated by SBERT should correspond more closely to the similarity between the actual texts. XLM-R is a multilingual language model so should be better able to deal with the mix of isiZulu and English documents in the FHYA.

The bi-encoder generates independent embeddings for mentions and entities. Mention embeddings are produced using a mention and its context separated by a special [ENT] token: *mention* [ENT] *context*. Similarly, entity embeddings were generated for each entity in the knowledge base using the entity description as the context. The similarity between a mention and entity is calculated using cosine similarity between their embeddings.

Since the entity embeddings are static they were pre-computed. Then when a mention needs to be linked a k-nearest neighbours (KNN) search is used used to find the 30 most similar entities to a mention based on their cosine similarity. The list is then filtered to remove candidates where the cosine similarity was less than 0.5 or until only the top six candidates remain. A value of 0.5 was chosen to ensure a high recall whilst reducing the number of candidates; based on earlier tests, correct candidates usually have high cosine similarity scores (> 0.9). However, at least the top six candidates were always passed on to the entity selection module to reduce the likelihood of the system incorrectly assigning a NIL entity. Since the cross-encoders, used in the entity selection module, tend to be more accurate than bi-encoders [12], used for candidate generation.

Entity Selection. Finally, the entity selection module used a cross-encoder to compare the entities from the candidate list with the mention. A cross-encoder takes two pieces of text as input and then tries to determine the extent to which they are similar. The cross-encoder was implemented using SBERT with the language model XLM-R. Since the cross-encoder cannot pre-compute embeddings it is slower than the bi-encoder [12]. Thus, the cross-encoder is only used for entity selection which requires far fewer comparisons since it only needs to compare each mention to the entities in the candiate list. The mention text is labelled with the ID from the knowledge base for the entity with which it has the highest similarity. The system dealt with out-of-KB entities using a threshold-based approach; if the similarity between a mention and the most similar entity was below the 0.03, a new entity was created and added to the knowledge base. The threshold was selected based on the results of training the cross-encoder.

5.2   Training

The Automatic NED system was evaluated using 10-fold cross-validation. Separate models were trained for each fold for each of the modules comprising the Automatic system to ensure that the test data used to evaluate the model had not been seen by any part of the system beforehand. The hyper-parameters were tuned separately for each module across all folds based on the training objective for the module. This meant that the evaluation set could have some affect on the module development but this should be negligible.

Training the Candidate Generation and Entity Selection. The bi-encoder and cross-encoder were trained using triples, which contained two mentions in context and an integer indicating if they referred to the same person. Triples are positive if both mention-context pairs refer to the same person; otherwise, they are classed as negative. The triples were created by selecting an entity from the knowledge base using a uniform random distribution. For each positive triple generated, a hard negative triple was created with a probability of 0.4, and an equal probability of the hard negative being created for the first or second mention in the triple. The hard negative examples were generated by finding mentions that refer to another entity but have a similar surface form. Incorporating hard negatives should prevent the models from only using the similarity between the entity name and mention.

6   Results

6.1   Evaluation

The precision, recall, F1-score and accuracy were calculated for each fold and then aggregated to produce the final score for each metric. The accuracy was calculated as the number of mentions detected by the system where the predicted entity and label were the same, divided by the total number of mentions

Table 1. A comparison of micro and macro precision, recall and F1 scores between the Automatic system (AT) and the baseline system (BL). MD indicates the system used for mention detection while EL indicates the system used for entity linking (candidate generation and entity selection)

Model	MD	EL	Macro-P	Macro-R	Macro-F1	Micro-P	Micro-R	Micro-F1
BL	Baseline	Baseline	0.316	0.264	0.287	0.410	0.307	0.350
BL-N	Automatic	Baseline	0.655	0.652	0.658	0.678	0.668	0.672
AT	Automatic	Automatic	**0.744**	**0.732**	**0.737**	**0.731**	**0.722**	**0.726**
BL-L	Label	Baseline	0.695	0.726	0.709	0.702	0.758	0.729
AT-L	Label	Automatic	**0.797**	**0.823**	**0.810**	**0.773**	**0.821**	**0.796**

detected. Thus, the accuracy did not consider mentions that the system failed to detect. This accuracy was used as it gives an easily interpretable metric for the performance of the candidate generation and entity selection modules.

6.2   Comparison with the Baseline

The Automatic NED system was compared with the baseline system (see Table 1). BL-N was also included as it used the same Mention Detection module as the Automatic NED system. Thus, any differences between the two models are due to the Entity Linking modules (Candidate Generation and Entity Selection). The Automatic NED system performs substantially better than the baseline model and at a similar level to BL-L. This indicates that the Automatic system can utilise the context around mentions.

6.3   Performance by Document Type

The document collection could be split into three broad categories: English, isiZulu, and metadata. In general, the metadata documents are shorter and more structured than the English or isiZulu documents as they follow a set template. Entities in the shorter more structured metadata seem to be easier to disambiguate as the F1-Score was highest for the metadata documents (0.88). The system still performed well on English documents, achieving an F1-score of 0.83. However, it performed significantly worse on the isiZulu documents with an F1-score of 0.74. The accuracy of the labels assigned to the detected mentions shows the same trends as the F1 scores, with the mentions found in metadata documents having the highest accuracy, followed by English documents and then isiZulu documents (see Fig. 3). It is interesting to note the differences in the performance of the candidate generation and entity selection systems for documents from the different categories (see Fig. 3). If the top candidate from the candidate generation module was selected as the final entity for the English and metadata documents, it would have performed close to or slightly better than the Entity Selection module. However, for the isiZulu documents, the candidate predicted by the Candidate Generation module was the correct candidate less

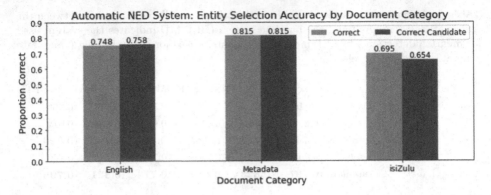

Fig. 3. Accuracy of entity labels assigned to mentions detected by the Automatic NED system split by language. *Correct* shows the proportion of mentions where the correct label was predicted while *Correct Candidate* shows the proportion of labels where the top candidate was the same as the label for the mention.

often than the entity selected by the Entity Selection module. This shows that the bi-encoder architecture is not as good at determining the similarity between mentions as the cross-encoder for isiZulu texts but has similar performances for English texts. This may be because the bi-encoder is used to produce embeddings that can be compared using cosine similarity, however, the cross-encoder acts as a classifier for a pair of mention-context pairs. Thus, the cross-encoder has access to each token from both mention-context pairs, potentially enabling better performances, with restricted access to training data.

7   Hybrid NED

Although the Automatic NED system performed reasonably well, the micro and macro F1-scores were lower for the isiZulu documents compared to the English documents. Previous research, such as by Hedderich et al. [19] and Wu, Liu, and Cohn [20], has shown that the inclusion of additional resources such as handcrafted rules can improve the performance of Named Entity Recognition (NER) systems. Thus, the Automatic NED system was augmented with handcrafted rules to form the Hybrid NED system, which was evaluated to answer the second research question – *"What are the effects of handcrafted features on the accuracy of the NED model?"*.

7.1   Mention Detection

The Mention Detection module performed reasonably well on all three document types. The largest difference was between the precision for the isiZulu and metadata documents, with the precision for the isiZulu documents 0.156 points lower. This indicates that the Mention Detection module identified spans of text that did not refer to a person as being a mention most often in the isiZulu documents. An analysis of the missed mentions showed that approximately 51% of the names that the system failed to detect had prefixes appearing before the name (e.g. uShaka). It is not entirely surprising that the system struggled to deal with the prefixes, since most of the data used during the training of the models did not attach prefixes to people's names.

In consultation with experts from the FHYA, the prefixes were analysed to determine which could be used as good indicators of whether a token was a person's name or not. Some prefixes such as u- are almost always used before a person's name. However, prefixes such as kwa- do not always precede a person's name. The prefixes were used to adjust the confidence scores from the NER model used by the mention detection modules. The tokens were then relabelled using the adjusted scores. Three methods were attempted to adjust the confidence scores from the NER model: handcrafted rules, a lookup table, and a combination of the handcrafted rules and lookup table.

All three methods were used to modify the probability the Mention Detection module assigned to the predicted label. For handcrafted rules, the adjustments were made by adding or subtracting an amount determined based on consultations with the experts from the FHYA. However, there may not be a handcrafted rule for all prefixes. Thus, the lookup table approach used a table of probabilities for the characters following a prefix (calculated based on the training data). At run time, if the token had a prefix and the prefix was in the lookup table, the score was calculated by interpolating the probability (P_M) predicted by the mention detection NED model and the probability from the lookup table (P_T). The new score is calculated as $\lambda P_M + (1 - \lambda)P_T$ where $\lambda \in [0,1]$ is a configurable parameter. The models were evaluated with $\lambda = 0.5$ and λ^* - calculated by optimising the F1-score for the Mention Detection module.

7.2   Entity Linking

The results from the cross-validation of the Automatic system were used to produce a list of mentions which were incorrectly linked. This showed that the system seemed to struggle when prefixes were added to a name -for example, predicting Zaka instead of zikaDingiwe. Thus, after consultation with experts from the FHYA, a set of simple linguistic rules was created for common prefixes indicating where they could be removed without changing whom the mention is referring to. For instance, kaSenzangakhona refers to a descendant of Senzangakhona and not Senzangakhona so the ka- prefix could not be removed. These rules were then applied to each mention before it was concatenated with its context used by the bi-encoder or cross-encoder. The prefix-removal rule was applied in both the candidate generation and entity selection modules. Applying the rule to only the entity selection module, the F1-score decreases thus the final models evaluated used either applied the rule to the candidate generation module or both the candidate generation and entity selection modules.

Table 2. A comparison of micro and macro precision, recall and F1-scores between the Automatic System. The CG (Candidate Generation) and ES (Entity Selection) columns indicate where hybrid approaches were used. λ^* is the value for λ determined by maximising the F1-score for the training data. †: The F1-score is significant at $p < 0.05$ compared to the Automatic system.

Mention Detection	λ	CG	ES	Macro-P	Macro-R	Macro-F1	Micro-P	Micro-R	Micro-F1
Lookup	0.5	✓		0.749	**0.734**	**0.741**†	0.743	0.727	0.734†
	0.5	✓	✓	0.747	0.733	0.739	0.744	**0.728**	0.735
Handcrafted + Lookup	λ^*	✓		**0.752**	0.731	0.741†	0.748	0.722	0.734†
	λ^*	✓	✓	0.751	0.729	0.739	**0.749**	0.724	**0.735**†
None		✓		**0.749**	**0.737**	**0.742**†	0.740	0.731	0.735†
		✓	✓	0.747	0.736	0.741†	**0.741**	**0.732**	**0.736**†
Automatic NED				0.744	0.732	0.737	0.731	0.722	0.726

7.3 Evaluation

The hybrid NED systems, which incorporated the handcrafted rules, were trained and evaluated using 10-fold cross-validation on the folds created in Sect. 3.3. The performance of the hybrid systems was evaluated using the same techniques and metrics as in the Automatic NED system. Additionally, a one-sided Wilcoxon signed-rank test was used to calculate the significance of the F1-scores for the hybrid NED systems compared to the Automatic NED system.

7.4 Comparison with Automatic NED System

The results presented in Table 2 show that all but one of the hybrid NED systems produced a significant increase in micro F1-score and only two systems failed to produce a significant increase in macro F1-score. The most significant increases were achieved by only adding rules to the entity linking module. The systems that applied rules to the mention detection module had significant increases in precision, however, these were countered by decreases in the recall, leading to marginal increases in the F1-score. This pattern held for both the micro and macro metrics.

8 Conclusion

An evaluation of the Automatic system compared to the baseline answered the first research question – *How accurate are transformer-based language models for NED applied to the text from the 500 Hundred Year Archive?* The improvement in the entity linking modules of the Automatic NED system compared to the baseline shows that the system could utilise the context within which the mentions appeared when linking them to an entity. These results indicate that transformer-based language models are effective for performing NED on documents from the FHYA. As expected, the performances were better when the documents were written in English (considered a high-resource language), compared to isiZulu. Thus, the Hybrid system was developed to explore the second research question – *What are the effects of handcrafted features on the accuracy of the NED model?* The evaluation of the Hybrid NED system indicates that hybrid rules can be used to improve the accuracy of NED models. However, it seems likely that more complex rules targeting specific cases will likely lead to larger improvements in the NED system whereas the relatively general rules applied to the hybrid models resulted in significant but small improvements.

References

1. Samuel, B.: Investigating entity knowledge in BERT with simple neural end-to-end entity linking. In: Proceedings of the 23rd Conference on Computational Natural Language Learning (CoNLL). Hong Kong, China: Association for Computational Linguistics, pp. 677–685 (2019). https://doi.org/10.18653/v1/K19-1063, https://www.aclweb.org/anthology/K19-1063
2. Louis, A., De Waal, A., Venter, C.: Named entity recognition in a South African context. In: Proceedings of the 2006 Annual Research Conference of the South African Institute of Computer Scientists and Information Technologists on IT Research in Developing Countries. SAICSIT 2006. Somerset West, South Africa: South African Institute for Computer Scientists and Information Technologists, pp. 170–179 (2006). isbn: 1595935673. https://doi.org/10.1145/1216262.1216281
3. Roald, E.: Government domain named entity recognition for South African languages. In: Proceedings of the Tenth International Conference on Language Resources and Evaluation (LREC 2016). Portorož, Slovenia: European Language Resources Association (ELRA), May 2016, pp. 3344–3348. https://aclanthology.org/L16-1533
4. Devlin, J., et al.: BERT: pre-training of deep bidirectional transformers for language understanding. In: Proceedings of the 2019 Conference of the North American Chapter of the Association for Computational Linguistics: Human Language Technologies, Volume 1 (Long and Short Papers). Minneapolis, Minnesota: Association for Computational Linguistics, June 2019, pp. 4171–4186. https://doi.org/10.18653/v1/N19-1423, https://www.aclweb.org/anthology/N19-1423
5. Olieman, A., et al.: Good applications for crummy entity linkers? the case of corpus selection in digital humanities. In: Proceedings of the 13th International Conference on Semantic Systems. Semantics 2017. Amsterdam, Netherlands: Association for Computing Machinery, pp. 81–88 (2017). isbn: 9781450352963. https://doi.org/10.1145/3132218.3132237
6. Heino, E., et al.: Named entity linking in a complex domain: case second world war history. In: Gracia, J., Bond, F., McCrae, J.P., Buitelaar, P., Chiarcos, C., Hellmann, S. (eds.) LDK 2017. LNCS (LNAI), vol. 10318, pp. 120–133. Springer, Cham (2017). https://doi.org/10.1007/978-3-319-59888-8_10
7. Boros, E., et al.: Robust named entity recognition and linking on historical multilingual documents. In: Conference and Labs of the Evaluation Forum (CLEF 2020), vol. 2696. Working Notes of CLEF 2020 - Conference and Labs of the Evaluation Forum. Thessaloniki, Greece: CEUR-WS Working Notes, pp. 1–17 (2020). https://hal.archives-ouvertes.fr/hal-03026969

8. Pruthi, D., Dhingra, B., Lipton, Z.C.: Combating adversarial misspellings with robust word recognition. In: Proceedings of the 57th Annual Meeting of the Association for Computational Linguistics. Florence, Italy: Association for Computational Linguistics, July 2019, pp. 5582–5591. https://doi.org/10.18653/v1/P19-1561, https://www.aclweb.org/anthology/P19-1561

9. Labusch, K., Neudecker, C.: Named entity disambiguation and linking historic newspaper OCR with BERT. In: Cappellato, L., et al. (eds.) Working Notes of CLEF 2020 - Conference and Labs of the Evaluation Forum, Thessaloniki, Greece, 22–25 September 2020, vol. 2696. CEURWorkshop Proceedings. CEUR-WS.org (2020). http://ceur-ws.org/Vol-2696/paper%5C_163.pdf

10. Vyas, Y., Ballesteros, M.: Linking entities to unseen knowledge bases with arbitrary schemas (2020). arXiv: 2010.11333 [cs.CL]

11. Logeswaran, L., et al.: Zero-shot entity linking by reading entity descriptions. In: Proceedings of the 57th Annual Meeting of the Association for Computational Linguistics. Florence, Italy: Association for Computational Linguistics, July 2019, pp. 3449–3460. https://doi.org/10.18653/v1/P19-1335, https://aclanthology.org/P19-1335

12. Wu, L., et al.: Scalable zero-shot entity linking with dense entity retrieval. In: Proceedings of the 2020 Conference on Empirical Methods in Natural Language Processing (EMNLP). Association for Computational Linguistics, pp. 6397–6407 (2020). https://doi.org/10.18653/v1/2020.emnlp-main.519, https://www.aclweb.org/anthology/2020.emnlp-main.519

13. Eiselen, R., Puttkammer, M.: Developing text resources for ten South African languages. In: Proceedings of the Ninth International Conference on Language Resources and Evaluation (LREC 2014). Reykjavik, Iceland: European Language Resources Association (ELRA), May 2014, pp. 3698–3703. http://www.lrec-conf.org/proceedings/lrec2014/pdf/1151_Paper.pdf

14. Hanslo, R.: Evaluation of neural network transformer models for named-entity recognition on low-resourced languages. In: 2021 16th Conference on Computer Science and Intelligence Systems (FedCSIS), pp. 115–119 (2021). https://doi.org/10.15439/2021F7

15. Ilievski, F., Vossen, P., Schlobach, S.: Systematic study of long tail phenomena in entity linking. In: Proceedings of the 27th International Conference on Computational Linguistics. Santa Fe, New Mexico, USA: Association for Computational Linguistics, pp. 664–674 (2018). https://www.aclweb.org/anthology/C18-1056

16. Sang, E.F., De Meulder, F.: Introduction to the CoNLL-2003 shared task: language-independent named entity recognition. In: Proceedings of the Seventh Conference on Natural Language Learning at HLT-NAACL 2003, pp. 142–147 (2003). https://www.aclweb.org/anthology/W03-0419

17. Reimers, N., Gurevych, I.: Sentence-BERT: sentence embeddings using Siamese BERT-networks. In: Proceedings of the 2019 Conference on Empirical Methods in Natural Language Processing and the 9th International Joint Conference on Natural Language Processing (EMNLP-IJCNLP). Hong Kong, China: Association for Computational Linguistics, pp. 3982–3992 (2019). https://doi.org/10.18653/v1/D19-1410, https://aclanthology.org/D19-1410

18. Conneau, A., et al.: Unsupervised cross-lingual representation learning at scale (2020). arXiv: 1911.02116 [cs.CL]

19. Hedderich, M.A., et al.: Transfer learning and distant supervision for multilingual transformer models: a study on African languages. In: Proceedings of the 2020 Conference on Empirical Methods in Natural Language Processing (EMNLP). Association for Computational Linguistics, pp. 2580–2591 (2020). https://doi.org/10.18653/v1/2020.emnlp-main.204, https://www.aclweb.org/anthology/2020.emnlp-main.204

20. Wu, M., Liu, F., Cohn, T.: Evaluating the utility of hand-crafted features in sequence labelling. In: Proceedings of the 2018 Conference on Empirical Methods in Natural Language Processing. Brussels, Belgium: Association for Computational Linguistics, Nov. 2018, pp. 2850–2856. https://doi.org/10.18653/v1/D18-1310, https://www.aclweb.org/anthology/D18-1310

Neural Speech Processing for Whale Call Detection

Edrich Fourie[1,3]($\boxtimes$) (iD), Marelie H. Davel[1,3,4] (iD), and Jaco Versfeld[2] (iD)

[1] Faculty of Engineering, North-West University, Potchefstroom, South Africa
edichfourie1@gmail.com
[2] Department of Electrical and Electronic Engineering, Stellenbosch University, Stellenbosch, South Africa
[3] Centre for Artificial Intelligence Research (CAIR), Cape Town, South Africa
[4] National Institute for Theoretical and Computational Sciences (NITheCS), Stellenbosch, South Africa

Abstract. Passive acoustic monitoring with hydrophones makes it possible to detect the presence of marine animals over large areas. For monitoring to be cost-effective, this process should be fully automated. We explore a new approach to detecting whale calls, using an end-to-end neural architecture and traditional speech features. We compare the results of the new approach with a convolutional neural network (CNN) applied to spectrograms, currently the standard approach to whale call detection. Experiments are conducted using the "Acoustic trends for the blue and fin whale library" from the Australian Antarctic Data Centre (AADC). We experiment with different types of speech features (mel frequency cepstral coefficients and filter banks) and different ways of framing the task. We demonstrate that a time delay neural network is a viable solution for whale call detection, with the additional benefit that spectrogram tuning – required to obtain high-quality spectrograms in challenging acoustic conditions – is no longer necessary. While the initial speech feature-based system (accuracy 96%) did not outperform the CNN (accuracy 98%) when trained on exactly the same dataset, it presents a viable approach to explore further.

Keywords: Convolutional neural network · Time delay neural network · Speech features · Whale call detection · Australian Antarctic Data Centre

1 Introduction

As human marine activities such as fishing and shipping have increased over the years, marine mammals' existence is increasingly being threatened [1]. Because animals are vocal and acoustics propagate well in water, passive acoustic monitoring of their movement can provide valuable information on marine mammal distribution [2].

Passive acoustic methods (PAM) use underwater microphones, also called hydrophones, to listen to and record signals of marine mammals for detection,

© The Author(s), under exclusive license to Springer Nature Switzerland AG 2022
A. Pillay et al. (Eds.): SACAIR 2022, CCIS 1734, pp. 276–290, 2022.
https://doi.org/10.1007/978-3-031-22321-1_19

classification and localisation. PAM performs better than visual observation as a monitoring method: it may be used at night, in bad weather, or under other conditions when visual observation is difficult [3].

Acoustic data collection and storage costs have decreased dramatically over the past decade. Nowadays, terabytes of data are collected in a single project [4]. With an increasing volume of acoustic data, extracting meaningful ecological information becomes more and more expensive [4]. Machine learning makes it possible to identify signals in large datasets at a relatively low cost and more consistently than human analysis [5].

Several successful machine learning approaches exist to whale call detection and classification, as discussed in more detail in Sect. 2. Currently, the most prominent approach uses convolutional neural networks (CNNs) together with spectrograms and has been applied successfully to several datasets [2,3,6,7]. However, a drawback of these spectrogram-based systems is that obtaining high-quality spectrograms over a wide range of underwater acoustic conditions can be challenging, and the process of training a model using spectrograms as features is computationally costly when datasets get large.

In this work, we explore an alternative approach to the task of whale call detection using a time delay neural network (TDNN)-based x-vector system, with traditional speech features as input. Specifically, we experiment with mel frequency cepstral coefficients (MFCCs) and filter banks (Fbanks).

We have two reasons for choosing this approach: In future work, we aim to utilise transfer learning to extend some of the smaller South African whale call datasets with internationally available data or models, an approach that an x-vector system is particularly good at. In addition, we aim to develop a system that does not require careful spectrogram calibration, especially when combining disparate datasets at a later point.

We, therefore, aim to demonstrate that a neural speech processing system based on a TDNN is a viable alternative technique for whale call detection. The TDNN we use is implemented using the Speechbrain toolkit [8]. We develop initial models using the "Acoustic trends for the blue and fin whale library" from the Australian Antarctic Data Centre (AADC) [9]. Results are compared with a spectrogram-based CNN developed using the same dataset.

The layout of this paper is as follows: Sect. 2 gives a brief overview of related work in the field. Background on neural speech processing and speech features are discussed in Sect. 3. In Sect. 4, we discuss the dataset, both its original version and how it was processed for experimentation. Section 5 discusses the CNN baseline In Sect. 6, we analyse whale call detection using Speechbrain. In Sect. 7, we analyse the results from the different systems. Finally, we discuss key findings in Sect. 8.

2   Related Work

Various studies use either speech features or spectrograms for whale call detection. While the most prominent model used is a convolutional neural network, other types of models are also used, as listed in Table 1. We are unaware of any studies using a TDNN and x-vectors for whale call detection.

Table 1. Significant whale call detection and classification studies.

Study	Dataset	Feature type	Model	Accuracy
Acoustic Detection of Humpback Whales Using a Convolutional Neural Network [1]	National Oceanic and Atmospheric Administration (NOAA)	Spectrograms	CNN	On the test set: 90% precision and 90% recall when a clip contains a humpback call
Blue whale calls classification using short-time Fourier and wavelet packet transforms and artificial neural network [10]	Saguenay-St. Lawrence Marine Park	Short-time Fourier Transform (STFT) and wavelet packet transform (WPT)	MLP	Classification performance: STFT/MLP obtained 86.25% and WPT/MLP obtained 84.22% on different blue whale calls
Whistle detection and classification for whales based on convolutional neural networks [2]	15 sound samples with a total duration of 120 min	Spectrograms	LeNet-5	97% correct detection rate for whistle calls and 95% correct classification rate on the testing set for distinguishing between killer whales and long-finned whales
Detection and classification of marine mammal sounds using AlexNet with transfer learning [11]	Watkins Marine Mammal Sound Database	Spectrograms	AlexNet	Detection accuracy of 99.96% for whale calls and Classification accuracy of 97.42% for distinguishing between killer whales and long-finned whales
Deep Machine Learning Techniques for the Detection and Classification of Sperm Whale Bioacoustics [7]	The Dominica coda dataset	Spectrograms	Detection model: CNN Classification model: LSTM and GRU RNN	Detection accuracy of 99.5% for sperm whale clicks
ORCA-SPOT: An Automatic Killer Whale Sound Detection Toolkit Using Deep Learning [6]	The Orchive and OrcaLab	Spectrograms	ResNet	Positive-predictive-value of 93.2% is achieved for detecting killer whale calls
Deep neural networks for automated detection of marine mammal species [4]	DCLDE 2013 workshop data, NOAA, Cornell University's Bioacoustics Research Program (BRP) and Kaggle data competition	Spectrograms	LeNet, ResNet, BirdNet and VGG	LeNet had the highest precision for a certain recall.
Classification of whale calls based on transfer learning and convolutional neural networks [12]	Whale FM	Spectrograms	CNN and transfer learning	Classification accuracy of 97.04% for two whale species and 91.47% classification accuracy for four whale sub-populations
Detection of baleen whale species using kernel dynamic mode decomposition-based feature extraction with a hidden Markov model [13]	Hydrophone recordings of SRW species found in the coastal water of False Bay, Western Cape, South Africa. Mobysound	Modified dynamic mode decomposition (mDMD). Kernel dynamic mode decomposition (kDMD)	Hidden Markov model (HMM)	The kDMD-HMM generally performed better than the mDMD-HMM and DMD-HMM detectors

3   Approach

We briefly describe neural approaches to different speech processing tasks, as well as the features we use in this work.

3.1   Neural Speech Processing

Several neural approaches produce state-of-the-art results for speech processing tasks such as speaker recognition and speech recognition. Speaker recognition refers to the process of identifying or confirming a person's identity given a segment of speech, while speech recognition refers to understanding the content of a speech utterance. We list a few.

Vaessen *et al.* [14] applied the wav2vec2 framework to speaker recognition and achieved on the test for the extended VoxCeleb1 a 1.88% Equal Error Rate (EER) compared to the ECAPA-TDNN with a 1.69% EER. (EER is defined as the point on the receiver operating characteristic (ROC) curve where the False Acceptance Rate (FAR) equals the False Rejection Rate (FRR). [15]) A study by Waibel *et al.* [16] used a TDNN for phoneme recognition which achieved a recognition rate of 98.5% compared to the Hidden Markov Models (HMM), which achieved 93.7%. Desplanques *et al.* [17] proposed the ECAPA-TDNN architecture that outperformed state-of-the-art TDNN-based system on the VoxCeleb. In speech, a TDNN is widely used for speech recognition and speaker identification. A TDNN with x-vectors is particularly successful for speaker identification [18,19].

SpeechBrain[1] is an open-source speech processing toolkit that implements various neural processing models. The model that we use is a TDNN with x-vectors. The inputs to the TDNN network are the frames of the extracted input features, either MFCCs or Fbanks.

The outputs are the probability distributions of each of the classes. In speech, we want to look at frames that are a little into the future or the past. These past or future frames are referred to as the context or delays. The process of looking at non-contiguous frames (skipping some of the frames) is referred to as dilation or subsampling. A TDNN consists of individual TDNN units, as illustrated in Fig. 1.

---

[1] https://speechbrain.github.io/.

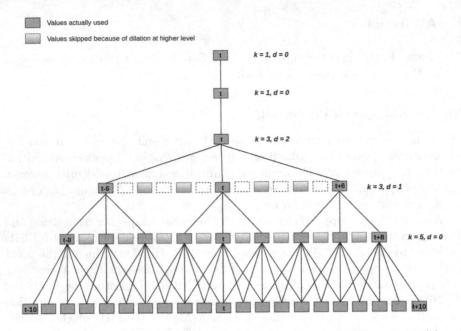

Fig. 1. The TDNN unit used in this study. Each layer has its own kernel size (k) and delay (d).

3.2   Speech Features

Three types of speech features are used in this study: spectrograms, Fbanks and MFCCs. While MFCCs and Fbanks are usually used to mimic human auditory processes, the frequency of the filters is adjusted to fall in the spectrum of whale sounds.

Spectrograms. In spectrograms, you can visually see the strength of a signal over time at different frequencies within a particular waveform [20]. Spectrograms are two-dimensional graphs with colours representing the third dimension [20]. The horizontal axis represents time, while the vertical axis represents frequency. The amplitude (or energy/loudness) of a particular frequency at a particular time is represented by the third dimension (colour) [20].

Filter Banks. A Fbank is a collection of bandpass filters used in image or signal processing to separate an input signal into a number of analysis signals, each of which carries a specific frequency sub-band from the original signal [21]. A filter bank is used to carry out a range of tasks, such as bandwidth reduction, sample rate modifications, combinations of the standard spectrum translation operations, and signal spectral composition and decomposition [21].

Mel Frequency Cepstral Coefficients (MFCCs). Different signal processing applications have extensively used MFCCs in feature extraction [22]. Image
identification, cetacean vocalisation and speech, gestures and drone sound recognition are just a few examples of such applications [22]. An MFCC is used to
extract distinctive features from a time-varying signal. To extract features from
the signal, it transforms the signal from the time domain into the Mel-scale
frequency domain using two types of filters (linear spaced filters and logarithm
spaced filters) [22].

4 Dataset

We first describe the original dataset (Sect. 4.1) before discussing our further
processing of the data in Sect. 4.2.

4.1 The AADC Dataset

The AADC library was created to enable automated algorithms to be applied to
a representative sample of recordings curated by the Acoustic Trends Working

Table 2. Classification and labelling system for Blue and Fin whale sounds in the
SORP Library of Annotated Recordings [9].

Call label	Call Type	Description
BmAnt-A	Antarctic Blue whale unit A	Tone with a constant frequency between 25 and 28 Hz (depending on the year) without other units. Usually repeated every 60-70 s
BmAnt-B	Antarctic Blue whale unit AB	The unit A tone follows an inter-tone down-sweep (unit B) for the Antarctic Blue whale
BmAnt-Z	Antarctic Blue whale z-call; (AKA 3 unit vocalisation)	An Antarctic Blue whale z-call with upper and lower tonal units A and C present (and down-swept unit B either present or absent).
Bm-D	Blue whale FM (AKA D-calls)	A Blue whale's down-swept frequency-modulated call. Typically, but not always, longer in duration and lower in frequency than FM calls from Fin and Minke whales.
Bp20Hz	Fin 20 Hz pulse	Fin whale pulse 20 Hz, with no significant energy at higher frequencies.
Bp20Plus	Fin 20 Hz pulse with energy at higher frequencies (e.g. 89 or 99 Hz components)	Pulse of a Fin whale 20 Hz, with secondary energy at higher frequencies (e.g. upper-frequency peak near 80-10 and Hz)
Bp-Downsweep	Fin whale FM calls (AKA' high frequency' down-sweep; AKA 40 Hz pulse)	Fin whales are thought to make frequency-modulated, normally down-swept calls. Blue whale calls are typically shorter in length and have a slightly higher frequency than FM calls from fin whales
Unidentified	Unidentifiable sounds	It includes any transient biological sound that cannot be confidently classified as one of the above. This included potentially biological sounds but significantly different from the categories above. It also included FM down-sweeps in which the analyst was uncertain whether they were Blue whale D sounds or Fin whale down-sweeps

282   E. Fourie et al.

Group of the International Whaling Commission's Southern Ocean Research Partnership (IWC-SORP) [9]. In the recorded data, there are four different call types for Antarctic Blue whales and three different call types for Antarctic Fin whales [9]. A detailed outline of calls that are annotated can be seen in Table 2 [9] in order to provide an indication of the diversity of calls. The first two letters correspond to the genus and species, where 'Bm' is for Blue whales, and 'Bp' is for Fin whales. The 'Ant' is for sub-species, in this case, Antarctic Blue whales. The remainder of the classification represents the particular call type.

The dataset consists of 1,880.25 h of audio across 11 different sites. The recording years were 2005, 2013, 2014, 2015 and 2017. In total, there were 105,161 annotations across all sites [9]. When analysing the data, it became clear that 3 of the 11 sites were better curated than the others. For 9 of the sites, some of the metadata contained inconsistent labels (the same call labelled with different annotations) and duration errors (calls starting before they end). Therefore, we select only three sites for experimentation: Elephant Island (2013 and 2014) and Balleny Island (2015). In Table 3, the three sites that we use can be seen together with the number of annotations per call type for the Blue and Fin whale calls.

Table 3. Number of annotations at each site by classification type for the Blue and Fin whale [9].

Site-Year	BmAnt-A	BmAnt-B	BmAnt-Z	Bm-D
Balleny Island 2015	923	44	31	47
Elephant Island 2013	2,625	1786	152	299
Elephant Island 2014	6,934	967	100	1,034
Total	10,482	2,797	283	1,380
Site-Year	Bp20Hz	Bp20Plus	Bp-Downsweep	Unidentified
Balleny Island 2015	951	148	78	18
Elephant Island 2013	3,662	1,859	1,042	22,927
Elephant Island 2014	4,940	2,912	3,660	890
Total	9,553	4,919	4,780	23,835

4.2 Data Processing and Event Selection

The audio files were recorded at different sampling rates. The highest recorded Blue and Fin whale call was observed at 135 Hz. In line with the Nyquist theorem, we resample all three sites 300 Hz.

We select two datasets for experimentation: a smaller and a larger one. The smaller set is selected for the training of the CNN (and a directly comparable SpeechBrain system) because it is computationally more expensive to run experiments. We selected as much data as we could process using the available infrastructure. The larger set is used for the training of additional Speechbrain systems because the training cycle is computationally less expensive.

The small and large sets are partitioned into a training, validation and test set each, using the ratio 70:10:20. The large set is used for the different machine learning (ML) tasks described in Sect. 6.1. The size, number of calls and overall duration of the two sets are shown in Table 4. Since more data is available than we can process, we construct a balanced dataset: we first select the available calls and then select background data ('no calls') of the same duration randomly from the available background data. A significant portion of background data is discarded in the process.

Table 4. Size of datasets used for experimentation.

Id	Set	Calls	No Calls	Total	Duration (hours)
Small	Training	8,069	8,069	16,138	17.93
	Validation	1,153	1,152	2,305	2.56
	Testing	2,306	2,305	4,611	5.12
Large	Training	29,782	29,782	59,564	109.26
	Validation	4,314	4,314	8,628	15.61
	Testing	8,470	8,469	16,939	31.22

5 CNN Baseline

This section describes the additional data processing required to develop the CNN baseline. We also describe the architecture used and the hyperparameter optimisation process.

5.1 Additional Data Processing

After resampling, we extracted the spectrograms using Python. The spectrograms are calculated with a Hanning window, a window length of 314, a hop length of 85 and a fast Fourier transform (FFT) with a size of 314. The spectrograms are divided into smaller chunks of 4 s in duration. The samples are divided into two classes: calls and no-calls. We use the annotations from the AADC to help us decide if the four-second chunk contains a call or not. If the chunk contains more than 40% of a call, it gets labelled as calls; otherwise, it gets labelled as no-calls. The no-call samples consist of background noise, other whales and biological sounds. Table 4 shows a summary of the dataset.

5.2 Architecture

The CNN architecture we use is a standard CNN, consisting of two hidden CNN layers and two fully-connected layers. The input for the first fully-connected layer varies depending on the architecture, and the output layer has two units that are used to indicate the predicted class for the last fully-connected layer. Each hidden CNN layer consists of a convolutional layer, batch normalisation, dropout, kernel with a size of two and max pooling layer.

The leaky rectified linear unit (LeakyReLU) is used as an activation function since it has been shown to speed up the learning of DNNs and tends to achieve good results in general [23]. The network is initialised with a uniform Kaiming scheme [24], which is the default choice for ReLU-activated networks. The CNN is optimised using either Adam or Stochastic Gradient Descent (SGD) as an optimiser. In addition to SGD, Adam is chosen for its adaptive estimates of lower-order moments [25].

5.3 Optimisation Protocol

The networks are trained on the spectrogram tensors from the Blue and Fin whale sounds for the CNN architecture of Sect. 5.2. The hyperparameter optimisation is done using the hyperparameter optimiser Optuna[2], which finds optimal hyperparameter values using different meta-optimisation algorithms. We use a Bayesian optimization algorithm referred to as a Tree-structured Parzen Estimator (TPE) [26]. The learning rate uses a step learning rate that decays the learning rate of each parameter group by gamma every n epochs. The batch size, learning rate, dropout level, the number of in and out channels of the convolution layers and the depth of the fully-connected layers are varied for each trial based on what Optuna selects. The number of epochs was kept at 30 because the training loss always converged before 30 epochs, and Optuna's pruning was implemented, which stopped an unpromising trial. Optuna is configured to minimise the validation loss. The hyperparameter ranges over which we optimised are shown in Table 5.

Table 5. Hyperparameter ranges searched over during optimisation.

Hyperparameters	Ranges
Batch size	32 and 64
Initial learning rate	0.1–0.0001
Dropout	0.1–0.5
Out Channel	32–128
Fully connected layer depth	100–1000
Optimiser	Adam, SGD
Epochs	30

The best accuracy for the CNN network was achieved with Adam as the optimiser and a learning rate of 0.0023. The optimal batch size was 32, with

---

[2] https://optuna.org/.

a dropout level of 0.3. The output channel for the convolutional layer was 32, with a 700-node fully-connected layer. Test accuracy of 97.92% is achieved, as discussed further in Sect. 7.

6   Whale Call Detection Using Speechbrain

In this section, we describe the different ML tasks and the additional data processing required for Speechbrain. We also describe the architecture used and the hyperparameter optimisation process.

6.1   Framing Whale Call Detection as Different Machine Learning Tasks

This section describes how the large dataset in Sect. 4.2 is used to create the data manifest files for the different ML tasks. In the data manifest, we tell Speechbrain where to find the audio data, the audio length, and the transcription.

- **Blue-fin-only:** In the blue-fin-only task, we consider two classes. The first class is 'Calls', which only contains the calls from Blue and Fin whales. The second class is 'No Calls', and this class contains only background noise.
- **All-whale-calls:** In the all-whale-calls task, we consider two classes. The first class is 'Calls', which contains the calls from any whale. The second class is 'No Calls', and this class contains only background noise.

In Table 6, the training, validation and testing sets for the two ML tasks are shown. The reason why we want to compare these different ML tasks is to see to what extent Speechbrain will perform well at identifying either specific whale calls or whale calls in general.

Table 6. Datasets used for the two Speechbrain tasks.

ML Task	Training			Validation			Testing		
ID	Size	Calls	No Call	Size	Calls	No Call	Size	Calls	No Call
blue-fin-only	59,022	29,511	29,511	8,550	4,275	4,275	16,762	8,381	8,381
all-whale-calls	59,564	29,782	29,782	8,628	4,314	4,314	16,939	8,470	8,469

6.2   Additional Data Processing

Only limited additional data processing was required. The resampled wav files were split into files containing calls and files that contained just background noise according to the tasks just mentioned. Calls were divided into segments and appropriately tagged.

6.3   Architecture

In this section, we motivate the decisions made, and principles followed to develop a whale call detection model using Speechbrain.

In the data augmentation and feature extraction below, the sample rate is changed 300 Hz (as previously discussed in Sect. 4.2). Part of Speechbrain's standard preprocessing pipeline is to add data augmentation before feature extraction. The data augmentation we use is a time-domain approximation of the SpecAugment algorithm [27]. In this data augmentation, we have three default speed perturbations (95, 100, 105) for Speechbrain to choose from.

We use Fbanks and MFCCs for feature extraction. In the Fbanks, we can use 'deltas', which will append derivatives and second derivatives to the features. When extracting MFFCs, the deltas are appended by default. The number of mel filters can also be adjusted in both Fbanks and MFCC; we will evaluate 23, 40 and 80 mel filters as these are the default values Speechbrain suggests. After the features are computed, we can apply mean and variance normalisation; we use the default where only mean normalisation is applied.

The TDNN constructed here has five time-delayed neural layers with the given dilation factors (0, 1, 2, 0, 0) and kernel size (5, 3, 3, 1, 1). We use the default values for dilations and kernel size. The LeakyReLU is used for the same reason as before. After each activation function, batch normalisation is used; this tends to improve the speaker identification performance [8]. After the convolutional layers, a statistic pooling layer is used to convert a tensor of variable length into a fixed-length tensor [8]. After the statistic pooling layer, a final linear transformation is performed using a linear layer with the number of neurons depending on the size of embedding required. The x-vectors are extracted here. Lastly, a multilayer perceptron (MLP) is implemented on top of the x-vector layer. The MLP consists of one fully-connected layer followed by the final softmax layer for classification. The number of input neurons in the layer depends on the size of the embeddings. The output neurons are the same as the number of classes that needs to be classified.

All the networks are trained with the Adam [28] optimiser. A scheduler with a linear annealing technique is implemented. We optimised the following hyperparameters using the validation set: embedding dimensions, TDNN channels, and the number of mel filter banks; we use the default Speechbrain ranges while we search over broader ranges for the learning rate, batch size, feature lobes and deltas, as shown in Table 7.

6.4   Optimisation Protocol

The algorithm we use is again TPE, which is a sequential model-based optimisation approach [29]. The hyperparameters that we searched over are shown in Table 7. Speechbrain comes with a convenience wrapper called 'hpopt', which can report objective values to Orion [30]. This facilitates the optimisation process.

We trained the model until convergence where, if the model's accuracy in the past 20% of the epochs was still increasing, another 20% of the past number of epochs is added to continue training. Minimum training is set at 10 epochs. If the model's accuracy did not increase in the past 20% of the epochs, training is stopped. We also implemented early stopping by selecting the model with the best validation accuracy in the training cycle and saving that model. For the hyperparameter sweep, we swept over the hyperparameters and their corresponding values, as seen in Table 7. The model sweeps over the selected hyperparameters with their corresponding values for ten trails, and after that, the model uses the best values from the sweep and does another hyperparameter sweep on only the learning rate (the most sensitive parameter) for final finetuning.

Table 7. Hyperparameters value ranges for the Orion search.

Hyperparameters	Search space
Learning rate	log_uniform (0.001, 0.1, precision = 2)
Embedding dimensions	[128, 256, 512, 768, 1024]
TDNN channels	[128, 256, 512, 768, 1024]
Batch size	[8, 16, 32]
Number mel filter banks	[23, 40, 80]
Feature lobes	[Fbank, MFCC]
Deltas	[True, False]

7   Analysis and Results

In this section, we describe the results obtained by both the CNN and Speechbrain in order to determine if Speechbrain is a feasible option for whale call detection.

7.1   CNN

The CNN performed well, resulting in a best validation accuracy of 97.98%. Only the best models obtained with SGD and Adam, respectively, were evaluated on the test set; reults are shown in Table 8. We can conclude from Table 8 that the model is not overfitting and has delivered good results. For the best trial in both cases, the hyperparameter values are not at the edges of the range provided.

7.2   Speechbrain's TDNN

The hyperparameter sweep is performed for 10 trials for each ML task. The hyperparameter values for the model with the best validation accuracy are then used to train the network again. After the network is trained, we test how the network performs on the test set. In both cases, MFCCs with deltas were the best-performing features (Results not shown here).

Table 8. CNN results obtained after hyperparameter optimisation with Optuna, using SGD and Adam as optimiser, respectively.

	SGD	Adam
Training loss	0.0573	0.0092
Training accuracy	98.07%	99.34%
Training recall positives	0.9660	0.9887
Training recall negatives	0.9955	1
Validation loss	0.0107	0.0001
Validation accuracy	97.31%	97.98%
Validation recall positives	0.9600	0.9606
Validation recall negatives	0.9900	0.9982
Testing accuracy	97.59%	97.92%
Testing recall positive	0.9626	0.9622
Testing recall negative	0.9895	0.9965

Table 9. TDNN results for the different ML tasks, after hyperparameter optimisation with Orion.

ML task	Set	Accuracy	Precision	Recall	Negative rate
blue-fin-only	Validation	93.36%	0.8827	1.0000	0.8671
	Test	95.89%	0.9382	0.9983	0.8976
all-whale-calls	Validation	97.79%	0.9638	0.9930	0.9627
	Test	97.67%	0.9624	0.9911	0.9614
cnn-compare	Validation	89.97%	0.8403	0.9235	0.8840
	Test	95.72%	0.9953	0.9186	0.9957

The result of the learning task 'cnn-compare' is directly comparable to the results of the CNN and the results for Speechbrain are less accurate than the CNN by more than 2% on the test set. This learning task is trained on the exact same data as used for the CNN. While the next results are not directly comparable (the CNN setup we used was restricted by the size of the dataset it could process) the Speechbrain system 'blue-fin-only' produced results that were slightly less accurate than the CNN but of the same order. However, the second machine learning task 'all-whale-calls' produced very competitive results. Finally, when evaluating all runs, it was clear that MFCCs with deltas outperformed Fbanks: each of the best systems selected by Optuna used MFCCs.

8   Conclusion

Our purpose with this study was to determine whether neural speech processing and specifically, a TDNN with x-vectors, could be a useful tool for whale call detection. We are specifically interested in this question since x-vectors lend themselves to data sharing across environments. This is important in the South African environment where local whale call databases are scarce. Since calls are split randomly between the training, validation and test sets, it is possible that some similar conditions may be observed within the data of a single site, even though conditions across sites could be very different. This should be taken into account when interpreting the high accuracies reported. In future work, we aim to determine how well these networks transfer to unseen sites, as well as the amount of further tuning that would be required.

In order to answer this question, we develop both a CNN-based and TDNN-based call detection system. The CNN-based system used spectrograms as features, and the TDNN-based system used either Fbanks or MFCCs. When performance is compared on a similar task (detecting Blue or Fin whales in background noise) the CNN performed better (97.92% vs 95.72% test accuracy). However, the results of the TDNN-based system were particularly promising when identifying a variety of whale calls in background noise (97.79% test accuracy). In both TDNN setups, MFCCs with deltas were the most successful features to use. We conclude that neural speech processing is a viable approach to whale call detection and one that we will be exploring further in future work.

References

1. Harvey, M.: Acoustic detection of humpback whales using a convolutional neural network (2018). https://ai.googleblog.com/2018/10/acoustic-detection-of-humpback-whales.html. Accessed 23 July 2021
2. Jiang, J.-J., Bu, L.-R., Duan, F.-J., et al.: Whistle detection and classification for whales based on convolutional neural networks. Appl. Acoust. **150**, 169–178 (2019)
3. Jiang, J.-J., Bu, L.-R., Wang, X.-Q., et al.: Clicks classification of sperm whale and long-finned pilot whale based on continuous wavelet transform and artificial neural network. Appl. Acoust. **141**, 26–34 (2018)
4. Shiu, Y., Palmer, K., Roch, M.A., et al.: Deep neural networks for automated detection of marine mammal species. Sci. Rep. **10**(1), 1–12 (2020)
5. Cireşan, D., Meier, U., Masci, J., Schmidhuber, J.: A committee of neural networks for traffic sign classification. In: The 2011 International Joint Conference on Neural Networks, IEEE, pp. 1918–1921 (2011)
6. Bergler, C., Schröter, H., Cheng, R.X., et al.: Orca-spot: an automatic killer whale sound detection toolkit using deep learning. Sci. Rep. **9**(1), 1–17 (2019)
7. Bermant, P.C., Bronstein, M.M., Wood, R.J., Gero, S., Gruber, D.F.: Deep machine learning techniques for the detection and classification of sperm whale bioacoustics. Sci. Rep. **9**(1), 1–10 (2019)
8. Ravanelli, M., Parcollet, T., Plantinga, P., et al.: SpeechBrain: a generalpurpose speech toolkit(2021). arXiv: 2106.04624 [eess.AS]
9. Miller, B.S., Balcazar, N., Nieukirk, S., et al.: An open access dataset for developing automated detectors of antarctic baleen whale sounds and performance evaluation of two commonly used detectors. Sci. Rep. **11**(1), 1–18 (2021)
10. Bahoura, M., Simard, Y.: Blue whale calls classification using shorttime fourier and wavelet packet transforms and artificial neural network. Digital Signal Process. **20**(4), 1256–1263 (2010)

11. Lu, T., Han, B., Yu, F.: Detection and classification of marine mammal sounds using alexnet with transfer learning. Ecol. Inf. **62**, 101277 (2021)

12. Yuea, H., Wanga, D., Zhanga, L., Wua, Y., Baoa, C., Wang, D.: Classification of whale calls based on transfer learning and convolutional neural network. In: 4th Underwater Acoustics Conference and Exhibition, pp. 537–544 (2017)

13. Usman, A.M., Versfeld, D.J.J.: Detection of baleen whale species using kernel dynamic mode decomposition-based feature extraction with a hidden Markov model. Ecol. Inf. **71**, 101766 (2022)

14. Vaessen, N., Van Leeuwen, D.A.: Fine-tuning wav2vec2 for speaker recognition. In: ICASSP 2022–2022 IEEE International Conference on Acoustics, Speech and Signal Processing (ICASSP), IEEE, pp. 7967–7971 (2022)

15. Singh, M., Pati, D.: Replay attack detection using excitation source and system features. In: Advances in Ubiquitous Computing, Elsevier, pp. 17–44 (2020)

16. Waibel, A., Hanazawa, T., Hinton, G., Shikano, K., Lang, K.: Phoneme recognition using time-delay neural networks. IEEE Trans. Acoust. Speech Signal Process. **37**(3), 328–339 (1989). https://doi.org/10.1109/29.21701

17. Desplanques, B., Thienpondt, J., Demuynck, K.: Ecapa-tdnn: emphasized channel attention, propagation and aggregation in tdnn based speaker verification, arXiv preprint arXiv:2005.07143 (2020)

18. Peddinti, V., Povey, D., Khudanpur, S.: A time delay neural network architecture for efficient modeling of long temporal contexts. In: Sixteenth Annual Conference of the International Speech Communication Association (2015)

19. Snyder, D., Garcia-Romero, D., Sell, G., Povey, D., Khudanpur, S.: Xvectors: robust DNN embeddings for speaker recognition. In: 2018 IEEE International Conference on Acoustics, Speech and Signal Processing (ICASSP), pp. 5329–5333 (2018). https://doi.org/10.1109/ICASSP.2018.8461375

20. Pacific Northwest Seismic Network. "What is a spectrogram?" (2012). https://pnsn.org/spectrograms/what-is-a-spectrogram. Accessed 23 July 2021

21. Electrical4U. "Filter bank: What is it? (dct, polyphase, gabor, mel and fbmc)." (2021). https://www.electrical4u.com/filter-bank/. Accessed on 25 Aug 2022

22. Ogundile, O.O., Usman, A.M., Babalola, O.P., Versfeld, D.J.: Dynamic mode decomposition: a feature extraction technique based hidden Markov model for detection of mysticetes' vocalisations. Ecol. Inf. **63**,101306 (2021)

23. Imoscopi, S.: Machine learning for text-independent speaker verification (2016)

24. He, K., Zhang, X., Ren, S., Sun, J.: Delving deep into rectifiers: surpassing human-level performance on imagenet classification. In: Proceedings of the IEEE International Conference on Computer Vision, pp. 1026–1034 (2015)

25. Goodfellow, I., Bengio, Y., Courville, A.: Deep Learning. MIT Press (2016). http://www.deeplearningbook.org

26. Lim, Y.: State-of-the-art machine learning hyperparameter optimization with optuna (2021). Accessed on 08 Apr 2022

27. Park, D.S., Chan, W., Zhang, Y., et al.: Specaugment: a simple data augmentation method for automatic speech recognition, arXiv preprint arXiv:1904.08779 (2019)

28. Kingma, D.P., Ba, J.: Adam: a method for stochastic optimization, arXiv preprint arXiv:1412.6980 (2014)

29. Ayuya, C.: Using random search to optimize hyperparameters (2022). https://www.section.io/engineering-education/random-search-hyperparameters/. Accessed 19 May 2022

30. Bouthillier, X., Tsirigotis, C., Corneau-Tremblay, F., et al.: Epistimio/orion: asynchronous Distributed Hyperparameter Optimization, version v0.2.4, May 2022. https://doi.org/10.5281/zenodo.3478592

Self-Supervised Text Style Transfer with Rationale Prediction and Pretrained Transformers

Neil Sinclair[iD] and Jan Buys[(✉)][iD]

Department of Computer Science, University of Cape Town, Cape Town, South Africa
sncnei001@myuct.ac.za, jbuys@cs.uct.ac.za

Abstract. Sentiment transfer involves changing the sentiment of a sentence, such as from a positive to negative sentiment, while maintaining the informational content. Given the dearth of parallel corpora in this domain, sentiment transfer and other text rewriting tasks have been posed as unsupervised learning problems. In this paper we propose a self-supervised approach to sentiment or text style transfer. First, sentiment words are identified through an interpretable text classifier based on the method of rationales. Second, a pretrained BART model is fine-tuned as a denoising autoencoder to autoregressively reconstruct sentences in which sentiment words are masked. Third, the model is used to generate a parallel corpus, filtered using a sentiment classifier, which is used to fine-tune the model further in a self-supervised manner. Human and automatic evaluations show that on the Yelp sentiment transfer dataset the performance of our self-supervised approach is close to the state-of-the-art while the BART model performs substantially better than a sequence-to-sequence baseline. On a second dataset of Amazon reviews our approach scores high on fluency but struggles more to modify sentiment while maintaining sentence content. Rationale-based sentiment word identification obtains similar performance to the saliency-based sentiment word identification baseline on Yelp but underperforms it on Amazon. Our main contribution is to demonstrate the advantages of self-supervised learning for unsupervised text rewriting.

Keywords: Text style transfer · Self-supervised learning · Transformers

1 Introduction

Advances in large language models have enabled the generation of high-quality open-ended text [12,20]. However, without fine-grained control, generated text has limited practical value. Despite some advances in controllable text generation, for example in writing text in a legal style [6], the transfer of text from one sentiment or style to another while maintaining the content of the source sentence [13] is still a challenging problem. Style transfer has a number of use-cases, including mitigating harmful content [22] in text, rewriting text in a more

A. Pillay et al. (Eds.): SACAIR 2022, CCIS 1734, pp. 291–305, 2022.
https://doi.org/10.1007/978-3-031-22321-1_20

Positive: the food here is **amazing** and the service **excellent!**

Negative: the food here is **terrible** and the service **appalling!**

Content: the food here is ... and the service ...!

Fig. 1. An example of a sentence rewritten from positive to negative sentiment, with the sentiment words in **bold**. To rewrite the text, sentiment words are identified and removed, leaving the sentence content (bottom). The content is then rewritten in the desired sentiment.

modern style [8], or de-formalising a piece of text [21]. However, one of the key challenges that differentiates sentiment and style transfer from other text transduction tasks is the lack of parallel corpora.

In this paper we extend previous unsupervised approaches to text sentiment transfer by utilizing self-supervised learning for rewriting sentences. Our approach extends one of the main paradigms to text sentiment and style transfer [13,23,29] that identifies and deletes sentiment-specific words and learns sentiment transfer through sentence reconstruction conditioned on the target sentiment (Fig. 1). Our work builds on previous approaches in three ways: First, we utilise the method of rationales [1,11], a neural network-based approach from the interpretability literature, to identify and mask sentiment words. This replaces the previous heuristic n-gram saliency approach [13,29]. Second, we fine-tune a pretrained BART [12] model to reconstruct masked sentences, which enables generating sentences with a different sentiment autoregressively. BART, pretrained with a denoising autoencoder (DAE) objective, is a natural fit for sentence reconstruction training. Third, we use self-supervised training to further improve the model's performance utilising its own generations: The fine-tuned model is used to generate a high-precision parallel corpus of sentences in opposite sentiments, on which BART is fine-tuned further to improve its style transfer accuracy.

We evaluate our approach on Yelp and Amazon review datasets [4]. We compare rationale-based sentiment word identification to the saliency-based approach, and BART to a sequence-to-sequence (Seq2Seq) [25] model with Long short-term memory (LSTM) [5]. Results using both automatic and human evaluations show that rationale-based sentiment word identification performs on par with the saliency-based approach on the Yelp dataset, but underperforms it on the Amazon dataset when used with the BART model. On the Amazon dataset BART obtained higher BLEU score but reduced sentiment transfer accuracy compared to the Seq2Seq model. Self-supervised training improves sentiment transfer accuracy on both datasets. Rationale-based sentiment word identification obtains similar performance to the saliency-based sentiment word identification baseline on the Yelp dataset while offering fine-grained control over the trade-off between content preservation and sentiment transfer accuracy. However it underperforms on Amazon due to its structure which makes it harder to identify style words. BART outputs were rated higher by human evaluators than the sequence-to-sequence models' on both datasets, in particular due to better fluency. The performance of our approach is close to that of state-of-the-art models [27,29] on Yelp, but lower on Amazon.

2   Background

The goal of text style transfer is to change some attribute of a sentence, such as its style or sentiment, while maintaining its content [13]. Due to the lack of parallel corpora across different styles in most domains [23], text style transfer is usually approached as an unsupervised learning problem, learning from non-parallel examples of text in different styles.

There are two main paradigms of approaches to sentiment and style transfer. In the first paradigm, sentences are seen as a combination of content and style elements. The style words are removed and the model is trained to reconstruct the full sentences from the content elements plus a token representing the style of the sentence. The model is effectively a semi-supervised denoising autoencoder (DAE), where the style words are removed based on some pre-defined criteria. The Delete Retrieve Generate (DRG) [13] model implements this paradigm by using a heuristic n-gram saliency-based approach to identify style-specific words. A sequence-to-sequence LSTM is trained with a DAE objective to reconstruct the original sentence. Additionally, the "retrieve" step retrieves relevant words or sentences in the target style, and conditions on this when generating sentences in the target style. We use the delete-only version of DRG as a baseline in this paper, as our BART-based model is trained similarly without a retrieval step. Figure 1 gives an example of sentiment transfer by deleting sentiment words and then rewriting the sentence in the target sentiment.

Subsequent work extends DRG to use a Transformer [26] instead of an LSTM while utilising the attention weights of the model to identify which words to remove [23], and uses a pretrained BERT [2] model that learns to fill in masked out style words [29]. This formulation simplifies learning but it limits the application of style transfer to be narrowly defined as deleting and replacing words.

The other paradigm to text style transfer involves encoding the input sentence in a latent representation, manipulating the encoded representation, and then decoding the sentence in another style. Wang et al. [27] utilise a Transformer to learn a latent representation of a sentence comprising both style and content, and then use a pretrained classifier to edit the entangled latent representation. Logeswaran et al. [14] and Lample et al. [10] approach style transfer as a combination of autoencoder (reconstruction) and back-translation objectives, while He et al. [3] apply variational inference to generalise these approaches.

3   Sentiment Word Identification

We review a widely-used approach to sentiment word identification based on n-gram saliency, and introduce our approach to apply an interpretable neural classifier to identify and mask or remove sentiment words.

Let $D = (x^{(1)}, c^{(1)}, ..., (x^{(m)}, c^{(m)})$ be the training set of sentences $x^{(j)}$ each annotated with sentiment marker $c^{(j)}$, where in our application the sentiments are restricted to $c^{(j)} \in \mathcal{C} = \{positive, negative\}$, and D_c denotes the subset of sentences in sentiment c.

3.1 Saliency Noising

DRG [13] uses a heuristic approach to identifying sentiment-specific words. This approach is conceptually similar to term frequency inverse document frequency (TF-IDF) where words that have higher discriminative power are weighted higher. The *salience* of word or n-gram w with respect to sentiment c is calculated as

$$s(w, c) = \frac{count(w, D_c) + \lambda}{(\sum_{c' \in \mathcal{C}, c' \neq c} count(w, D_{c'})) + \lambda}, \tag{1}$$

where λ is a smoothing parameter. The method identifies w as a sentiment marker for sentiment c if $s(w, c) > \gamma$, where γ is the saliency threshold.

3.2 Rationales Noising

Motivation. Interpretable neural network-based text classifiers identify which words are most important for a classifier's classification decision for a given input. Gradient-based methods such as Integrated Gradients [24] sum the gradients across input features in order to understand which parts of the input are most sensitive to changes in the output. In this manner it uncovers the contribution of each feature to the output, enabling identifying the features or rules the model uses to make decisions.

Another approach for interpretable neural networks is the method of rationales [11]. This method aims to understand how a neural network-based classification model reaches its prediction by identifying the words or phrases that are essential for the model to achieve the correct sentence classification. The model learns a binary mask over the input sentence that learns to mask words which do not contribute to the rationale. Bastings et al. [1] extends this work using the Hard Kumaraswamy distribution to enable the differentiable binary masking of tokens. In contrast to Integrated Gradients, which provides a real number representing the contribution of each feature to a model's output, the method of rationales identifies a discrete set of tokens most impacting the classification.

In the context of sentences written in a particular sentiment, the words that are most important in classifying its sentiment will most likely be the sentiment words of a sentence. We therefore propose a novel approach to identifying sentiment words based on interpretable classifiers.

Method. Given an input sentence x, the method of rationales [1,11] defines a latent binary variable Z_i corresponding to each sentence token x_i, indicating whether x_i is included in the rationale for the classification decision. For each sentence, the tokens in x that do not form part of the rationale are masked out, while the unmasked words are used as input to a trainable classifier that predicts the sentiment marker c. Masking can be formulated as taking the element-wise Hadamard product of the latent variable and the sentence tokens, i.e., $z \odot x$.

The rationales (masks) Z_i can each be seen to be derived from a Bernoulli distribution, and the sentence classification variable C from a categorical distribution:

$$Z_i | x \sim \text{Bern}(g_i(x; \phi)), \tag{2}$$
$$C | x, z \sim \text{Cat}(f(x \odot z; \theta)). \tag{3}$$

Functions $g_i(x; \phi)$ and $f(x \odot z; \theta)$ are neural networks based on the LSTM [5] architecture and are parameterised by ϕ and θ respectively.

To enable differentiable training with respect to a discrete set of masks, a modified version of the Kumaraswamy distribution [9] called the Hard Kumaraswamy distribution is employed [1]. The reparameterization trick [7] is utilised during training. The per-example loss function is defined as

$$- \mathbb{E}_{P(z|x,\phi)}[\log P(c|x, z, \theta)] + \lambda \sum_{i=1}^{n} z_i, \tag{4}$$

where λ is a hyper-parameter. The first part of the loss is the lower bound on data log-likelihood $\log P(c|x)$, the likelihood of the sentiment label given the sentence x. The second part is a sparsity loss to prevent the model from choosing the entire sequence as the rationale. We do not include an additional loss used by [1] that encourages choosing longer contiguous stretches of words.

4   Self-supervised Text Sentiment Transfer

The training pipeline for our approach comprises five steps: training a classifier, training and applying the rationales sentence noiser, training the BART DAE, generating a high-precision parallel corpus, and finally self-supervised training. The training process is visualized in Fig. 2.

Classifier Training. A BART encoder is fine-tuned to classify the sentiment of sentences in the training corpus (step 1 in Fig. 2). This classifier is used for self-supervised training.

Sentence Noising. The rationales model [1] is trained as a (second) sentiment classifier. The extracted rationales are represented as a vector of binary masks for each sentence. Each token that is part of the rationale is replaced with a <mask> token or removed in the case of the sequence-to-sequence model. This process is shown in step 2 of Fig. 2. We refer to the masking or removal of sentiment-specific words from the sentences used as input to the sentiment transfer models as *noising*, as both the BART and seq2seq models are trained as DAEs. We train rationale extraction models with different noising levels, and as sentiment word identification baseline we also use saliency-based noising.

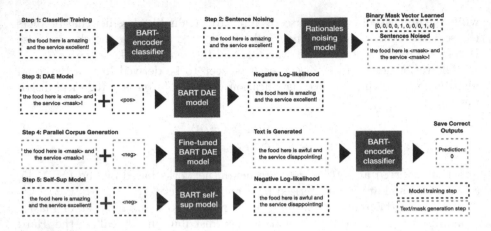

Fig. 2. Overview of classifier pretraining, sentence noising, and DAE training in the BART training pipeline.

DAE Model Training. BART [12] is a pretrained encoder-decoder Transformer that can be seen as a generalisation of BERT [2] as the encoder and GPT [19] as the decoder. BART is trained with a DAE objective in which encoder input tokens are randomly masked (and permuted) and the input is reconstructed autoregressively by the decoder.

In our approach the BART encoder-decoder is fine-tuned as a DAE to reconstruct the sentences in the training corpus in which the sentiment words have been masked with a `<mask>` token, using either the rationale-based or saliency-based method. No word permutation is done. The sentiment token (`<pos>` or `<neg>`) is appended as the first input token in the decoder. This is shown as step 3 in Fig. 2.

The model is trained to optimise the negative log-likelihood of each of the sentences that it aims to reconstruct:

$$L_{\text{DAE}} = -\sum_{j=1}^{m} \log p(x^{(j)}|\text{Mask}(x^{(j)})), \qquad (5)$$

where $\text{Mask}(x)$ represents the sentence with the sentiment words masked out.

Early stopping is performed based on the accuracy of validation sentences translated to the opposite sentiment, as measured by the classifier trained in step 1. Every fixed number of batches 1 000 sentences from the development set are sampled and masked and fed to the partially fine-tuned BART model to translate by seeding the decoder with the sentiment token opposite of the original sentence. Sentiment transfer accuracy is measured by feeding the generated output sentences to the classifier trained in step 1.

Parallel Corpus Generation. We apply the fine-tuned BART model to generate a parallel training set of sentences in opposite sentiments by translating all the (masked) sentences from the training set. This is shown as step 4 in Fig. 2. Sentences are generated auto-regressively using greedy decoding, where the next word with the highest probability of occurring is chosen deterministically. Minor formatting corrections such as removing double spaces and some excess punctuation is also necessary. The generated sentences are filtered so that only sentences classified as having been accurately transferred into the target sentiment (using the classifier from step 1) are kept.

Self-supervised Training. Finally, the BART model is fine-tuned further to translate sentences from one sentiment to another with self-supervised training (step 5 in Fig. 2). Here, we use the term self-supervised as the parallel training set was generated and filtered entirely by the model. In this training step original (unmasked) sentences are used as inputs, as we found that the model obtains higher sentiment transfer accuracy in such a setting. The same setup is followed at test time.

The self-supervised training loss is defined as

$$L_{SS} = - \sum_{(x,y,c,c') \in D'} \log P(y|x,c'), \tag{6}$$

where D' represents the new dataset of generated parallel sentences with y as the generated sentence translated from sentiment c to sentiment c'. The model is trained until the early stopping criterion is met where translation accuracy does not improve over three validation checks of 250 batches each.

5 Experimental Setup

5.1 Data

We use two common sentiment transfer datasets, which are truncated versions of Yelp and Amazon reviews that have been categorised as either positive or negative [4]. The training sets have 440k and 555k training examples, and an average sentence length of 7.9 and 13.8 tokens, respectively. The test sets include human-annotated reference translations which are used to calculate the automatic evaluation Bilingual Evaluation Understudy (BLEU) scores [15]. These scores measure the reliability of the content of the generated sentences in the new sentiment with respect to a gold standard set of sentences.

5.2 Sentiment Word Identification

We use the original implementation of the rationales model [1] to learn sentiment token identification.[1] To test the impact of different levels of token masking or removal on sentiment transfer accuracy and content consistency of the generated sentences, five levels of rationales-based noising are used, namely 15%, 20%, 30%,

---

[1] https://github.com/bastings/interpretable_predictions.

40% and 50%. There is some variance in the actual level of noising achieved per sentence, and also some discrepancy between the mean level of noising constraint fed to the model and the mean level of noising achieved.

As a sentiment word identification baseline we use a PyTorch implementation of the saliency-based method [18].[2] Saliency noising achieves a mean level of token masking or removal of 32.4% on Yelp and 31.1% on Amazon, so it is closest to the rationales 30% noising scheme. However its inter-sentence variance is higher than that of rationales noising.

The rationales model [1] is trained for 20 epochs with an Adam optimizer with learning rate set to 2e–4, and a batch size of 128. Pretrained GLoVE embeddings [16] are used with an embedding size of 300. For saliency noising the smoothing parameter, λ is set to 1; spans of up to four words are considered as attribute markers; and γ, the threshold for the saliency score, is set to 15 and 5.5 for Yelp and Amazon respectively. These hyperparameters follow [13], in which they were chosen through tuning on the development set.

5.3   Sentiment Transfer Models

We use the HuggingFace Transformers [28] pretrained implementation of BART-Base with a language modelling head. Due to a degree of stochasticity in the results each version of the model was trained three times and the final accuracy and BLEU results of the three runs averaged on the test set for reporting. During validation and testing, all sentences are generated using greedy decoding. This lead to better performance than sampling-based decoding, where each word is sampled based on the next word probability distribution.

BART-Base consists of six encoder and decoder blocks in the encoder and decoder with 139 million parameters in the model. The dimensionality of the model is $d_{model} = 768$. The weights are optimized using the Adam optimizer with a learning rate of 2e–5 which was empirically found to achieve the best results. We use early stopping if the translation accuracy fails to improve for 3 consecutive validations, which are conducted every 250 batches. This is to prevent model over-fitting and is used due to the model appearing to learn very quickly from the data. A batch size of 64 is used for training due to memory constraints. All the parameters are fine-tuned except the positional embeddings.

As a non-pretrained baseline we use the delete-only Seq2Seq model of [13], utilising the PyTorch implementation of [18]. This model is also trained with various levels of token masking using the rationales model, in addition to using the saliency method. This model uses word-based tokenization with a vocabulary size of 16,000, in contrast to the word-piece tokenization used in BART. The model is trained utilising an Adam optimizer with learning rate of 2e–4. The model is trained for 70 epochs for both datasets. The embedding dimension is kept at the default of 128 and hidden dimension of the LSTM encoder and decoder is kept as 512 as per the original paper.

---

[2] https://github.com/rpryzant/delete_retrieve_generate.

Table 1. Classification accuracy and BLEU score of BART and Seq2Seq models for different levels of token masking or removal for the Yelp and Amazon datasets.

Noising	Yelp				Amazon			
	BART		Seq2Seq		BART		Seq2Seq	
	Accuracy	BLEU	Accuracy	BLEU	Accuracy	BLEU	Accuracy	BLEU
15%	70.3	29.4	49.0	24.3	28.5	37.4	39.2	19.5
20%	74.5	27.7	69.2	16.1	31.4	33.5	40.0	29.6
30%	78.6	26.3	72.0	17.8	35.2	27.2	42.4	22.3
40%	83.5	21.1	81.2	13.6	48.1	23.1	51.7	16.4
50%	93.7	13.6	92.0	8.74	44.7	18.1	58.8	11.0
Saliency	74.2	26.2	82.2	15.1	53.6	37.0	47.4	21.3
Wu et al. (2019)	97.3	14.4	–	–	84.5	28.5	–	–
Wang et al. (2019)	95.4	22.6	–	–	85.3	34.1	–	–

6 Results

6.1 Automatic Evaluation

We evaluate our approach based on the accuracy of rewriting sentences from one sentiment to another as assessed by the pretrained classifier, as well as by the BLEU score [15] comparing generated sentences with a set of reference human translated sentences.[3] BLEU represents the outputs' content preservation, and to a lesser extend their fluency. For both models sentiment word identification is performed with rationale-based noising with different noising levels as well as with the saliency-based sentiment word identification baseline.

Table 1 gives the results of both our BART-based self-supervised model and our replication of the delete-only Seq2Seq DAE model of [13]. The table reports BART test set results with the full training pipeline including self-supervised learning. Additionally we include the results of Wu et al. [29] and Wang et al. [27], which are state-of-the-art on the Yelp and Amazon datasets, respectively. As an ablation experiment we found that BART with only DAE training obtains style transfer accuracies that are between 4.3% and 20.4% lower than including self-supervised training (across different levels of rationales noising on both datasets). This highlights how effective the models are at learning the sentiment of a sentence conditioned on a sentiment token. This is especially evident as sentence noising or token masking increases. However, it also highlights the limitations of training without the paired synthetically generated self-supervised training set. These gains in accuracy are likely due to the model learning an explicit mapping from one sentiment to another when using the synthetically generated training set. This is in contrast to learning to simply replace masked words in a sentence conditioned on a sentiment token.

The results on the Yelp dataset shows that our BART-based model achieves higher accuracy and higher BLEU scores than the Seq2Seq model for the same

---

[3] We use SACREBLEU [17] with default settings to calculate the BLEU score.

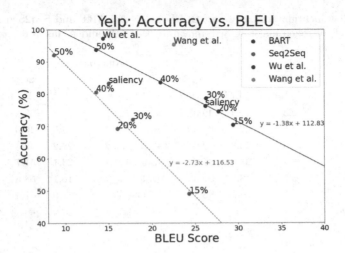

Fig. 3. Test set sentiment transfer vs. BLEU score for BART (solid blue line) and Seq2Seq (striped red line) on the Yelp dataset. (Color figure online)

level of rationale-based noising. Saliency-based noising leads to a higher BLEU score but lower classification accuracy on BART compared to the Seq2Seq baseline. On the Amazon dataset BART achieves higher BLEU but lower accuracy than Seq2Seq for all the levels of rationale-based noising. Saliency-based noising leads to the highest overall trade-off between accuracy and BLEU score, as can be seen in Fig. 4. Seq2Seq with 50% rationale-based noising, however, obtains the highest overall classification accuracy.

The trade-off between sentiment transfer accuracy and content preservation as measured by BLEU is visualised in Figs. 3 and 4. The higher the level of noising, the better a model is able to transfer from one sentiment to another but the less faithful the translation is to the content of the original sentence. The existence of this trade-off is consistent with previous sentiment transfer research [13,29]. With rationale-based noising BART achieves a better trade-off than the Seq2Seq baseline on the Yelp dataset. The performance of saliency-based noising is similar to that of rationale-based noising with a similar BLEU score on both models, although on Seq2Seq saliency noising is slightly preferable.

On the Amazon dataset the BART model's trade-off across different noising levels is almost identical to that of the Seq2Seq model. BART's high performance with saliency noising is an outlier here compared to the rationale-based models. The combination of BART's ability to generate more fluent sentences and the saliency-based method for masking or removing n-grams of up to four tokens may explain the better performance of that configuration. The results show that all of our models obtain relatively low sentiment transfer accuracies compared to the state-of-the-art, suggesting that the Amazon dataset may be less suitable to sentiment transfer based on identifying individual sentiment words.

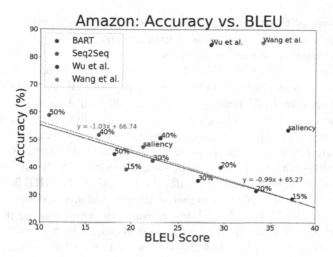

Fig. 4. Test set sentiment transfer vs. BLEU score for BART (solid blue line) and Seq2Seq (striped red line) on the Amazon dataset. (Color figure online)

Table 2. Human evaluation results (scores scale 1 to 5, higher is better) on the Yelp and Amazon datasets.

Model	Yelp			Amazon		
	Content	Sentiment	Fluency	Content	Sentiment	Fluency
BART 30%	–	–	–	3.4	3.4	3.9
BART 40%	3.9	4.4	4.1	–	–	–
BART 50%	3.2	4.5	4.1	2.8	3.8	3.9
Seq2Seq Saliency	3.4	3.9	3.4	3.3	3.2	3.5
Wu et al. (2019)	4.0	4.4	4.2	4.1	4.0	4.0
Wang et al. (2019)	3.5	3.6	3.8	4.2	4.0	4.1

On Yelp our best model's performance is close to that of state-of-the-art approaches, although both [27,29] obtain a slightly better trade-off between accuracy and BLEU. This is shown by these models having accuracy and BLEU scores that lie above the average trade-off, represented by the solid blue line in 3. The simpler structure of examples in the Yelp dataset makes it particularly suitable for the word-based "Mask and Infill" approach [29].

6.2   Human Evaluation

Due to the limitations of automatic evaluation in text generation tasks such as sentiment and style transfer, we performed a human evaluation, broadly following the methodology of [13] and subsequent work on style transfer. Amazon Mechanical Turk crowd-workers were asked to rate a sample of 100 sentiment transferred generations for each of the models included in the evaluation. The evaluation assesses the fluency, sentiment transfer accuracy and content preservation of the generations, each scored using a Likert scale from 1 to 5. Each output was evaluated by three evaluators.

The results are shown in Table 2. We include the rationale-based BART models with the best trade-off between accuracy and BLEU according to the automatic evaluation (40% noising on Yelp and 30% on Amazon) and the highest overall accuracy (50% noising), as well as the baseline Seq2Seq model with saliency noising. The human evaluation results of Wu et al. [29] and Wang et al. [27], as reported in the original papers, are given as an additional comparison.

The results show that the BART-based models clearly outperform the Seq2Seq baseline on both datasets, with the exception of BART 50%'s content preservation. BART 50% obtains higher sentiment transfer accuracy and similar fluency than the BART models with less noising, but lower content preservation. On Yelp and Amazon the BART-based models obtain higher sentiment transfer accuracy than the Seq2Seq baseline on the human evaluation despite similar or lower automatic classification accuracies. Content preservation is broadly in line with the BLEU scores. The increased fluency of the BART models can be attributed to its pretraining compared to the Seq2Seq baseline which is trained from scratch. This may also explain why BART achieves higher BLEU scores than the Seq2Seq model for the same level of noising.

Compared to state-of-the-art models, the BART 40% model performs better on Yelp than [27] and on par or within 0.1 evaluation points of [29]. Our approach therefore obtains state-of-the-art performance despite performing slightly worse on the automatic evaluation. On Amazon the BART 50% model obtains sentiment and fluency scores within 0.2 points of the state-of-the-art (compared to the automatic classifier which showed a much greater gap in sentiment transfer accuracy), but much worse content preservation. BART 30%'s content preservation is higher but still far below that of the state-of-the-art, and at the expense of lower sentiment transfer accuracy.

In spite of these mixed automatic evaluations, the higher human evaluations still show a clear benefit to using the pretrained model for generation. This suggests that fluency ratings may have an inadvertent impact on other human ratings and that more fluent model outputs are preferred by human evaluators. This is the case even if sentiment translation accuracy, as measured by automatic evaluations, does not match that of a less fluent model.

6.3   Qualitative Analysis

A selection of model outputs on the test data, along with the original sentences and reference human sentiment translations, are shown in Tables 3 and 4. In the first and last Yelp examples the Seq2Seq model manages to translate the sentence, but fluency is worse when compared with the source sentence. While the BART 40% model manages to preserve the content of the sentences, the BART 50% and Seq2Seq models are less successful.

Although the translation of the first Amazon sentence seems to be relatively obvious, the correct translation of the second and third examples appears less

Table 3. Original, reference translation and model sentiment transfer output examples on the Yelp test set.

Model	Output
Original	But was very disappointed with what actually arrived
Reference	Was very happy with what arrived
BART 40%/30%	Everything was very good with what actually arrived
BART 50%	But was very happy with what they did
Seq2Seq Saliency	But it was very nice and easy to say that was dessert.
Original	I do not like the size of the dance floor
Reference	I love the size of this dance floor!
BART 40%/30%	I also really like the size of the dance floor
BART 50%	I also really liked the feel of the dance floor
Seq2Seq Saliency	I like the size of the dance floor
Original	Its quiet and nice people are here
Reference	Nice people are here, but it is too quiet and boring
BART 40%/30%	Its dirty and rude people are here
BART 50%	Its not like they are very busy
Seq2Seq Saliency	Its quiet people are here

Table 4. Original, reference translation and model sentiment transfer output examples on the Amazon test set.

Model	Output
Original	The cookbook that comes with it is adequate
Reference	The cookbook that comes with it is terrible
BART 40%/30%	The cookbook that comes with it is terrible
BART 50%	the cookbook that came with it was terrible
Seq2Seq Saliency	The only good thing i ve used it is with it is adequate
Original	It s almost like putting the phone into a high end pair of socks
Reference	It s almost like putting the phone into a low end pair of socks
BART 40%/30%	It s not like putting the phone into a high end pair of shoes
BART 50%	It s not even worth the effort into a single pair of socks
Seq2Seq Saliency	It s almost like they are not into a high end pair of socks
Original	So not that great for leaving on at night
Reference	Perfect for night
BART 40%/30%	So far that works for leaving on at night. great product
BART 50%	So far that works for me on at work
Seq2Seq Saliency	So not that great for leaving a timer on

obvious. This is predominantly due to the ambiguity of the sentiment of the original sentences in these examples. These examples also show how the sentence content can become corrupted during sentiment transfer, both with the Seq2Seq examples and some of the BART 50% examples.

7   Conclusion

We proposed a self-supervised training pipeline for text style or sentiment transfer. We diverge from previous work by using an interpretable classifier to identify which sentiment words to mask or remove, and by fine-tuning a pretrained encoder-decoder Transformer, first as a DAE and then with self-supervised learning. The outputs of our models are preferred by human evaluators over a non-pretrained baseline with saliency-based sentiment word identification. The BART model also obtains a better trade-off between sentiment transfer accuracy and content preservation according to automatic evaluation. On Yelp the performance of our approach is comparable to the state-of-the-art, although on the Amazon dataset the approach performs less well. Self-supervised learning improves performance over a model that is already pre-trained and fine-tuned as a denoising autoencoder. As future research this method could be extended to text style transfer for low resource languages, given that parallel sentences are not required.

References

1. Bastings, J., Aziz, W., Titov, I.: Interpretable neural predictions with differentiable binary variables. In: ACL, no. 1, pp. 2963–2977 (2019)
2. Devlin, J., Chang, M., Lee, K., Toutanova, K.: BERT: pre-training of deep bidirectional transformers for language understanding. In: NAACL-HLT, no. 1, pp. 4171–4186 (2019)
3. He, J., Wang, X., Neubig, G., Berg-Kirkpatrick, T.: A probabilistic formulation of unsupervised text style transfer. In: ICLR (2020)
4. He, R., McAuley, J.: Ups and downs: modeling the visual evolution of fashion trends with one-class collaborative filtering. In: Proceedings of the 25th International Conference on World Wide Web, pp. 507–517 (2016)
5. Hochreiter, S., Schmidhuber, J.: Long short-term memory. Neural Comput. 9(8), 1735–1780 (1997)
6. Keskar, N.S., McCann, B., Varshney, L.R., Xiong, C., Socher, R.: Ctrl: a conditional transformer language model for controllable generation (2019)
7. Kingma, D.P., Welling, M.: Auto-encoding variational Bayes. In: ICLR (2014)
8. Krishna, K., Wieting, J., Iyyer, M.: Reformulating unsupervised style transfer as paraphrase generation. In: EMNLP, pp. 737–762 (2020)
9. Kumaraswamy, P.: A generalized probability density function for double-bounded random processes. J. Hydrol. 46(1–2), 79–88 (1980)
10. Lample, G., Subramanian, S., Smith, E., Denoyer, L., Ranzato, M., Boureau, Y.L.: Multiple-attribute text rewriting. In: ICML (2019). https://openreview.net/forum?id=H1g2NhC5KQ
11. Lei, T., Barzilay, R., Jaakkola, T.S.: Rationalizing neural predictions. In: EMNLP, pp. 107–117 (2016)
12. Lewis, M., et al.: BART: denoising sequence-to-sequence pre-training for natural language generation, translation, and comprehension. In: ACL, pp. 7871–7880 (2020)
13. Li, J., Jia, R., He, H., Liang, P.: Delete, retrieve, generate: a simple approach to sentiment and style transfer. In: NAACL-HLT, pp. 1865–1874 (2018)

14. Logeswaran, L., Lee, H., Bengio, S.: Content preserving text generation with attribute controls. In: Advances in Neural Information Processing Systems, pp. 5108–5118 (2018)
15. Papineni, K., Roukos, S., Ward, T., Zhu, W.J.: BLEU: a method for automatic evaluation of machine translation. In: Proceedings of the 40th Annual Meeting of the Association for Computational Linguistics, pp. 311–318 (2002)
16. Pennington, J., Socher, R., Manning, C.D.: Glove: global vectors for word representation. In: EMNLP, pp. 1532–1543 (2014)
17. Post, M.: A call for clarity in reporting BLEU scores. In: WMT, pp. 186–191 (2018)
18. Pryzant, R., Richard, D.M., Dass, N., Kurohashi, S., Jurafsky, D., Yang, D.: Automatically neutralizing subjective bias in text. In: AAAI (2020)
19. Radford, A., Narasimhan, K., Salimans, T., Sutskever, I.: Improving language understanding by generative pre-training (2018)
20. Radford, A., Wu, J., Child, R., Luan, D., Amodei, D., Sutskever, I.: Language models are unsupervised multitask learners (2019)
21. Rao, S., Tetreault, J.R.: Dear sir or madam, may I introduce the GYAFC dataset: corpus, benchmarks and metrics for formality style transfer. In: NAACL-HLT, pp. 129–140 (2018)
22. Nogueira dos Santos, C., Melnyk, I., Padhi, I.: Fighting offensive language on social media with unsupervised text style transfer. In: Proceedings of the 56th Annual Meeting of the Association for Computational Linguistics (Volume 2: Short Papers), pp. 189–194. Melbourne, Australia (2018). https://doi.org/10.18653/v1/P18-2031, https://aclanthology.org/P18-2031
23. Sudhakar, A., Upadhyay, B., Maheswaran, A.: Transforming delete, retrieve, generate approach for controlled text style transfer. In: EMNLP/IJCNLP, no. 1, pp. 3267–3277 (2019)
24. Sundararajan, M., Taly, A., Yan, Q.: Axiomatic attribution for deep networks. In: ICML, pp. 3319–3328. PMLR (2017)
25. Sutskever, I., Vinyals, O., Le, Q.V.: Sequence to sequence learning with neural networks. In: Advances in Neural Information Processing Systems, pp. 3104–3112 (2014)
26. Vaswani, A., et al.: Attention is all you need. In: Advances in Neural Information Processing, pp. 5998–6008 (2017)
27. Wang, K., Hua, H., Wan, X.: Controllable unsupervised text attribute transfer via editing entangled latent representation. In: Advances in Neural Information Processing Systems, pp. 11034–11044 (2019)
28. Wolf, T., et al.: Transformers: state-of-the-art natural language processing. In: EMNLP (Demos), pp. 38–45 (2020)
29. Wu, X., Zhang, T., Zang, L., Han, J., Hu, S.: Mask and infill: applying masked language model for sentiment transfer. In: IJCAI, pp. 5271–5277 (2019). https://doi.org/10.24963/ijcai.2019/732

Socio-Technical and Human-Centered AI

AI for Social Good: Sentiment Analysis to Detect Social Challenges in South Africa

Koena Ronny Mabokela[1][(✉)] ⓘ and Tim Schlippe[2]

[1] University of Johannesburg, Johannesburg, South Africa
`krmabokela@gmail.com`
[2] IU International University of Applied Sciences, Erfurt, Germany
`tim.schlippe@iu.org`

Abstract. Sentiment analysis has the potential to help analyse people's opinions and emotions on social issues [1]. We believe that in multilingual communities sentiment analysis systems should be even used to quickly discover what social challenges exist. This would help government departments target those issues more precisely and effectively. Consequently, in this paper we describe our experiments to apply cross-lingual sentiment analysis on South African tweets to detect social challenges described in English, Sepedi (i.e. Northern Sotho) and Setswana tweets. We investigated the polarities of the 10 most emerging topics in the tweets that fall within jurisdictional areas of 10 South African government departments. Our AI-driven systems indicate that the topics of *employment*, *police service*, *education*, and *health* are particularly problematic for the investigated multilingual communities since more than 50% of the tweets are categorised as *negative*, whereas the mood regarding the topics of *agriculture* and *rural development* is rather *positive*. Our developed systems can be easily extended to other topics and languages.

Keywords: AI for Social Good · Sentiment Analysis · Natural Language Processing · South Africa

1 Introduction

AI for Social Good (AI4SG) is an emerging subject of research that focuses on applying artificial intelligence (AI) to address significant social, environmental and public health issues that are facing society [2]. One can think of AI4SG as the intersection of AI with social sciences, health sciences as well as environmental sciences [3]. The field is focused on delivering positive social impact in accordance with the priorities outlined in the United Nations' Sustainable Development Goals [4] which are adopted by and applicable to South Africa. According to the 2019 Country Report, the National Development Plan has a 74% convergence with the Sustainable Development Goals and prioritises job creation, elimination of poverty, reduction of inequality and growing an inclusive economy [5]. South

A. Pillay et al. (Eds.): SACAIR 2022, CCIS 1734, pp. 309–322, 2022.
https://doi.org/10.1007/978-3-031-22321-1_21

Africa established a national coordinating mechanism to strengthen implementation of development policies. Therefore, South African national government departments such as *Health, Employment and Labour, Water and Sanitation, Rural Development, Education*, and *Police Service* have a mandate to support the development of these Sustainable Development Goals [5].

Sentiment analysis is the process of automatically detecting a sentiment from textual information and then classifying the information into classes such as *negative, neutral* or *positive* [6]. It is a growing research area of natural language processing (NLP) with proven track record using Twitter as a source for sentiment-related text data [7–10]. Many of the sentiment analysis applications are developed for English [11]. However, English is only spoken by 19% of the world population [12] and in multilingual communities other languages are also relevant. South Africa with its 11 official languages has a lot of multicultural and multilingual communities. However, its Niger-Congo Bantu languages are classified as low-resource languages except for English and Afrikaans [13] due to the lack of digital language resources. But South African citizen also communicate and express emotions on social media platforms such as Twitter in the Bantu languages. Consequently, a sentiment analysis application that can detect the sentiments from texts in the South African languages to address social challenges would be extremely beneficial.

As in [14], we define a social challenge or problem as "any condition or behavior that has negative consequences for large numbers of people and that is generally recognis(z)ed as a condition or behavior that needs to be addressed". Due to South Africa's unemployment rate of 33.9% [15], poor conditions in healthcare facilities, lack of educational tools as well as lack of water, electricity and sanitation [16,17] and other social challenges, for this study we decided to automatically analyse the following government departments related topics with the help of AI: *employment, sanitation, police service, education, health, small business, transport, home affair, rural development*, and *agriculture*. In this paper, we present a cross-lingual sentiment analysis approach to detect the social challenges experienced in South Africa.

Our systems employ Google's Bidirectional Encoder Representations from Transformers (BERT) model [18] to classify tweets into the 3 categories *negative, neutral* and *positive*. BERT is a pre-trained language model which was originally trained on over 100 languages using Wikipedia text corpora. We fine-tuned BERT to perform the sentiment analysis task with the help of the *SAfriSenti* corpus [19]. *SAfriSenti* is a multilingual sentiment corpus for South African languages. For our study, we additionally collected a new set of tweets, over 16,000 tweets in three languages containing our government departments related topics: the *SAGovTopicTweets* corpus. Our contributions are:

- With the help of Twitter's Academic API, we collected a new corpus of tweets in 3 South African languages covering 10 South African government departments related topics—the *SAGovTopicTweets* corpus.
- We leveraged the *SAfriSenti* corpus to fine-tune the BERT model for sentiment classification.

- We utilised Sepedi-English and Setswana-English machine translation to enable cross-lingual sentiment analysis systems for Sepedi and Setswana.
- Based on the fine-tuned BERT model and the machine translation, we classified the tweet in our *SAGovTopicTweets* corpus into *negative, neutral* and *positive* and analysed the polarity of the topics and for each topic the need to take action.
- Our results can be used as recommendations for the South African government departments to improve the social challenges identified on Twitter.

In the next section we will describe related work. The experimental setup of our collection and sentiment analysis of tweets in English, Sepedi and Setswana will be presented in Sect. 3. In Sect. 4 we will demonstrate the results of our experiments. Finally, we will summarise our work and indicate possible future steps.

2 Related Work

AI4SG is an emerging field of study with recent successful areas such as the development of AI interventions to improve community's well-being [20]. [21] provide a detailed analysis of approaches, use cases and examples in AI4SG. Numerous AI4SG applications use learning, reasoning, heuristic search, and problem-solving algorithms [21]. These algorithms are used by the majority of organisations and economic sectors [22]. The demand for AI applications that benefit society is great and may be used to solve several difficulties [23].

In the area of NLP for social good, [24] used sentiment analysis to automatically examine gender and race bias. [1] investigated sentiment analysis methods to classify the five major social issues of corruption, women violence, poverty, child abuse, and illiteracy. They collected English tweets and applied machine learning algorithms. The analysed text data is retrieved from microblogging services like Twitter since these sources share situational information, cover a lot of topics and contain negative, neutral and positive tweets [25,26]. Several studies investigated different data collection methods for tweets [7,8,25–27]. [25] explored methods to collect millions of annotated tweets from various places, hours, and writers. Other studies used emoticons and keywords [7,8] to extract and build Twitter-based corpora via distant supervision. [8]'s and [10]'s corpora contain code-switched tweets. To ensure a correct labeling, [10] let the tweets label by three annotators following the *SentiStrength* [28] strategy. [19] use the distant supervised methods with emoticons and keywords together with a word frequency based language identification to collect tweets in Sepedi, Setswana and English.

For automatic sentiment analysis, different machine learning algorithms like support vector machines, decision trees, random forests, multilayer perceptrons and long short-term memories were analysed [29–32]. [11] demonstrated that the Transformer models BERT [18] and RoBERTa [33] (Robustly Optimized BERT Pretraining Approach) usually outperformed the other machine learning algorithms. Lexicon-based approaches were also investigated, e.g. in [34,35], but machine learning algorithms usually outperform the lexicon-based approaches.

Some researchers propose cross-lingual NLP approaches to solve the problems of low-resource languages by benefiting from rich-resource languages like English [11,28,29,36,37]. For sentiment analysis, they usually translate the comments from the original low-resource language to English. This allows to do the classification task of sentiment analysis with well-performing models trained with a lot of English resources.

In our work we used the pre-trained BERT model [18] to build an English sentiment analysis system that is an essential component of our monolingual English and our two cross-lingual Sepedi and Setswana sentiment analysis systems.

3   Experimental Setup

In this section, we will first describe how we used the Twitter API and a word frequency based language identification to gather South African tweets in the languages English, Sepedi and Setswana which contain the 10 most emerging topics that fall within jurisdictional areas of 10 South African government departments. Then we will present the data set which we used to train our English sentiment analysis system that is an essential component of our monolingual and our two cross-lingual sentiment analysis systems to classify the collected tweets into *negative*, *neutral* and *positive*.

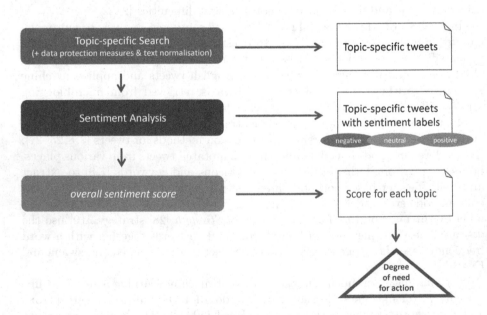

Fig. 1. Pipeline of topic-specific search, sentiment analysis and scoring.

3.1   Overview of Our Systems

Figure 1 shows the pipeline of our systems. In the first step, topic-specific tweets are collected with the help of search terms and data protection measures and text normalisation steps are applied. Then a sentiment analysis system classifies the tweets into *negative*, *neutral* and *positive*. In the last step, an overall sentiment score is computed for each topic that indicates the degree of need for action.

3.2   Collection of SAGovTopicTweets

For our study the goal was to collect South African tweets in English, Setswana and Sepedi covering the topics *employment, sanitation, police service, education, health, small business, transport, home affair, rural development,* and *agriculture*—the *SAGovTopicTweets* corpus. We focused on these 10 government departments related topics since they were highlighted in the State of the Nation Address for 2021 as key government issues to strengthen the economy [38].

Over the past years, Twitter has taken steps to expand and improve the Twitter API functions [39]. For example, Twitter's Academic API[1] has recently been released with the ability to get historical tweets dated back to 2006, a cap of 10 million tweets per month and more advanced filter functionality to collect relevant data [39]. The API offers the capability to collect a large pool of real-time and historic tweets for free.

Consequently, we used the Twitter API for Academic Research to crawl tweets for 24 h which were posted between January and August 2022. To find exclusively tweets which cover our 10 government departments related topics, we searched for tweets in the following two ways:

1. We used the government departments' names[2] *Employment and Labour, Education, Police Service, Rural Development, Health, Small Business, Transport, Home Affairs, Water and Sanitation* and *Agriculture* as search terms in English and their Sepedi and Setswana translations.
2. We collected tweets which are comments on the tweets of the 10 government departments which can be found on the departments' Twitter handles.

Using Twitter's geolocation feature, we ensured that we only collected tweets from South Africa. With the help of a language identification based on word frequency [19], we also made sure that we exclusively collected tweets from our three target languages. Based on the search term that led to its download, each tweet was tagged with its topic. Once downloaded, all tweets in *SAGovTopicTweets* were subjected to our strict preprocessing steps, which we also apply in our *SAfriSenti* corpus collection. These steps are described in [19] in detail and contain data protection measures and text normalisation.

Table 1 shows three examples of government departments related South African tweets which are classified as *negative*. The first tweet is in English and

---

[1] https://developer.twitter.com/en/use-cases/do-research/academic-research.
[2] https://www.gov.za/about-government/government-system/national-departments.

Table 1. Examples of government topic related South African tweets.

English tweet [home affairs] [negative]	The queue in home affairs at Wynberg is very slow One cannot get anything without waiting the service is bad
Sepedi tweet [sanitation] [negative]	Ke neng re lla ka mohlagase le makhura tsa go tura
English translation	I was crying about expensive electricity and fuel
Setswana tweet [health] [negative]	Puso ya rena ya ANC ga e re hlokomele ka tsa maphelo
English translation	Our ANC government does not take care of our health

covers the topic of *home affairs*. The words "very slow" and "bad" indicate the negative sentiment. The second tweet is in Sepedi and contains information on the topic of *sanitation*. The Sepedi words "re lla" (crying) and "go tura" (expensive) reflect the negative mood. The third tweet is in Setswana and belongs to the topic of *health*. The Setswana word sequence "ga e re hlokomele" (not take care) describes the negative attitude.

3.3   The SAfriSenti Corpus

The *SAfriSenti* corpus is to date the largest sentiment dataset available for South African languages with 64.3% of monolingual tweets in English, Sepedi and Setswana and 36.6% of code-switched tweets between these languages [19].

For our experiments, we trained an English sentiment analysis system that is used to classify the collected English tweets plus the Sepedi and Setswana tweets translated into English. For that, we used a subset of the English part of *SAfriSenti* with 5,998 tweets that were labeled in a semi-automatic process with strict annotation guidelines [19]. To evaluate the sentiment analysis systems' performances on high-qualified labeled data which will be described in Sect. 3.4, we used 1,499 different English tweets, 2,106 Sepedi and 1,053 Setswana tweets from *SAfriSenti*.

3.4   Sentiment Analysis

As visualised in Fig. 2, we used one monolingual and two cross-lingual sentiment analysis systems to classify the collected tweets into *negative, neutral* and *positive*. In all three systems, an English sentiment analysis system (*English Sentiment Analysis*) was the essential component, which was trained using our

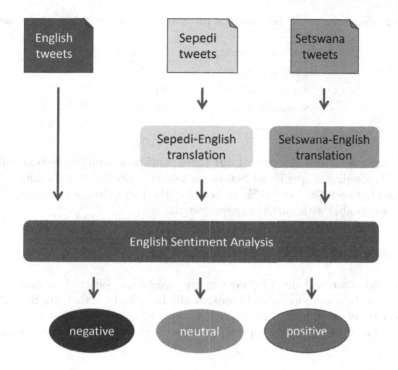

Fig. 2. Sentiment analysis for english, sepedi and setswana.

5,998 English labeled tweets from the *SAfriSenti* corpus. In the monolingual system, the English tweets were directly classified with the English sentiment analysis system. In contrast, in the cross-lingual systems the tweets were machine-translated from Sepedi and Setswana to English and then classified with the English sentiment analysis system. For the Sepedi-English machine translation task, we used Google's Neural Machine Translation System [40]. An overview of the system's BLEU scores over languages is given in [41]. For the Setswana-English machine translation task, we used the Autshumato Machine Translation Web Service[3] [42] as Setswana-English machine translation was not available in Google's Neural Machine Translation System. We randomly checked the quality of the English output and can report that it was acceptable.

Our sentiment analysis system is based on the English BERT[4]. We trained our BERT models with 4 epochs and a batch size of 16 using the AdamW optimizer [43] with an initial learning rate of 2e−5. Furthermore, we used a dropout layer for some regularisation and a fully-connected layer for our output. To get the predicted probabilities from our model, we applied a softmax function to the output. For the implementation, we used Google Colab[5].

---

[3] https://mt.nwu.ac.za.

[4] https://huggingface.co/bert-base-uncased.

[5] https://colab.research.google.com.

Table 2. Performances of the sentiment analysis systems (on SAfriSenti test sets).

System	Accuracy (%)	F-score (%)
English	86.36	86.01
Sepedi	84.19	84.03
Setswana	83.26	82.74

The accuracies and F-scores of our three sentiment analysis systems on the *SAfriSenti* English, Sepedi and Setswana test sets are listed in Table 2. With accuracies between 83% and 86%, we see that the prediction of the three classes is quite acceptable with our AI-driven systems.

4   Experiments and Results

In this section we will describe how many tweets we collected to analyse the social issues. Furthermore, we will look at the benefits of classifying Sepedi and Setswana tweets in addition to English tweets. Finally, we will investigate the polarity of each investigated topic and draw conclusions about the urgency to act.

4.1   Language Distributions of Collected Tweets

Figure 3 illustrates the distribution of our collected tweets over languages and topics. In total, we collected 16,787 tweets with the Twitter API covering the topics of *employment, police service, education, health, small business, transport, home affair, rural development,* and *agriculture*. We see that for some topics such as *transport* (2,172 tweets), *home affair* (2,232 tweets) and *sanitation* (2,361 tweets), we found significantly more tweets, while for other topics such as *rural development* (621 tweets) and *agriculture* (1,011 tweets) much less tweets existed. Moreover, most tweets were in English (12,102 tweets, i.e. 72.1%), second most in Sepedi (3,088 tweets, i.e. 18.4%) and least in Setswana (1,597 tweets, i.e. 9.5%). We expected to see significantly more tweets in English, as many Sepedi and Setswana speakers communicate in English on the Internet. Nevertheless, we wanted to find out how strong the impact of the 27.9% tweets from Sepedi and Setswana is on the average polarities of the individual topics.

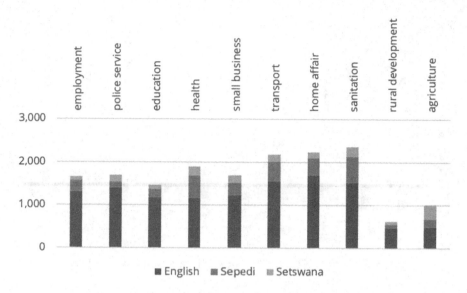

Fig. 3. Numbers of collected tweets over languages and topics.

4.2 Sentiment Analysis to Detect Social Challenges

To represent the distribution of the classified tweets with only one score, we defined an *overall sentiment score* in the following way:

$$overall\ sentiment\ score = \frac{\#negative * (-1) + 0 * \#neutral * (+1) * \#positive}{\#allsentiments}$$

The *overall sentiment score* lies between −1 and +1, where −1 expresses a completely negative sentiment and +1 a completely positive sentiment. The benefit of this score is that it gives a clear tendency in only one score and makes it easier to compare the topics. Based on the *overall sentiment score*, the governmental institutions can strategically decide according to which priority they will tackle the problems identified from the tweets. This lays a foundation for a recommender system that automatically analyses the polarity in text data on the internet and makes recommendations based on the value of the score where action is needed. Of course, our formula can be extended or its factors be adapted in case of more sentiment classes (e.g., very negative, negative, neutral, positive, very positive) or if neutral tweets—which are not weights in our case— should be given more weight.

Figure 4 shows the *overall sentiment score* distribution of English tweets (*sentiment score_EN*) compared to adding Sepedi and Setswana tweets (*sentiment score_all3*) over our 10 investigated topics based on the sentiment classification of our systems. We see that the *overall sentiment scores* of 8 of the 10 topics are in the negative number range. The topics *employment* and *police service* are particularly problematic for the investigated multilingual communities since their scores are less than -0.3, whereas with positive scores the mood regarding the

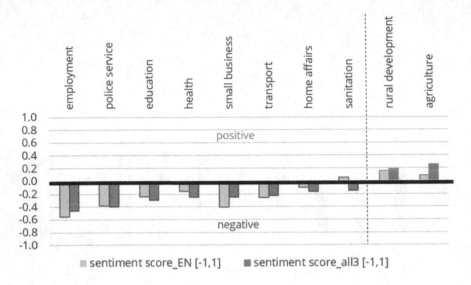

Fig. 4. Overall sentiment scores of the investigated topics.

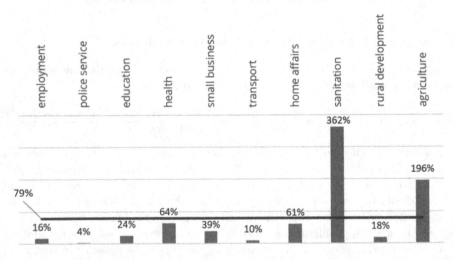

Fig. 5. Relative change in overall sentiment score with *all3* compared to *EN*.

topics of *agriculture* and *rural development* is rather *positive*. It can also be seen that the scores differ greatly in some cases if only English tweets are considered (*EN*) and if tweets in Sepedi and Setswana are also taken into account (*all3*).

Figure 5 visualises the relative change in *overall sentiment scores* with *all3* compared to *EN*. We see that there are differences between 4% and 362%. The average deviation is 79%. This shows that it is very important to add tweets in other languages besides English to ensure a fair representative analysis.

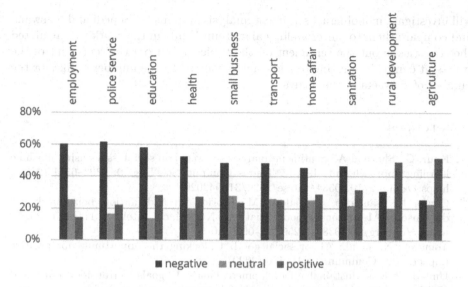

Fig. 6. Sentiment distribution of the investigated topics.

Figure 6 shows the sentiment distribution of the classes *negative, neutral* and *positive* over the 10 topics analysed. Here, too, we see what the *overall sentiment scores* in Fig. 4 have already indicated: The topics *employment, police service, education,* and *health* are particularly problematic for the investigated multilingual communities—more than 50% of the tweets are categorised as *negative*. The mood regarding the topics of *agriculture* and *rural development* is rather *positive*.

5 Conclusion and Future Work

With a lack of service delivery and enormous social challenges in South Africa [16], we require technologies which can assist the government to make informed decisions based on the perceptions from the citizens. Consequently, in this paper we have demonstrated that multilingual sentiment analysis is able to detect social challenges in South Africa. We investigated the polarities of the 10 most emerging topics in the tweets that fall within jurisdictional areas of 10 South African government departments. Our AI-driven systems indicate that the topics of *employment, police service, education,* and *health* are particularly problematic for the investigated multilingual communities since more than 50% of the tweets are categorised as *negative*, whereas the mood regarding the topics of *agriculture* and *rural development* is rather *positive*.

To further understand the details of the social challenges, our next goal is to retrieve further information from the tweets with the help of information extraction algorithms. Moreover, we plan to extend our developed systems and data collections to other topics, languages and sentiment classes. Additionally, we will investigate monolingual sentiment analysis systems for Sepedi and Setswana and compare them to our cross-lingual system. While in this work we calculated the statistics about the sentiment on the topics only from the sentiment of the tweets, it can be interesting to add other features like the number of likes or the number of retweets in the analysis.

References

1. Kaur, C., Sharma, A.: Sentiment analysis of tweets on social issues using machine learning approach. Int. J. Adv. Trends Comput. Sci. Eng. **9**, 6303–6311 (2020). https://doi.org/10.30534/ijatcse/2020/310942020
2. Cowls, J., Tsamados, A., Taddeo, M., Floridi, L.: A definition, benchmark and database of AI for social good initiatives. Nat. Mach. Intell. **3**, 111–115 (2021). https://doi.org/10.1038/s42256-021-00296-0
3. Tomašev, N., et al.: AI for social good: Unlocking the opportunity for positive impact. Nat. Commun. **11**(1), 1–6 (2020)
4. United Nations: Sustainable Development Goals: 17 goals to transform our world (2022). https://www.un.org/sustainabledevelopment/sustainabledevelopment-goals. Accessed Aug 2022
5. Sustainable Development Goals: Country report 2019 – South Africa. Technical report ISBN 978-0-621-47619-4, Statistics South Africa (2019)
6. Wankhade, M., Rao, A., Kulkarni, C.: A survey on sentiment analysis methods, applications, and challenges. Artif. Intell. Rev. 1–50 (2022). https://doi.org/10.1007/s10462-022-10144-1
7. Pak, A., Paroubek, P.: Twitter as a corpus for sentiment analysis and opinion mining. In: The 7th Edition of the Language Resources and Evaluation Conference (LREC 2010), pp. 1320–1326 (2010)
8. Nakov, P., Ritter, A., Rosenthal, S., Sebastiani, F., Stoyanov, V.: SemEval-2016 task 4: Sentiment analysis in Twitter. In: International Workshop on Semantic Evaluation (SemEval) (2016)
9. Nguyen, H., Nguyen, M.-L.: A deep neural architecture for sentence-level sentiment classification in Twitter social networking. In: Hasida, K., Pa, W.P. (eds.) PACLING 2017. CCIS, vol. 781, pp. 15–27. Springer, Singapore (2018). https://doi.org/10.1007/978-981-10-8438-6_2
10. Vilares, D., Alonso, M.A., Gómez-Rodríguez, C.: Sentiment analysis on monolingual, multilingual and code-switching Twitter corpora. In: The 6th Workshop on Computational Approaches to Subjectivity, Sentiment and Social Media Analysis, pp. 2–8 (2015)
11. Rakhmanov, O., Schlippe, T.: Sentiment analysis for Hausa: Classifying students' comments. In: The 1st Annual Meeting of the ELRA/ISCA Special Interest Group on Under-Resourced Languages (SIGUL 2022). Marseille, France (2022)
12. Statista: The most spoken languages worldwide in 2022 (2022). https://www.statista.com/statistics/266808/the-most-spoken-languages-worldwide. Accessed Aug 2022
13. Duvenhage, B.: Short text language identification for under resourced languages. In: South African Forum for Artificial Intelligence Research (FAIR 2019) (2019)
14. University of Minnesota: Social problems (2022). https://open.lib.umn.edu/socialproblems/chapter/1-1-what-is-a-social-problem. Accessed Oct 2022

15. Quarterly Labour Force Survey - Quarter 2: 2022 (2022). https://www.statssa.gov.
    za/publications/P0211/P02112ndQuarter2022.pdf
16. Lorraine, M., Molapo, R.: South Africa's challenges of realising her socio-
    economic rights. Mediterr. J. Soc. Sci. **5** (2014). https://doi.org/10.5901/mjss.2014.
    v5n27p900
17. Nkomo, S.: Public service delivery in South Africa councillors and citizens critical
    links in overcoming persistent inequities. Technical report 42, Afrobarometer (2017)
18. Devlin, J., Chang, M.W., Lee, K., Toutanova, K.: BERT: Pre-training of deep
    bidirectional transformers for language understanding. In: NAACL (2019)
19. Mabokela, K.R., Schlippe, T.: A sentiment corpus for South African under-
    resourced languages in a multilingual context. In: The 1st Annual Meeting of
    the ELRA/ISCA Special Interest Group on Under-Resourced Languages (SIGUL
    2022), pp. 70–77 (06 2022)
20. Musikanski, L., Rakova, B., Bradbury, J., Phillips, R., Manson, M.: Artificial intel-
    ligence and community well-being: A proposal for an emerging area of research.
    Int. J. Commun. Well-Being **3**(1), 39–55 (2020). https://doi.org/10.1007/s42413-
    019-00054-6
21. Shi, Z.R., Wang, C., Fang, F.: Artificial intelligence for social good: A survey.
    CoRR abs/2001.01818 (2020)
22. Bjola, C.: AI for development: Implications for theory and practice. Oxford Dev.
    Stud. **50**(1), 78–90 (2022). https://doi.org/10.1080/13600818.2021.1960960
23. Hager, G., et al.: Artificial intelligence for social good. ArXiv abs/1901.05406
    (2019)
24. Kiritchenko, S., Mohammad, S.M.: Examining gender and race bias in two hundred
    sentiment analysis systems. ArXiv abs/1805.04508 (2018)
25. Go, A., Bhayani, R., Huang, L.: Twitter sentiment classification using distant
    supervision. Processing **150** (2009)
26. Indriani, D., Nasution, A.H., Monika, W., Nasution, S.: Towards a sentiment ana-
    lyzer for low-resource languages. CoRR abs/2011.06382 (2020)
27. Agarwal, A., Sabharwal, J.S.: End-to-end sentiment analysis of Twitter data. In:
    Conference: Proceedings of the Workshop on Information Extraction and Entity
    Analytics on Social Media Data (2012)
28. Vilares, D., Alonso Pardo, M., Gómez-Rodríguez, C.: Supervised sentiment anal-
    ysis in multilingual environments. Inf. Process. Manage. **53**(3), 595–607 (2017).
    https://doi.org/10.1016/j.ipm.2017.01.004
29. Balahur, A., Turchi, M.: Comparative experiments using supervised learning and
    machine translation for multilingual sentiment analysis. Comput. Speech Lang. **28**,
    56–75 (2014)
30. Nguyen, P.X.V., Hong, T.V.T., Nguyen, K.V., Nguyen, N.L.T.: Deep learning ver-
    sus traditional classifiers on Vietnamese students' feedback corpus. In: The 5th
    NAFOSTED Conference on Information and Computer Science (NICS) (2018)
31. Kumar, A., Sharan, A.: Deep learning-based frameworks for aspect-based senti-
    ment analysis. In: Agarwal, B., Nayak, R., Mittal, N., Patnaik, S. (eds.) Deep
    Learning-Based Approaches for Sentiment Analysis. AIS, pp. 139–158. Springer,
    Singapore (2020). https://doi.org/10.1007/978-981-15-1216-2_6
32. Rakhmanov, O.: A comparative study on vectorization and classification techniques
    in sentiment analysis to classify student-lecturer comments. Procedia Comput. Sci.
    178, 194–204 (2020)
33. Liu, Y., et al.: RoBERTa: A robustly optimized BERT pretraining approach (2019)

34. Kolchyna, O., Souza, T.T.P., Treleaven, P.C., Aste, T.: Twitter sentiment analysis: Lexicon method. computation and language, machine learning method and their combination. arXiv (2015)
35. Kotelnikova, A., Paschenko, D., Bochenina, K., Kotelnikov, E.: Lexicon-based methods vs. AIST, BERT for text sentiment analysis. In: International Conference on Analysis of Images. Social Networks and Texts, pp. 71–83. Springer, Cham (2021)
36. Lin, Z., Jin, X., Xu, X., Wang, Y., Tan, S., Cheng, X.: Make it possible: Multilingual sentiment analysis without much prior knowledge. In: IEEE/WIC/ACM International Joint Conferences on Web Intelligence (WI) and Intelligent Agent Technologies (IAT), vol. 2, pp. 79–86 (2014). https://doi.org/10.1109/WI-IAT.2014.83
37. Can, E.F., Ezen-Can, A., Can, F.: Multilingual sentiment analysis: An RNN-based framework for limited data. In: ACM SIGIR 2018 Workshop on Learning from Limited or Noisy Data (2018)
38. Ramaphosa, C.: State of the nation address (2021). https://www.stateofthenation.gov.za/assets/2021/SONA%202021.pdf. Accessed Aug 2022
39. Twitter Developers: A Python wrapper around the Twitter API (2021). https://python-twitter.readthedocs.io/en/latest. Accessed Aug 2022
40. Wu, Y., et al.: Google's neural machine translation system: Bridging the gap between human and machine translation (2016). https://arxiv.org/abs/1609.08144
41. Aiken, M.W.: An updated evaluation of Google translate accuracy. Stud. Linguist. Lit. (2019)
42. Biljon, E.V., Pretorius, A., Kreutzer, J.: On optimal transformer depth for low-resource language translation. In: The International Conference on Learning Representations (ICLR 2020) (2020)
43. Kingma, D.P., Ba, J.: Adam: A method for stochastic optimization (2014). https://doi.org/10.48550/ARXIV.1412.6980

A Model for Biometric Selection in Public Services Sector

Mapula Elisa Maeko and Dustin van der Haar[(✉)]

University of Johannesburg, Auckland Park, PO Box 524, Johannesburg 2006, South Africa
dvanderhaar@uj.ac.za

Abstract. The need to authenticate people using their biometric attributes and tighten information security in organisations significantly increased over the years and public services are no exception. Selecting suitable, robust, relevant and beneficial multimodal biometric attributes in public services environment for person authentication and access control is essential. The major challenge is deploying the wrong multimodal biometric technology in the organisation, which results in failed system deployment. Artificial intelligence (AI) has the potential to significantly drive the adoption and deployment of multimodal biometric authentication in public services. The study recommends a multimodal biometrics selection model for authentication to prevent fraudulent and invalid documents for identification. This study focuses on the human factor elements of public awareness, acceptance, perception and usability relevant to multimodal biometric deployment success. The formalised model proposed in the study could be of value to public services that need to deploy multimodal biometric authentication technologies, thereby minimising future failed deployments.

Keywords: Artificial intelligence · Multimodal biometrics · Acceptance · Usability · Deployment

1 Introduction

Authentication of persons through their biometric attributes to access applications, systems, and the physical environment is a fast-growing technology among individuals and organisations. It is a business need and an individual need to protect personal or confidential data from unauthorised access [5]. Knowing which biometric method to choose for an organisation and protecting its valuable information is complex. There is still some confusion on which attributes to consider when selecting the appropriate biometric solution for the licencing centre environment. Public services have been around for many years and are used for a person's identification and verification. The driving licence card is primarily used as a form of identification to indicate that one is licenced to drive a vehicle. This information connects to the available database information about the person. The issue of operating a motor vehicle with fake, invalid driver's licence documents or unlicensed driving is the main contributor to road accidents and fatal crashes in many countries worldwide [21, 25, 27]. The issue of road crashes causes immeasurable psychological and physical trauma, and they negatively impact the world economy

A. Pillay et al. (Eds.): SACAIR 2022, CCIS 1734, pp. 323–334, 2022.
https://doi.org/10.1007/978-3-031-22321-1_22

with costs of over a billion US dollars [21, 26]. A study conducted on modelling and simulation of the Nigerian Airspace Management Agency Billing System using python simulation packages highlights that the high percentage of unlicenced/unauthorised drive vehicles is likely a widespread concern worldwide [11, 20]. The problem is made worse by issuing fake/counterfeit and illegal driving licences, which lead to scams, fraud and corruption that occur [28, 30].

The use of multimodal biometric technologies can reduce corruption, eradicate fraud-ulent driving documents and identity fraud and improve the integrity and reliability of the driver's licence document [7, 10]. However, choosing the suitable biometric modality or correct combination of biometric modalities is difficult considering the varying com-plexities in each environment, along with their user base. Selecting multimodal biometric authentication technology for use in a large-scale environment such as public services depend on its needs, the budget cost, technical infrastructure and user acceptance of the technology [14, 15]. The study argues that driver licencing centres could benefit from multimodal biometric access control that focuses on user awareness, positive per-ception and usability assessment of multimodal biometric technologies. We propose a model that promotes selecting appropriate multimodal biometric attributes for biomet-ric authentication solutions in driver licencing centres. The entire paper is organised as follows; the next section is Sect. 2, which discusses the problem background. Section 3 discusses related work on technology acceptance, Sect. 4 is about the Evaluation of biometric authentication methods, and Sect. 5 is about A model discussion for multi-modal biometric selection and its components. Finally, Sect. 6 concludes the paper with recommendations.

2   Problem Background

This section reviews similar works in the literature to understand the factors that affect the selection of multimodal biometric attributes for implementing a multimodal bio-metric authentication system in public services. The factors affecting biometric attribute selection are an essential tool for deploying a suitable technology for the driver licencing and testing centres (DLTCs) environment and can be leveraged in biometric selection for a successful deployment or the adoption of multimodal biometric solutions in an organisation [22]. A more systematic biometric selection approach of an authentica-tion system could benefit the organisation by ensuring the highest accuracy, improved security detection and reduced duplicate identities to help ensure an appropriate bio-metric system is deployed [13, 15]. Sharma [29] highlighted the potential benefits of the Aadhaar system to the people who cannot prove their identities at each step they take. Thus, underprivileged people can benefit from available social welfare schemes and access to the public's resource distribution systems. However, some limitations of the Aadhar system are duplicate numbers, lack of awareness, failures in authentication, identity theft, fraud, and bribes since its inception [30]. Multimodal biometrics in this study for fingerprint and iris are used to uniquely identify a person, ensuring one Aadhar unique number for one beneficiary. Fingerprint and iris multimodal biometrics were in this study used to enable them to identify fake identities, allowing a fair distribution of social welfare schemes [17]. However, challenges were identified, such as rolling out the project on a large scale, which requires vast infrastructure and extensive databases for a large population [14, 29].

Whitman [34] conducted a study investigating the ease of using biometric security solutions in a public environment (Telecare industry). According to the author, it is significant to consider ease of use through human authentication while selecting a biometric solution. They revealed that a three-dimensional choice rubric could demonstrate the effectiveness of choosing a multimodal biometric solution that ensures ease of using the technology among users, organisations, and the technological perspective. In this study, more careful consideration of the stakeholder's perspective on the ease of use of the multimodal biometric solution for selection is required focusing on a broader view. The purpose is to close the gap in the digital divide by including the public in accessing public services and promoting users to learn how to use the technological solution effectively. The Aadhar management system proposed using biometrics relating to public services to address challenges posed by information security breaches and maintain data quality by selecting the right technology to manage a high or massive volume of public demographical data and biometric data. Like the Aadhar system, the driver licensing environment processes large volumes of general demographics. The authors propose multimodal biometric technology to achieve better quality data and a high level of accuracy and security. Omolara [20] reviewed the importance of the benefits of multimodal biometric systems, their applications and their performance. The review of multimodal biometric systems is significant in ensuring a reliable and secure method. It shows a need to increase security in government organisations, financial institutions, and border control in a developing country such as West Africa, similar to security needs in any developed country. The author highlights that combining one or more biometric traits for human identification promises robust and high accuracy at a decision level. Moreover, Over the years, the public sector has seen an increase in the deployment of AI to render services to the public. The study refers to the work conducted by Mikalef et al. [16] highlights the influencing factors of enabling AI deployment in government services. Based on the discussed literature, we can see that the biometric selection of an authentication system can benefit the organisation by ensuring the highest accuracy, improved security and detecting duplicate identities [35]. The authors suggest that choosing the right biometric combination will result in a better mechanism to authenticate persons in public service to identify a person and can help ensure that only authorised users access driver's licence systems.

3 Related Work on Technology Acceptance

Several factors affect the selection of biometric attributes and the acceptance of biometrics authentication systems. Organisations require a deeper analysis of the type of environment and infrastructure, cultural, legal aspects and stakeholders to ensure they mitigate the risk of deploying unsuitable and acceptable technology. It shows that for successful biometric selection, the organisation must have the plan to manage change and public perception, which results in positive acceptability. The technology acceptance frameworks identified in this study form a significant part of multimodal biometric attributes selection and the implementation of the authentication system. We explore the three-technology acceptance framework, which plays a critical role in selecting suitable biometric attributes. The following models were identified for biometric attribute selection the Technology acceptance model (TAM), the Unified Theory of acceptance of user technology model (UTAUT), and the Theory of planned behaviour (TPB). TAM is applicable for assisting the user in using and accepting the new technology; in this instance, TAM may assist in the decision to select the right combination of biometric attributes. The TAM model helps ensure a better understanding of attitudes towards the biometric system, in which attitudes negatively or positively affect selection. However, TAM has limitations that do not address technology acceptance and successful application for a largescale environment such as the driver licencing environment. However, TAM could be useful for individual users' applications as influenced by the intention of the user to adopt the technology [2].

UTAUT model applies to the perception and the intention to use biometric systems, and it ensures the selection of suitable biometric attributes for acceptance among the users. It further focuses on public perception, social influence and facilitating conditions for user acceptance [33]. The model helps better understand the importance of perception on a person's intention towards using a biometric system. Though many researchers have cited UTAUT as an excellent method to measure the intention for technology adoption, the Theory is not extensively used in the practical environment, influencing the intention to adopt the technology. UTAUT theory has been used to support arguments and is limited in evaluating the factors influencing the technology's underlying consequences.

The Theory of Planned Behaviour (TPB) shows significant relevance in the study as attention is drawn to the crucial factors of selecting the right biometric attributes for a suitable multimodal biometric authentication system. This model supports the prediction of an individual's behavioural intention to accept and the intention to use the biometric system. In biometric attributes selection, the model ensures that predicting behaviour shows the impact on the user's expectation and behaviour regarding the current and future use of the biometric system. Planned behaviour is influenced further by the individual's knowledge and understanding of the biometric authentication system, which allows selecting between the various biometric attributes [2]. However, TPB has some limitations in that the Theory does not consider other external factors, such as the users' behaviour, to evaluate the intention to adopt the technology. The model fails to control the perception of behaviour and belief control [9].

Moreover, the biometric selection also needs to consider the system's reliability and convenience to achieve its purpose: security and the system's ability to reduce security incidents and prevent attacks. Information security is at the core, providing value for the organisation's information and resources. The organisation must consider tackling security concerns to ensure the privacy of the information supplied and stored in the future [6, 19, 31]. Information security ensures confidentiality, integrity, and accountability and influences the perception and usability of biometric technology. The selected multimodal biometric technology should consider its feasibility factor for the organisation's future while maintaining ample security for its users [10, 17].

4 Evaluation of Biometric Authentication Methods

Pooe and Labuschagne et al. [22] investigated the factors impacting the implementation of biometric technology by banks, and their study shows that gaps exist in the successful adoption of technology. They investigated user acceptance, perception, the technology, culture, reliability, security and privacy of authentication technology to enhance information security among banks despite the biometric authentication technology not being broadly implemented. The study concluded that the challenges in integrating technology and lack of awareness might be why a biometric authentication system is not entirely implemented in the banks.

Wolf et al. [35] highlighted the challenges faced by expected and non-expect users on new biometric authentication, where both experienced reliability and dissatisfaction with the reliability of technology, resulting in poor adoption of the biometric technology, even though a higher level of security explained. The risk of adopting a technology with low confidence levels increases the concerns about the reliability of the biometric technology. Therefore, careful considerations of the implementation of the biometric authentication technology, the people who use the technology and the environment are necessary, thereby further motivating the need to include biometric attribute selection at the initial stage.

A study on biometric identification by Sailaja et al. [26] states that driver verification using the Aadhaar database provides a solution to the issues of checking a vehicle driver using the fingerprint biometric. The advantages of the study are that, for example, when vehicle drivers get reviewed, should the driver not be found in the database, the driver must then pay a fine. The disadvantage is that a driver is not successfully verified in the event of an error rate or false match [29]. Therefore, the law enforcement agency should act against the driver due to a false match, and the disadvantage results in low user buy-in [21]. It is a complex task to select a suitable solution for a large-scale environment considering usability, users' perspective, and a cost-effective solution to ensure deployment will not fail. Biometric-based technological authentication systems are not without limitations; for example, if an individual's identity is compromised, the biometric traits cannot be changed and potentially make other systems that use that same biometric vulnerable [13, 14, 19].

No single study based on the highlighted literature and observations in this study draws attention to the impact of biometric selection for deployment, whether for a developing or a developed country. Biometric selection mechanisms used for biometric deployment on a large scale are lacking [14]. Biometric attribute selection of an authentication system could benefit the organisation in ensuring the highest accuracy and improved security detection of duplicate identities. The primary focus is on various aspects of biometric implementation.

This study's primary focus is on the factors affecting the biometric attributes selection considering the issue of usability of the technology, the security of collected information, awareness, perception, acceptability from the stakeholders, and public satisfaction. This study focuses on addressing the identified gaps. We hope selecting the appropriate multimodal biometric solution will help prevent fake driver's licenses, invalid driver's licenses, fraud and corruption, and reduce the issue of inadequate driver authentication and illegitimate driver applicants in the driver licencing centres [17]. Organisations can deploy almost any biometric technology; however, successful deployment only depends on selecting the appropriate biometric attributes for a multimodal biometric authentication system for the stakeholders using the solution [23]. In the various studies, we lack biometric selection guidelines, such as selecting and deploying a particular multimodal biometric solution, identifying the required functions, and the expectation from the solution. In this instance, deployment becomes a failure, raising negative cost implications [30]. Selection guidelines could prevent the problem of the risk of loss and the cost implications due to deployment failures.

5   A Model Discussion for Multimodal Biometric Selection

Multimodal biometric selection allows an organisation to select an appropriate biometric solution according to its requirements, and there are various factors to consider when choosing a suitable biometric technology [18]. An organisation needs to compare and evaluate the appropriateness of the solution regarding infrastructure considerations, usability, robustness, performance, accuracy, security and cost. We propose a more systematic approach to choose the most appropriate biometric modality to fully benefit the driving licence centres. In addressing the identified gaps, the multimodal biometric selection model focuses exclusively on the human factor element for directing public awareness, acceptance, perception and the usability of the authentication system. It is at the initial stage of the system design process that the organisation may use the model aiming for high levels of technology acceptance [16].

Moreover, organisations must consider ethical, legal, religious and cultural aspects that may arise during biometric selection. Efficiency and effectiveness are critical factors for ensuring user satisfaction with the biometric system so that users can accomplish tasks efficiently and on time [18, 28]. Figure 1 below shows the proposed biometric selection model based on public awareness, usability, perception, acceptance and satisfaction. The biometric selection incorporates these factors to select a suitable, robust and acceptable multimodal biometric authentication technology. The following section discusses usability as a significant factor in the biometric attribute selection model.

5.1   Usability

The significant element of usability measures the impact of usability on multimodal biometric attribute selection. The usability factor concerns the user's ability to perform tasks. Social and cultural beliefs could influence and impact the usage of multimodal biometric technology for authentication [1, 3, 23]. A difficult-to-use kind of technology could deter the public from using technology and negatively impacts public satisfaction. The biometric selection also needs to consider the system's effectiveness, efficiency, reliability and convenience to achieve its purpose and effort. The attribute selection of current biometric technology (fingerprint, face, iris and signature) for authentication in public services affects the adoption/implementation of biometric technology.

Furthermore, factors such as environment, design requirements, operations and post-deployment outlined will also impact the usage of the selected current biometric systems

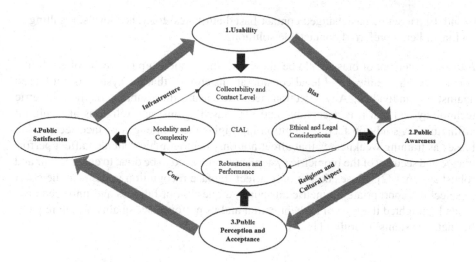

Fig. 1. The proposed multimodal biometric selection model

[22]. Usability is essential in adopting biometric systems in various sectors [1, 19]. Furthermore, the model for multimodal biometric selection based on usability considers the role of infrastructure requirements on the collectability and contact level of the system, ensuring that it is easy to collect the person's biometric attributes used to authenticate a person. We discuss cost, infrastructure, bias, and religious and cultural aspects within the usability factor. Cost: Cost in organisations is another aspect of the cost of infrastructure and deployment, which impact selecting biometric attributes for a multimodal biometric authentication system. Choosing multimodal biometric authentication technology for use in a large-scale environment such as the driver licencing environment depends on its needs and the cost of implementation [14]. The organisation needs to know the cost impact of selecting and deploying a particular modality type. Awareness of cost implications ensures long-term sustainability and maintenance of the biometric system. There is a broader need to raise awareness of the cost of maintaining the implemented technology to ensure sustainable long-term usage of the system.

Infrastructure: Infrastructure requirements ensure enough support in terms of hardware, software, capacity, and bandwidth should unique technology be required to meet demands [4, 35]. While selecting a biometric solution for the organisation, ethical concerns around the misuse of acquired biometric samples are considered paramount [6]. All stakeholders need to consent to the acquisition of the biometric samples without violating the privacy of persons. Some people may be cynical about their biometric samples and wonder whether the organisation use biometric data for other purposes and situations other than those intended for [8, 35].

Religious, Cultural Aspect: The model examines religion and culture as factors critical to public acceptance of biometric technology. The multimodal biometric system should be user-friendly for all computer-literate or illiterate users and not be complicated. The

Covid-19 pandemic has changed contact-based attributes' tolerance levels, resulting in no longer being preferred contactable solutions.

Bias: The element of bias should be alleviated; the relationship between collectability, contact level and ethical and legal considerations ensures that the system is not biased against any individual. Any selection needs to ensure the usability of the biometric technology and that it is acceptable and not biased against any religious and cultural affiliations. A study on face recognition highlights the challenges of face recognition biases as training weaknesses that affect not only demographics but also affect performance and security of the biometric system, but more need to be done to ensure a fair and robust system [32]. Biases do not only affect the face recognition system. In the quest to select the appropriate biometric attributes, a question of fairness and unbiased approach highlighted the type of authentication and its performance quality which impacts biometric systems' usability [1, 24].

5.1.1 Biometric Modality and Complexity

Biometric attributes change over time, such as fingerprint patterns deteriorating with overuse of chemicals is different from eye deterioration because of surgery; therefore, it is necessary to consider the various factors during selection to ensure the biometric system's successful implementation. The biometric attribute changes get affected due to environmental changes, age, and health, with each biometric feature being affected differently. It is, therefore, essential to consider the various changes in the design of biometric authentication systems since data modalities matter. Over time, the changes affect the accuracy, effectiveness and reliability of implemented multimodal biometric systems [24].

5.1.2 Collectability and Contact Level

The collected biometric attributes need to safeguard against unauthorised access. Therefore, from an information security point of view, the organisation should analyse the biometric solution for its ability to withstand risk and security threats. For the implementation of biometrics in a sensitive environment that requires high-level security, the selected biometric needs to have high-level security standards [14]. Information security is at the core of biometric authentication, providing value for the organisation's information and resources. The organisation needs to consider tackling current security concerns and, in the future, the privacy of data stored at their disposal [4, 17, 31]. Information security ensures trustworthiness and influences the perception and usability of biometric technology. The selected multimodal biometric technology should consider the technology's long-term feasibility factor in the organisation's future [2, 22]. There is an increased lack of trust, and confidence in the biometric system and unsatisfied users, which has a direct negative impact on public satisfaction for implementation, where the biometric data that the organisation collected for authentication are used for other purposes [19].

Furthermore, contact level determines the convenience, effectiveness and simplicity of using the multimodal biometric authentication system. The look and feel of multimodal biometric authentication systems impact the public acceptance of biometric systems. Therefore, at the contact level, the system's ease of use must be well thought out when selecting multimodal biometric modalities. Social and cultural beliefs could also influence and impact the usage of multimodal biometric technology for authentication.

5.2 Public Awareness

The human factor element of awareness plays a significant role in biometric attribute selection for a desired biometric system. In this instance, awareness measures the impact of biometric attribute selection based on understanding, knowledge, comfort with using biometric technologies, and usefulness for biometric attribute selection. Some users have not experienced the use of biometric technology; therefore, awareness could help the public with a good understanding of the technology. Awareness of biometric technology plays a significant role in the biometric attributes selection for a suitable biometric authentication system in public services. Therefore, awareness needs to form part of the selection to assist users in overcoming their technological fears and uncertainties. The public needs to know the multimodal biometric technology for adoption and understand its benefits [12]. Therefore, awareness of the biometric systems among the people is crucial before deploying multimodal biometric technology.

5.3 Public Perception and Acceptability

The effect of public perception concerning multimodal biometric authentication technology requires careful observation before biometric selection and exposure to the technology. Public perception and acceptance focus on a person's attitude towards biometric technology and whether others' opinions positively influence acceptance and use of biometric technology for authentication purposes [3]. Furthermore, public satisfaction ensures decision-makers understand the type of biometric modalities selected and the underlying complexities for the organisation. The objective is to ensure that challenges get addressed at the beginning of the planning phase. Public acceptability measures the impact of acceptability on current biometric attribute selection, multimodal attribute selection and esoteric biometric attribute selection for adopting or implementing biometric systems in the driver licensing environment. Acceptability is measured in terms of the observed variable of readiness to use, acceptance of risk, trust and confidence to use the biometric system.

5.4 Public Satisfaction

The primary factor of acceptability is a measure of the feedback and information received on the aspect of performance which impacts satisfaction. Confidence and trust in biometric authentication systems mainly depend upon the users' experience [28]. The element of public satisfaction explores robustness and performance, ensuring the public can perform a task without challenges or errors. The model will guide requirements extraction and allow the context to be better understood.

5.5 Confidentiality, Integrity, Availability and Learnability (CIAL)

Increased security is the main reason for implementing multimodal authentication technology in various organisations, including law enforcement. Information security is becoming critical as more people, computers, and networks become interconnected [26]. At this point, organisations should safeguard their information assets against security breaches. The model's central points, confidentiality, integrity, availability and learnability (CIAL), ensure that information maintains its value, accuracy, and reliability and is free from unauthorised modifications. The model explores the element of CIAL from the organisation's perspective when selecting biometric attributes for multimodal biometric authentication systems. In the context of driver licencing centres, confidentiality ensures the protection of information from the threat of unauthorised access, accidental information breach, and information misuse, should the data fall into the wrong hands. Selecting the appropriate multimodal biometric attributes also involves the importance of information integrity. Information integrity ensures the protection of information against unauthorised modifications and tampering, and that the data is consistent and reliable. Finally, the element of availability should be considered for biometric attribute selection. Information availability ensures that the required information is always available. It can impact the organisation negatively should the request for information from authorised users be denied. Learnability ensures that a person can quickly learn to use a biometric authentication system, which is the element that impacts the critical aspect of usability for continuous use of biometric authentication systems [23]. Therefore, CIAL impacts deployment success by providing a technique for solid security and preserving organisational information against unauthorised access following applicable policies, regulations and laws. There must be adequate support, to ensure the organisation attain confidence with the multimodal biometric system for deployment.

6 Conclusion

This study focused on a multimodal biometric attribute selection model in driver licencing centres and determined if it is a viable solution to deploying a biometric system unsuitable for the particular environment. Biometrics, as a subcategory of artificial intelligence, impacts the radical transformation of communities. The proposed model argues that driver licencing centres could benefit from selecting an appropriate biometric attribute for a suitable and robust multimodal biometric authentication system. Furthermore, to derive the benefit from the model's effectiveness and successful deployment, it is essential to consider the type of environment, the infrastructure, usability, and public acceptance.

Therefore, biometric deployment will pave the way for successful implementation. This study indicates that the challenges of selecting the wrong kind of biometric attributes negatively impact the deployment and continuous usage of the multimodal biometric authentication system. Knowing which biometric method to choose in an organisation to protect its valuable information is complicated. Selecting an appropriate multimodal authentication method in such a large-scale environment guided by some framework is significant for its successful deployment and long-term sustainability in terms of continuous support and maintenance.

References

1. Ahmad, M., et al.: Security, usability, and biometric authentication scheme for electronic voting using multiple keys. Int. J. Distrib. Sens. Netw. **16**(7), 1550147720944025 (2020)
2. Ajibade, P.: Technology acceptance model limitations and criticisms: exploring the practical applications and use in technology-related studies, mixed-method, and qualitative researches. Libr. Philos. Pract. **9**, 197–255 (2018)
3. Ajzen, I., Fishbein, M., Lohmann, S., Albarracín, D.: The influence of attitudes on behaviour. In: The Handbook of Attitudes, pp. 197–255 (2018)
4. Ani, U.D., Watson, J.M., Nurse, J.R., Cook, A., Maples, C.: A review of critical infrastructure protection approaches: improving security through responsiveness to the dynamic modelling landscape (2019)
5. Beigi, H.: Access control through multifactor authentication with multimodal biometrics. United States patent US 10,042,993 [BCPRS13] Bolle RM, Connell JH, Pankanti NK (2018)
6. Boult, T.E., Woodworth, R.: Privacy and security enhancements in biometrics. A multimodal biometric system for secure identification. Asian J. Technol. Manag. Res. **5**(02), 423–445 (2015)
7. Chanukya, P.S.V.V.N., Thivakaran, T.K.: Multimodal biometric cryptosystem for human authentication using fingerprint and ear. Multimed. Tools Appl. **79**(1), 659–673 (2020)
8. Datta, P., Bhardwaj, S., Panda, S.N., Tanwar, S., Badotra, S.: Survey of security and privacy issues on biometric system. In: Gupta, B.B., Perez, G.M., Agrawal, D.P., Gupta, D. (eds.) Handbook of Computer Networks and Cyber Security, pp. 763–776. Springer, Cham (2020). https://doi.org/10.1007/978-3-030-22277-2_30
9. Dwivedi, Y.K., Rana, N.P., Jeyaraj, A., Clement, M., Williams, M.D.: Re-examining the unified theory of acceptance and use of technology (UTAUT): towards a revised theoretical model. Inf. Syst. Front. **21**(3), 719–734 (2019)
10. El-Abed, M.: Usability assessment of keystroke dynamics systems. Int. J. Comput. Appl. Technol. **55**(3), 222–231 (2017)
11. Farrell, S.: Biometrics in air transport: no flight of fancy. Biometr. Technol. Today **2019**(1), 5–7 (2019)
12. Furnell, S., Evangelatos, K.: Public awareness and perceptions of biometrics. Comput. Fraud Secur. **2007**(1), 8–13 (2007)
13. Genovese, A., Muñoz, E., Piuri, V., Scotti, F.: Advanced biometric technologies: emerging scenarios and research trends. In: Samarati, P., Ray, I., Ray, I. (eds.) From Database to Cyber Security. LNCS, vol. 11170, pp. 324–352. Springer, Cham (2018). https://doi.org/10.1007/978-3-030-04834-1_17
14. Goldstein, J., Angeletti, R., Holzbach, M., Konrad, D., Snijder, M.: Large-scale biometrics deployment in Europe: identifying challenges and threats. JRC Scientific and Technical Reports (2008)
15. Kannammal, A., Prasanalakshmi, B.: Analysing security measures with unimodal and multimodal biometrics. In: International Conference on Sensors, Security, Software and Intelligent Systems, Beijing (2009)
16. Mikalef, P., et al.: Enabling AI capabilities in government agencies: a study of determinants for European municipalities. Gov. Inf. Q. **39**, 101596 (2021)
17. Myers, D., Willingham, M., Stewart, K.: Law enforcement guide to false identification. Pacific Institute for Research and Evaluation, Beltsville, MD (2001)
18. Nag, A.K., Dasgupta, D., Deb, K.: An adaptive approach for active multi-factor authentication. In: 9th Annual Symposium on Information Assurance (ASIA14), vol. 39 (2014)
19. Oh, J., Lee, U., Lee, K.: Usability evaluation model for biometric system considering privacy concern based on MCDM model. In: Security and Communication Networks (2019)

20. Omolara, O.O., Reginald, A.C., Oyedepo, O.M., Wasiat, E.A., Taiwo, O.A.: Modelling and simulation of Nigerian airspace management agency billing system using Python simulation packages. Am. J. Math. Comput. Model. 5(4), 109–115 (2020)
21. Plaatjies, P.: National training framework for road traffic management. In: SATC 2010 (2010)
22. Pooe, A., Labuschagne, L.: Factors impacting on the adoption of biometric technology by South African banks: an empirical investigation. South. Afr. Bus. Rev. 15(1), 20 (2011)
23. Ramakumar, N., Reddy, P.S.N., Naik, R.N., Jilani, S.A.K.: Authentication based systematic driving license issuing system. In: 2017 International Conference on Intelligent Computing and Control Systems (ICICCS), pp. 1327–1331. IEEE (2017)
24. Rommetveit, K.: Introducing biometrics in the European Union: practice and imagination. In: Delgado, A. (ed.) Technoscience and Citizenship: Ethics and Governance in the Digital Society. TILELT, vol. 17, pp. 113–126. Springer, Cham (2016). https://doi.org/10.1007/978-3-319-32414-2_8
25. Sagberg, F.: Characteristics of fatal road crashes involving unlicensed drivers or riders: implications for countermeasures. Accid. Anal. Prev. 117, 270–275 (2018)
26. Sailaja, D., Navya Sri, M., Samuel, P.: An biometric-based embedded system for e-verification of vehicle users for RTO. In: Bhateja, V., Satapathy, S., Zhang, Y.D., Aradhya, V. (eds.) ICICC 2019. AISC, vol. 1034, pp. 711–718. Springer, Singapore (2019). https://doi.org/10.1007/978-981-15-1084-7_69
27. Sam, D., Velanganni, C., Evangelin, T.E.: A vehicle control system using a time synchronized Hybrid VANET to reduce road accidents caused by human error. Veh. Commun. 6, 17–28 (2016)
28. Sathye, M., Dugdale, A.: Fraud in e-government transactions risks and remedies. In: Business evaluation of Government Conference and e-Government, pp. 41–52. Australian Government/IPAA (2004)
29. Sharma, V.: Aadhaar-a unique identification number: opportunities and challenges ahead. Res. Cell: Int. J. Eng. Sci. 4(2), 169–176 (2011)
30. Smith, M., Mann, M., Urbas, G.: Biometrics, Crime and Security. Routledge, Abingdon (2018)
31. Sujithra, M., Padmavathi, G.: A survey of biometric iris recognition: security, techniques and metrics. Asian J. Inf. Technol. 14(6), 192–199 (2015)
32. Thakur, K., Vyas, P.: Social impact of biometric technology: myth and implications of biometrics: issues and challenges. In: Sinha, G.R. (ed.) Advances in Biometrics, pp. 129–155. Springer, Cham (2019). https://doi.org/10.1007/978-3-030-30436-2_7
33. Venkatesh, V., Morris, M.G., Davis, G.B., Davis, F.D.: User acceptance of information technology: toward a unified view. MIS Q. 27, 425–478 (2003)
34. Whitman, R.: Information security: a study on biometric security solutions for telecare medical information systems. Doctoral dissertation. University of New Hampshire (2016)
35. Wolf, F., Kuber, R., Aviv, A.J.: An empirical study examining the perceptions and behaviours of security-conscious users of mobile authentication. Behav. Inf. Technol. 37(4), 320–334 (2018)

Technology Days: An AI Democratisation Journey Begins with a Single Step

Danie Smit[1]([⊠])(ⓘ), Sunet Eybers[1](ⓘ), Nhlanhla Sibanyoni[1], and Alta de Waal[2,3]

[1] Department of Informatics, University of Pretoria, Pretoria, South Africa
d5mit@pm.me
[2] Department of Statistics, University of Pretoria, Pretoria, South Africa
[3] Centre for Artificial Intelligence Research (CAIR), Pretoria, South Africa

Abstract. Due to AI's numerous potential benefits, embedding AI as part of an organisation's analytics portfolio has become essential. Also, organisations want to avoid a situation where only a few business areas reap the benefits of using AI. Some organisations are setting up central AI teams. However, a central AI adoption approach is not always feasible, as domain knowledge, specific to each business unit, is often required. An alternative approach would be to democratise AI and position citizen data scientists across the organisation. These citizen data scientists do not necessarily need the same level of skills as a central AI expert team; however, they require a certain level of AI-related knowledge. Technology Days are popular events that provide people insight into today's latest technologies and can be used for marketing, recruitment of experts, or knowledge transfer. The paper investigates the effectiveness of an auto-motive organisation's use of Technology Days to support the democratisation of AI by creating the intent among employees to become citizen data scientists. Based on the constructs of the technology acceptance model, a survey was used to gather feedback from the Technology Days attendees. A single case study contextualised the organisational setting and suggested that Technology Days can be used to showcase the effectiveness of AI and the ease of use of AI tools. Technology Days can help create a positive attitude and create the intent of employees in the organisation to become citizen data scientists. However, Technology Days remains only one of many incremental steps towards democratised AI.

Keywords: Democratisation · Artificial intelligence · Technology day · Knowledge transfer · Technology acceptance model · Citizen data scientist

1 Introduction

Due to AI's numerous potential benefits, embedding AI as part of an organisation's analytics portfolio has become essential [8,13,21]. As a result of this strong business case, organisations want to avoid a situation where only a few business areas reap the benefits of using AI. Furthermore, a central AI adoption

A. Pillay et al. (Eds.): SACAIR 2022, CCIS 1734, pp. 335–347, 2022.
https://doi.org/10.1007/978-3-031-22321-1_23

approach is not always feasible, as domain knowledge, specific to each business unit, is often required in each unit. An alternative approach to grouping AI experts centrally, would be to democratise AI and position citizen data scientists across the organisation. Citizen data scientists are employees who create or generate models that leverage predictive or prescriptive analytics, but whose primary job function is outside of the field of statistics and analytics [33]. Even though these citizen data scientists do not necessarily require the same level of skills as a central AI expert group, who formally received training as specialized data scientists, they need a certain level of AI-related knowledge. As a result, there is a requirement to up-skill employees and embed citizen data scientists in various business units across the organisation.

Unfortunately, many social and technical factors often hinder the development of citizen data scientists. Despite the focus on the technological challenges [2], social barriers such as trust in AI and lack of skills is causing AI projects to remain in somewhat experimental state [7]. If these barriers are not addressed, the intent to become a citizen data scientist might be limited. For example, the lack of knowledge can hinder the intent of people to become citizen data scientists, because the benefits might not be clear. Furthermore, in literature there is a lack of understanding of effective training methods for using AI in an organisational setting [17]. To complicate the situation more, traditional organisations might have little, or no experience in AI and are now required to embed AI as part of their analytics portfolio [36]. Although organisations can undoubtedly use methods such as formal training, forums, and workshops to facilitate the increase of knowledge [39], this paper argues that Technology Days can be used as an alternative method. A formal definition for Technology Days is lacking; however, in industry, it is understood as an event that provides people with insight into today's latest technologies[1]. It is focused on the area of computing, science, and technology[2]. A Technology Day can include presentations, training sessions, technical expertise networking opportunities[3] and can be used for marketing, recruitment of experts, or knowledge transfer. The question arises: *how effective are Technology Days in supporting the democratisation of AI within organisations by creating the intent among employees to become citizen data scientists?* This question has rarely been scrutinised in information systems research. A case study, in which an organisation used Technology Days, was analysed to answer the question. In the case study, the objective of the Technology Days was to encourage the intent among employees to become citizen data scientists. A survey was used to gather qualitative and quantitative data and distributed to Technology Day attendees. The data was analysed using descriptive statistics. The study occurred in South Africa, at a global automotive manufacturer's IT

---

[1] Techday Headcourter [website], https://techdayhq.com/new-york, (accessed 21 August 2022).

[2] Techday with Google [website], https://techday.withgoogle.com/, (accessed 21 August 2022).

[3] Microsoft Techday [website], https://www.microsoft.com/ro-ro/techday, (accessed 21 August 2022).

hub of an automotive manufacturer in South Africa. To measure the effectiveness of Technology Days to create the intent among IT people to become citizen data scientists, the technology acceptance model (TAM) was used as a theoretical framework [14]. The TAM is a technology adoption theory used to explain how users come to adopt and use technology, with behavioural intention as the main factor that leads to the use of technology [14].

The rest of this paper is structured as follows: the background section provides an overview of the democratisation of AI, technology acceptance, citizen data scientists, and democratisation in a South African context. The case study is described thereafter, followed by a summary of findings and a conclusion.

2   Background

The use of AI as part of organisations' analytics portfolio is still relatively new, and research shows that early adopters should expect and manage both social and technical challenges [24]. Certainly, the challenge of democratising AI is not to be underestimated. Even the meaning of AI democratisation is a much-debated topic and often differently understood and interpreted among industry practitioners. For example: does the democratisation of AI imply that all employees should become citizen data scientists, or does it simply suggest that all employees should have access to AI tools, or the benefits reaped from AI implementations? The following section addresses these questions.

2.1   Democratisation

The term democracy refers to 'rule by the people' [42] and is characterised by certain democratic concepts, procedures, and operative principles. For example, on a conceptual level, a democracy refers to an operation where citizens hold rulers accountable for their actions through competition using procedural processes such as elections [45]. To empower citizens to do this, a democratisation process involves training and educating people. Additionally, a democracy is not to be understood as a predetermined end-state but as an imaginary future of what ought to be [42]. As a result, the democratisation process is moving towards an outcome that is neither predetermined nor fully stable [45], where citizens' knowledge is also continuously increased.

Drawing from this basic understanding of 'democracy', one can deduce that the democratisation of AI relates to concepts, procedures, and operative principles that allows for the inclusively of AI initiatives [37]. This can be achieved by providing people access to the required tools and platforms [39], while increasing the knowledge about AI in the organisation. The knowledge required does not only relate to the technical, but also the AI principles and fundamental knowledge that will support access to the benefits of AI [11] and empower people to ensure accountability within the organisation. Moreover, because a democracy is an imaginary future (as previously mentioned), the democratisation of AI should cater for an iterative approach with incremental improvement [45] based on a philosophical method of what the future should be. Lastly, because democracy is about the 'rule by the people' [42] and not the 'rule by the machines', the democratisation of AI should include avoiding an oppressive future [22].

Similar to the democratisation of nations, the question of identifying the elements of persuasion to support the true democratisation of AI arises [45]. Both democracy-related [42] and innovation related research [31] can provide a basis for understanding. In democracy-related research, specifically, the transition towards a democracy model [42], the transition from an authoritarian rule towards democracy is a process that includes a preparatory phase, a decision phase, and a consolidation phase [42]. Likewise, the diffusion of innovation (DOI) theory, which is concerned with technology adoption [31], has an awareness and persuasion stage (as preparation for adoption) and then a decision stage. As part of the preparation for the democratisation of AI in organisations, organisations must address the AI awareness and knowledge about AI requirements [38]. However, developing skills and knowledge in AI is challenging for traditional organisations [37]. It requires a process of knowledge transfer, which involves an entity that has the knowledge, another entity that does not have the required knowledge. Furthermore, it requires a communication channel which connects the two units [31]. This study focuses on using Technology Days as a method that connects knowledgeable AI experts with less knowledgeable employees to support the democratisation of AI within organisations.

2.2 Democratisation of AI in the Context of a Developing Country

The developments and advancements in AI have attracted both academic and public attention as it has the potential to disrupt and transform socio-economic activities. Research shows that many jobs to be impacted in the process are in Africa, while the European countries benefit more from AI as they are better equipped and prepared [19]. Governments and businesses are also positioning themselves to understand and analyse the potential of AI to transform societies and its disruptive impact across industries.

South African organisations are feeling the pressure of scaling the implementation of AI, which has the potential to encourage worker productivity at a trade-off accidental or design risk. In other words, designing AI applications without appropriate training could introduce dangerous processes or social inequalities [20]. However, to successfully democratise and disseminate AI to the masses, people, technology, and data, must be considered together to promote a digital revolution. Data Science Africa (DSA) is a non-governmental organisation established in 2013, based in Kenya. DSA encourages AI practitioners and researchers across Africa to use their platform to discuss and share knowledge of the usefulness and developments of AI via workshops (Data Science Africa, 2022). GovTech[4] is an annual conference organised and hosted by the South African State IT Agency (SITA[5]) to showcase innovations that will shape 4IR and enable cit-

---

[4] GovTech 2022 [website], https://www.govtech.gov.za/, (accessed 21 August 2022).
[5] SITA [website], https://www.sita.co.za/content/govtech-0, (accessed 21 August 2022).

izens into the future. Most noticeably, GovTech aims to bring together thought leaders in ICT to engage and share experiences and solutions and identify new and creative ways to showcase technology's ability to improve sustainable development, grow inclusive economies and ensure a positive impact on the lives of citizens.

2.3   Democratisation and Technology Acceptance

To evaluate the effectiveness of Technology Days as a method to support the democratisation of AI, the factors that influence technology acceptance as postulated by Davis' TAM are used as a theoretical basis [14]. Even though other technology adoption theories exist [16, 43, 44], we argue that the TAM is specifically applicable to this study, due to its simplicity and its centredness around intention. According to the TAM, behavioural intention is a factor that leads people to use technology, whereas behavioural intention is influenced by the attitude toward the general impression of the technology. The perceived usefulness and the perceived ease of use are notable factors influencing attitude. Even though the TAM is a dominant theory for technology adoption, it has some limitations. One is that the adoption of the technology depends on self-reported usage, and another is that the TAM provides a limited investigation of the full range of essential consequences of technology adoption [15]. But, as this study's scope is limited to the relevance and possible contribution of Technology Days to support AI adoption and AI democratisation, the limitations (of not measuring the actual usage or considering all the other impacts of the technology adoption) are not in question. Figure 1 is a graphical depiction of the research landscape. It depicts how TAM was used in this study as a theoretical framework to evaluate the effectiveness of Technology Days as a method to support the democratisation of AI.

The first influence of the TAM is the *external variables* [14] which constitutes the context of the study [42]. For example, a certain level of data-driven maturity is needed before AI can be generalised [17]. Data-driven maturity refers to the level of technical aspects such as access to data, automation capabilities and AI platforms [12, 47]. In addition, social aspects such as skills, organisational culture, top management support, employee-AI trust, AI strategy, and compatibility with existing systems are of importance [17, 40, 41].

Perceived ease of use is the second TAM influencing factor [14]. Due to its inherent arcane nature, AI is a rather complex technology to understand [34]. To support AI democratisation, organisations will have to make investments to ensure that AI in the organisation is perceived as user-friendly, understandable, and customised for each target user group [37]. To enable organisations to deal with the complexity, organisations can increase their process and technology knowledge [41]. One major enabler is the open-source movement that allows for the availability of free coursework, which everyone can use to educate themselves on the fundamentals of AI [9].

The third influencing factor of the TAM is *perceived usefulness* [14]. On an organisational level, the usefulness is related to the importance of a strong

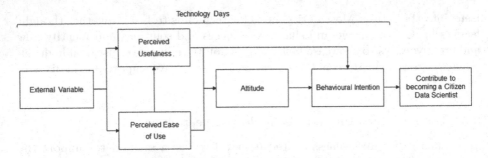

Fig. 1. Applying the TAM to evaluate the effectiveness of Technology Days as a communication channel to support the democratisation of AI (Adapted from [14])

business case for implementing and adopting AI [10]. Two of the main differences between AI and other traditional technologies is that AI has the ability to learn and act autonomously [27]. AI has the potential to automate repetitive tasks and may provide employees more opportunities to focus on work which requires their core competencies [17]. However, as AI can automate informed decisions, organisations will not only have to know how to build and apply algorithms but also learn how to build solutions where AI acts on behalf of humans [25]. Furthermore, for managers and product owners to be able to identify business use cases for the application of AI, they should have a certain level of understanding of the technology [37].

Attitude towards technology is the fourth influencing factor of the TAM [14]. Behavioural intent is influenced by the attitude and perceived usefulness of technology. The perceived ease of use, on the other hand, influences attitude. Also, according to the TAM, external variables, such as social perceptions, determine the attitude. The idea of democratising AI is to make AI technologies available to a greater number of employees. To create a positive attitude towards AI, as well as better citizen data scientists, employees need to be empowered by open datasets, no-code or auto machine learning solutions and cloud-based platforms [23]. As managers are required to make decisions based on business value and the benefits of what the technology offers, they will need to have knowledge about AI [8]. Also, if AI can make predictions and decisions in an automated and inexpensive way, managers need to know how to implement such functionality [17] responsibly. Moreover, if there are any trust issues related to AI adoption [40] or that it is perceived that AI contrasts with any of the subjective norms of the organisation [5], the attitude to AI technology adoption will not be positive.

During the decision stage of adopting technology, individuals weigh the advantages and disadvantages and, as a result, form an intent to adopt or reject it [31]. The *behavioural intention* in return leads to the use of technology [14]. Even though intention is not the only factor that leads to the use of technology, this study is interested in measuring the effectiveness of Technology Days in creating intent among employees to become citizen data scientists. However, as the democratisation of AI within an organisation takes place within a complex environment [46], it is worth noting that the context of the study is important.

3   Case Study

A case study research approach was followed to gain a deeper understanding of the role Technology Days in supporting the democratisation of AI within organizations by creating the intent among employees to become citizen data scientists. This case study is part of a larger study that focuses on identifying the enabling factors for organisational AI adoption. In this study, the function of AI is to support analytics. When AI is referred to in this study, it is to support analytics, and not to achieve artificial general intelligence [7].

3.1   The Study Setting

The study was conducted at an IT hub, which acts as an IT service department to a global head office, located in South Africa. The organisation includes more than 1700 IT consultants. Just over 300 of these consultants are in the data analytics domain; however, less than 20 are involved in AI products. The organisation is viewed as a leader in industrial digital transformation [4] and actively encourages its employees to up-skill and stay updated with the latest technologies. In order to support the democratisation of AI, a week of Technology Days were hosted. The organisation referred to this event as a Tech Indaba. An Indaba is a meeting to discuss a serious topic [1]; thus, the name represents the importance of the democratisation of AI to the organisation. During the Tech Indaba, experts in the field of AI presented sessions which covered topics such as an introduction to AI principles and techniques, an overview of a managed machine learning service in the cloud and practical tutorials. The sessions exposed the attendees to Amazon SageMaker, a cloud machine-learning platform[6]. Also, a gamification technique was used in the form of an AWS DeepRacer competition[7]. This provided attendees with an introduction to reinforcement learning regardless of their prior knowledge of machine learning. More than 200 people attended the Tech Indaba, which followed a hybrid attendance model, with around 80% of the joining the sessions virtually. As the Tech Indaba comes at a cost, in the form of time invested by the participants and the presenters, the organisation's managers were interested in determining if it contributed to creating more citizen data scientists.

3.2   Data Collection

To evaluate the effectiveness of the sessions, a survey was created that included questions focused on determining the intention of the participants to pursue data

---

[6] Amazon SageMaker [website], https://aws.amazon.com/pm/sagemaker/, (accessed 27 August 2022).

[7] AWS DeepRacer [website], https://aws.amazon.com/deepracer/, (accessed 27 August 2022).

science topic after attending one or more sessions of the indaba. The survey was distributed online to the attendees of the Tech Indaba and resulted in a combination of quantitative and qualitative data. A total of 48 people completed the questionnaires of which only two rated themselves as experienced data scientists using AI. However, more than half of the respondents have experience in descriptive analytics which makes them a relevant target group for this study: they have some data skills not related to AI but have room for improvement in their current skill set. Furthermore, the expectation of participants, as indicated on the questionnaires, was to learn and better understand AI's capabilities.

3.3   Data Analysis

Descriptive statistics were used to visualise and understand the data. The main component of the survey was based on the TAM, and asked the respondents how much they agreed in terms of the following five questions: *'did the AI Technology Day sessions address some of the external variables that prevents you from becoming a citizen data scientist?'*, *'did the AI Technology Day sessions make it in any way easier for you to be a citizen data scientist?'*, *'do you perceive AI to be more useful after attending the AI Technology Day sessions?'*, *'do you feel more positive about being or becoming a citizen data scientists after attending the AI Technology Day sessions?'* and *'do you intent to use (or continue to use) AI in the future as a result of the Technology Day sessions?'*. The respondents could choose from a 5-point Likert scale: *'Strongly agree'* (5), *'Agree'* (4), *'Neither agree nor disagree'* (3), *'Disagree'* (2) and *'Strongly disagree'* (1). To gain more insight into the reason for the rating, the respondents were also asked to supply a reason for their choice using a comment box why. The other questions in the survey related to respondents' opinion of Technology Days in general and their current experience in data science. Table 1 summarises the results of the questionnaire.

Table 1. Summary statistics for external variables, perceived usefulness, perceived ease of use, attitude, and behavioural intention.

	External	Ease of use	Usefulness	Attitude	Intention
Count	48	48	48	48	48
Mean	3.3542	3.8125	3.9167	3.8750	3.8542
Std	0.8627	0.7043	0.8711	0.64	0.7987

Most of the elements had a rather high average rating, of *'Agree'* or *'Strongly agree'*. However, the rating of Technology Days and its influence on addressing external variables was lower than the other elements. Figure 2 visually depicts the feedback of these questions and how the elements might impact each other. It shows the flow of the TAM and each element depicting the rating of the people that attended the sessions. From this flow, one can deduce that the intention to become a citizen data scientist is likely to increase because of a Technology Day. However, it is also evident not addressing external impediments will hamper the adoption.

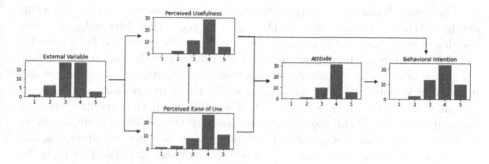

Fig. 2. TAM flow with ratings on addressing: external variables; perceived usefulness; perceived ease of use; attitude; behavioural intention

Moreover, more than 85% of the respondents believed that Technology Day sessions can be used to assist in creating more citizen data scientists. When analysing the qualitative data, it was clear that the respondents believe that Technology Days are not only valuable to increase the knowledge of AI within the organisation, but also to communicate the benefits of AI. Furthermore, when asking the participants what elements of the sessions were particularly useful the feedback was that the Technology Days was useful to provide a basic understanding of AI. Also, it gave the people hands-on practical experience with real-world examples and some insight into the ease of use of the platform and its tools. Other feedback included: *'AI has always been a bit intimidating. Today's session has shown that it is very valuable'*, *'boosted confidence'*, *'I will start with small steps to first gain more knowledge'*, *'the myth around AI were broken down'*, *'useful tools were demonstrated'* and *'the enormous value that AI can add'*. From this feedback, one can presume that the Technology Day contributed to showing the benefit of using AI and providing insight into the tools, techniques, and principles knowledge, which is in line with the theory of diffusion of innovation [31]. Although the Technology Day sessions were perceived as positive, there was still room for improvement. For example: feedback included comments such as: 'training or resource links must be given for beginners', 'opportunities are needed to become a citizen data scientist', 'still need to understand more', and even 'still no clue how to start'.

4   Discussion

Previous research shows that multiple communications channels are used to support knowledge transfer [31]. This is also true in the case of increasing organisations' knowledge of AI [3,18]. Even though the use of general training methods is a well-researched topic, there is a need to develop innovative methods to help employees learn through newer channels [35]. Furthermore, advanced knowledge on effective training methods for using AI in an organisational setting is lacking [17]. Also, the use of Technology Days for knowledge transfer and democratisation is not yet understood.

The study draws upon the transition towards a democracy model [42], a theory in another discipline, and the DOI [30], to argue that knowledge is key to the democratisation of AI. The overlap and similarities between the transition towards a democracy model and the DOI are useful to highlight the sociotechnical nature of AI democratisation. Taking from the concepts of transition towards a democracy model [42], this study argues that the democratisation of AI is not an end state that can be achieved, it is a continuum, that requires continuous improvement. Technology Days can support the continuous improvement process, by addressing some of the attitude and knowledge requirements, and thereby act as an innovative method to help employees learn [35]. The TAM [14] was used as a theoretical model to illustrate that Technology Days can be used to support the democratisation of AI by creating the intent of people within the organisation to become citizen data scientists. The findings contribute to practice by providing organisations with an example that a Technology Day is valuable and worth the investment to support the democratisation of AI. Also, the study helps organisations understand that the use of Technology Days is only one of many communication channels that should be used. Furthermore, Technology Days is one of many incremental steps towards democratised AI, which is an outcome that is neither predetermined nor fully stable [45], where citizens' knowledge is also continuously increased. As an outcome of this study, the organisation where the case study took place acknowledged that Technology Days were valuable to demystify AI [28] and contributed to the democratisation of AI in the organisation. However, the organisation also acknowledged that Technology Days are undoubtedly not the only intervention the organisation will have to implement to be successful in the democratisation of AI. Lastly, this study argues that democratising AI through Technology Days will ensure that employees are kept in the loop. Furthermore, as Technology Days can empower people with the required knowledge to keep organisations accountable, Technology Days can provide a simple and practical contribution toward actionable AI ethics [32].

5   Conclusion

AI is a complex social-technical system that has the ability to make decisions, learn and interact appropriately, quicker than humans [6,26]. Human-centred AI research focuses on empowering and supporting employees towards self-efficacy, controlling technologies, providing clarity to division of labour (between AI and employees) and promoting creativity [29]. Additionally, we argue that democratized AI is 'rule by the people' [42] and not 'rule by the machines'.

This study aimed to contribute toward expanding Human-centred AI research by analysing how, using a real-life case study and TAM as a framework, Technology Days can be used as a method to democratise AI and increase knowledge. This paper makes multiple contributions to human-centered AI research. It shows Technology Days can be used to showcase the usefulness of AI. Furthermore, it can help demonstrate how easy it can be to use AI tools. Moreover, Technology Days can be helpful in creating a positive attitude towards the use of AI. Following the logic of the TAM, Technology Days can also be used to increase the intent of people in the organisation to become citizen data scientists. We conclude that Technology days can effectively support AI's democratisation by creating the intent among employees to become citizen data scientists. However, Technology Days remains only one of many incremental steps towards democratised AI.

This study has some limitations. The case study focused on Technology Days at one organisation. Even though the Technology Days were found to be useful, it is unclear what makes a Technology Day successful. Therefore, we suggest future research to investigate what contributes to the success of Technology Days. This is significant, as hosting Technology Days seems to be an important step in an organisation's AI democratisation journey.

References

1. Collins english dictionary (2022). https://www.collinsdictionary.com/us/dictionary/english/indaba
2. Ågerfalk, P.J., et al.: Artificial intelligence in information systems: state of the art and research roadmap. Commun. Assoc. Inf. Syst. **50**(1), 420–438 (2022). https://doi.org/10.17705/1CAIS.05017, https://aisel.aisnet.org/cais/vol50/iss1/21/
3. Alfaro, E., Bressan, M., Girardin, F., Murillo, J., Someh, I., Wixom, B.H.: BBVA's data monetization journey. MIS Q. Exec. **18**(2), 117–128 (2019)
4. ARC advisory group: industrial digital transformation top 25. ARC Special Report (2022). https://www.arcweb.com
5. Awa, H.O., Ojiabo, O.U., Orokor, L.E.: Integrated technology-organization-environment (TOE) taxonomies for technology adoption. J. Enterp. Inf. Manag. **30**(6), 893–921 (2017)
6. Baxter, G., Sommerville, I.: Socio-technical systems: from design methods to systems engineering. Interact. Comput. **23**(1), 4–17 (2011). https://doi.org/10.1016/j.intcom.2010.07.003
7. Benbya, H., Davenport, T.H.: Artificial intelligence in organizations: current state and future opportunities. MIS Q. Executive **19**(4) (2020)
8. Berente, N., Gu, B., Recker, J., Santhanam, R.: Managing artificial intelligence. MIS Q. **45**(3), 1433–1450 (2021)
9. Campbell, S.D., Jenkins, R.P., O'Connor, P.J., Werner, D.: The explosion of artificial intelligence in antennas and propagation: how deep learning is advancing our state of the art. IEEE Antennas Propag. Mag. **63**(3), 16–27 (2021). https://doi.org/10.1109/MAP.2020.3021433
10. Chui, M.: Artificial intelligence the next digital frontier. McKinsey Company Glob. Inst. **47** (2017)

11. Clyde, A.: AI for science and global citizens. Patterns **3**(2), 100446 (2022). https://doi.org/10.1016/j.patter.2022.100446
12. Davenport, T.H.: The AI Advantage: How to Put the Artificial Intelligence Revolution to Work. MIT Press, Cambridge (2018)
13. Davenport, T.H., Harris, J.G.: Competing on Analytics: the New Science of Winning. Harvard Business School Press, Boston (2007)
14. Davis, F.D.: User Acceptance of Computer technology: a comparison of two theoretical models. Manage. Sci. **35**(8), 982–1003 (1989)
15. Dwivedi, Y.K., Wade, M.R., Scheberger, S.L.: Information Systems Theory. Explaining and predicting our digital society, vol. 1. Springer, New York (2012). https://doi.org/10.1007/978-1-4419-6108-2
16. Fishbein, M., Ajzen, I.: Belief, attitude, intention and behavior: an introduction to theory and research. Philos. Rhetoric **10**(2) (1977). https://doi.org/10.2307/2065853
17. Giraud, L., Zaher, A., Hernandez, S., Akram, A.A.: The impacts of artificial intelligence on managerial skills. J. Decis. Syst. (2022). https://doi.org/10.1080/12460125.2022.2069537
18. Gust, G., Sthroehle, P., Flath, C.M., Neumann, D., Brandt, T.: How a traditional company seeded new analytics capabilities. MIS Q. Exec. **16**(3), 123–139 (2017)
19. Gwagwa, A., Kraemer-Mbula, E., Rizk, N., Rutenberg, I., De Beer, J.: Artificial intelligence (AI) deployments in Africa: benefits, challenges and policy dimensions. Afr. J. Inf. Commun. **26**, 1–28 (2020). https://doi.org/10.23962/10539/30361
20. Hagerty, A., Rubinov, I.: Global AI ethics: a review of the social impacts and ethical implications of artificial intelligence. arXiv preprint arXiv:1907.07892 (2019)
21. Johnson, D.S., Muzellec, L., Sihi, D., Zahay, D.: The marketing organization's journey to become data-driven. J. Res. Interac. Mark. (2019)
22. Kane, G.C., Young, A.G., Majchrzak, A., Ransbotham, S.: Avoiding an oppressive future of machine learning: a design theory for emancipatory assistants. MIS Q.: Manag. Inf. Syst. **45**(1), 371–396 (2021)
23. Korot, E., Gonçalves, M.B., Khan, S.M., Struyven, R., Wagner, S.K., Keane, P.A.: Clinician-driven artificial intelligence in ophthalmology: resources enabling democratization. Curr. Opin. Ophthalmol. **32**, 445–451 (2021). https://doi.org/10.1097/ICU.0000000000000785
24. Lacity, M.C., Willcocks, L.P.: Becoming strategic with intelligent automation. MIS Q. Exec. **20**(2), 1–14 (2021)
25. Leyer, M., Oberlaender, A., Dootson, P., Kowalkiewicz, M.: Decision-making with artificial intelligence: towards a novel conceptualization of patterns. In: PACIS 2020 Proceedings (2020). https://aisel.aisnet.org/pacis2020
26. Makarius, E.E., Mukherjee, D., Fox, J.D., Fox, A.K.: Rising with the machines: a sociotechnical framework for bringing artificial intelligence into the organization. J. Bus. Res. **120**, 262–273 (2020). https://doi.org/10.1016/j.jbusres.2020.07.045
27. Mayer, A.S., Haimerl, A., Strich, F., Marina, F.: How corporations encourage the implementation of AI ethics. In: ECIS 2021 Research Papers (2021)
28. Natarajan, P., et al.: Demystifying AI for the Enterprise?: A Playbook for Business Value and Digital Transformation. Productivity Press, New York (2022)
29. Oosthuizen, R., Van't Wout, M.C.: Sociotechnical system perspective on artificial intelligence implementation for a modern intelligence system (2019). https://researchspace.csir.co.za/dspace/handle/10204/11347
30. Rogers, E.M.: New product adoption and diffusion. J. Consum. Res. **2**, 290–301 (1976). https://0-web-b-ebscohost-com.pugwash.lib.warwick.ac.uk/bsi/pdfviewer/pdfviewer?vid=1&sid=69d38d76-f5ce-4740-bc3c-66ccee9413a1

31. Rogers, E.M.: Diffusion of Innovations, 4th edn. The Free Press, New York (1995)
32. Ruttkamp-Bloem, E.: The quest for actionable AI ethics. In: Gerber, A. (ed.) SACAIR 2021. CCIS, vol. 1342, pp. 34–50. Springer, Cham (2020). https://doi.org/10.1007/978-3-030-66151-9_3
33. Sakpal, M.: How to use citizen data scientists to maximize your D& A strategy (2022). https://www.gartner.com/smarterwithgartner/use-data-and-analytics-to-tell-a-
34. Salam, A.F., Pervez, S., Nahar, S.: Trust in AI and intelligent systems: Central core of the design of intelligent systems. In: AMCIS 2021 Proceedings (2021)
35. Santhanam, R., Yi, M.Y., Sasidharan, S., Park, S.H.: Toward an integrative understanding of information technology training research across information systems and human-computer interaction: a comprehensive review. AIS Trans. Hum.-Comput. Interact. **5**(3), 134–156 (2013). https://aisel.aisnet.org/thci
36. Simoudis, E.: The Big Data Opportunity in Our Driverless Future. Corporate Innovators, Menlo Park (2017)
37. Sjödin, D., Parida, V., Palmié, M., Wincent, J.: How AI capabilities enable business model innovation: scaling AI through co-evolutionary processes and feedback loops. J. Bus. Res. **134**, 574–587 (2021). https://doi.org/10.1016/j.jbusres.2021.05.009
38. Smit, D., Eybers, S.: Towards a socio-specific artificial intelligence adoption framework. In: Proceedings of 43rd Conference of the South African Institute of Computer Scientists and Information Technologists, vol. 85, pp. 270–282 (2022)
39. Smit, D., Eybers, S., De Waal, A.: A data analytics organisation's perspective on the technical enabling factors for organisational AI adoption. In: AMCIS 2022 Proceedings, p. 11 (2022). https://aisel.aisnet.org/amcis2022/sig_dsa/sig_dsa/11
40. Smit, D., Eybers, S., Smith, J.: A data analytics organisation's perspective on trust and ai adoption. In: Jembere, E., Gerber, A.J., Viriri, S., Pillay, A. (eds.) SACAIR 2021. CCIS, vol. 1551, pp. 47–60. Springer, Cham (2022). https://doi.org/10.1007/978-3-030-95070-5_4
41. Smit, D., Eybers, S., de Waal, A., Wies, R.: The quest to become a data-driven entity: identification of socio-enabling factors of AI adoption. In: Rocha, A., Adeli, H., Dzemyda, G., Moreira, F. (eds.) WorldCIST 2022. Lecture Notes in Networks and Systems, vol. 468, pp. 589–599. Springer, Cham (2022). https://doi.org/10.1007/978-3-031-04826-5_58
42. Sorensen, G.: Democracy and Democratization: Processes and Prospects in a Changing World, 3rd edn. Taylor & Francis Group (2007)
43. Taylor, S., Todd, P.A.: Understanding information technology usage: a test of competing models. Inf. Syst. Res. **6**(2), 144–176 (1995)
44. Tornatzky, L.G., Fleischer, M.: The Processes of Technological Innovation. Lexington Books, Lexington (1990)
45. Whitehead, L.: Democratization: Theory and Experience. Oxford University Press on Demand, Oxford (2002)
46. Wihlborg, E., Söderholm, K.: Mediators in action: organizing sociotechnical system change. Technol. Soc. **35**(4), 267–275 (2013)
47. Wixom, B.H., Owens, L., Beath, C.: Data is everybody's business. MIT Sloan Center for Information Systems Research (2021)

The Preparation of South African Companies for the Impact of Artificial Intelligence

Tiaan Taljaard[1] and Aurona Gerber[1,2]($\boxtimes$)

[1] University of Pretoria, Pretoria, South Africa
aurona.gerber@up.ac.za
[2] The Center for AI Research (CAIR), Pretoria, South Africa

Abstract. Within conversations about strategic interventions that businesses should embrace given technological advancements, Artificial Intelligence (AI) features prominently. Thought leaders such as Gartner publish reports regularly with predictions that AI will affect the future job market and business competitiveness. In this paper we report on a survey that analysed the preparation of South African companies for the impact of AI. The survey had 120 respondents across all provinces and industries within South Africa. The results indicate that more than 80% of participants believe that AI will be crucial to compete in markets and that they need to prepare for AI's impact to stay relevant. However, more than 50% of all participants have not started initiatives to upskill staff or create initiatives to explore and deploy AI technology. The main impact of AI technology is believed to be the automation of mundane tasks, which could provide opportunities to prepare for the adoption of AI technology in a company. The results may be of value for researchers who aim to understand how to assist business within South Africa with the implementation of interventions for AI adoption.

Keywords: Artificial Intelligence adoption · AI for business relevance

1 Introduction

The current business landscape includes many conversations about the impact of the current rapid technology developments on business and business strategy. Many organisations worldwide are acknowledging the need to prepare for the impact of new technologies on business models as well as the associated workforce skills required to generate competitive advantages [3, 12, 19]. However, in spite of the so-called urgency, the changes triggered by developments such as the fourth industrial revolution with the associated artificial intelligence (AI) technologies may take years to realise [8, 17, 21].

This research study focuses on AI specifically and literature acknowledges the increasing impact of AI on business and the necessity to adopt AI strategically in order to remain competitive in the transformed economy [1, 7, 9, 11, 14, 16, 18]. Literature on the future impact of AI technologies on business is divided between excitement and concerns. One of the biggest discussion points is the impact on jobs where employees either fear that they will not gain the skills required to compete against robots in the workplace or are excited to learn new skills to prepare for future jobs [5, 10, 15, 26].

Given the stated importance to adopt AI within business in order to stay competitive, this study surveyed South African companies to understand how they view AI, and how they are preparing for the impact of AI on their business. The remainder of the paper is structured as follows: Sect. 2 provides background on AI and AI adoption within business. Section 3 discusses the research approach, Sect. 4 presents the results, a discussion is provided in Sect. 5 and Sect. 6 concludes.

2 Background and Related Work

For the purpose of this study, the broad definition of Nilsson for artificial intelligence (AI) namely "...the ability of a system to act intelligently and to do so in ever wider regions" is adopted as starting point [21]. This wide definition allows for the inclusion of the business perspective on AI because this study is specifically concerned with how business view AI, and how they prepare for the adoption of AI.

Guggenheimer categorises the use of AI within business and the workplace into four emerging patterns namely Virtual Agents, Ambient Intelligence, AI-Assisting Professionals and Autonomous systems [12]. In the first place, virtual agents are concerned with creating intelligent applications that can interact with employees and customers on behalf of the organisation. Virtual agents can be programmed to help answer questions, provide support and, in the future, become a proactive representative of an organisation for client and employee interactions [12]. The second pattern concerns ambient intelligence that tracks people and objects in a physical space such as a factory floor with objects, machines, and workers. Ambient intelligence is where AI can overlay a digital world over the factory to measure out-of-place objects, possible machine failures and signal a warning to employees to react and reduce possible incidents [12]. Thirdly, AI-Assisting Professionals could assist any professional to be more productive with the decisions they have to make. Applications using machine learning and other cognitive services allows a machine to generate and deliver insights to professionals to equip them to make more informed decisions with their daily activities [12]. The last pattern entails autonomous systems including systems that function without any human interaction such as systems that identify and remove possible threats that could harm an organisation's network [12].

The patterns for the use of AI in business identified support the case that AI might influence all business functions eventually. Furthermore, AI is predicted to change the workplace substantially [1, 7, 10, 14, 18]. Some studies indicate that AI will create more jobs than the number of jobs it replaces [3]. MGI predicts that AI investments could lead to adding 20 to 50 million new jobs globally by 2030 [4]. However, having vacancies created does not necessarily mean the skills will be available to perform these new roles [2]. There will be a significant gap between the skills employers are looking for and the potential skills required by roles in the future. Even if the markets create more jobs than it eliminates, several people will still be unemployed due to skills shortages [6].

In summary, countries are listing AI as a top priority from the government level down to large and small organisations [7, 13, 24, 25]. AI has different levels of complexity, and AI suppliers create and enhance AI services every day to make it easier to start using AI through low/no-code services [3, 20, 22]. There is already a skills gap between AI users and AI suppliers, and this gap challenges organisations with how they could equip employees with in-depth technical skills to create AI products or to use AI services offered by leading AI suppliers effectively [26]. The skills required to be AI competent should be considered and organisations should analyse their people's current skill set and plan to upskill people to perform the task required for AI deployment. Lastly, there is no best practice guidance on how to prepare for AI or how to implement AI. Leading AI suppliers offer guidance, but there is still more research and deployments required before there could be a guiding list of considerations. The wide range of considerations for businesses when preparing to adopt AI provides the motivation for this study, namely to analyse how South African businesses are preparing for the impact of AI, including the business strategy, changed job descriptions and required skills.

3 Research Approach

The study aimed to determine if South African businesses are preparing for the impact AI through a survey gathering information from senior IT leaders in South African businesses. These IT leader roles include CIOs, IT heads, and business executives in the digital space. The data was gathered from companies across South Africa: companies of all sizes and operating in various industries. These respondents were included to ensure a large sample that should include different views. Using purposeful sampling (i.e. when the study targets individual participants that possess the required knowledge, experience and skills required for the study), the survey was distributed using a database of more than 2000 IT leaders from more than 1000 companies in South Africa. The aim was to obtain at least one hundred responses from senior IT leaders. Ethical clearance was granted and the survey included a consent letter. Responses from one hundred different companies, varying in size and industry were considered to be sufficient for an indication of whether South African companies are preparing for AI's impact on their businesses based on recommendations from quantitative methods specifying how many survey responses are sufficient [27]. For an experimental study, the recommendation is to obtain a minimum of 10% or at least 100 responses [27]. The study is cross-sectional and used online surveys as it is the best option to simultaneously reach out to many company leaders and collect information from all industries [27]. The survey questions were tested by doing pilot tests with five senior leaders and experts in data and AI technologies. The survey used in this study included a combination of nominal and ordinal data [23], and the survey was developed and distributed through email and on LinkedIn using Qualtrix. The survey questions are summarised in the appendix. Out of 1923 email requests, 80 participants responded, and through LinkedIn, 32 out of 800 requests responded. In total 112 valid responses were received.

4 Results

4.1 Demographics

Responses were received from business operating across all provinces in South Africa. However, the majority of responses were from Gauteng (35.93%), Western Cape (23.35%) and KwaZulu-Natal (12.57%). The top three provinces comprised 71.86% of all responses. The responses include participants from different company size and company age (Fig. 1). The majority of the responses are from companies operating for more than 20 years.

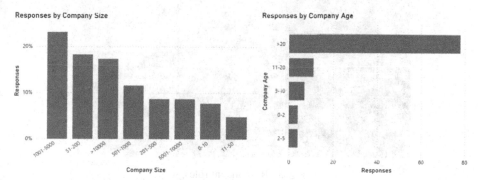

Fig. 1. Respondents' company size and age.

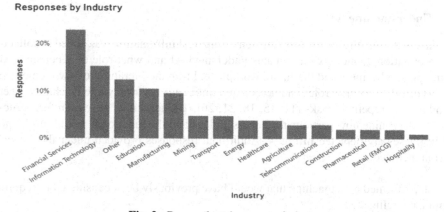

Fig. 2. Respondents' company industry.

The collected data includes responses from companies in all industries as depicted in Fig. 2. Most of the respondents are from Financial Services (22.86%) and the Information Technology industry (15.24%).

The last component of respondents' demographics was included to determine whether the respondents are familiar with AI and their business' strategy related to implementing AI technology was. Figure 3 indicates that most respondents are IT managers, followed by Other roles. Other roles listed are IT specialist, solutions architect, systems engineer and CEO or founder. With regards to the follow up question asking about experience with AI, 28.48% of the respondents indicated they do not have AI experience, resulting in 71.52% of the respondents indicating they have previously used, developed, deployed, or have done training on AI technologies.

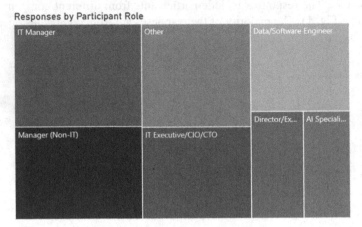

Fig. 3. Respondent roles.

4.2   Understanding AI

The first section in the survey after the demographic information was aimed at collecting information about how participants understand AI and what role they perceive AI would play in business and the future workplace. From the literature review it emerged that AI could create new roles and tasks that humans can perform while machines cover mundane and repetitive tasks [11, 15, 18, 21, 26]. The questions were therefore structured to determine how participants understand AI and the function it would fulfil, and the responses are depicted in Fig. 4. Additional responses provided under the 'other' section included:

- Task performed by a machine that would have previously been considered to require human intelligence.
- AI is programs, written by humans, that run on machines. AI is basically algorithms run off big servers/super computers. My own opinion is that we are very far away from computers actually thinking.
- Both option 2 & 3, some task that repeated over and over loses human interest while AI routines will not, replacing base tasks with AI and only passing exception for human intervention is the solution to many problems in our environment;
- Computers mimicking human intelligence.
- Computers mimicking human intelligence in processing.

- Computers returning outputs produced by running statistical models rather than as imperative instructions.

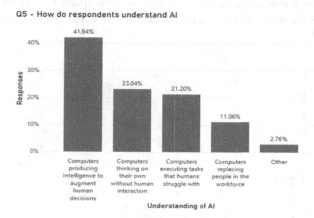

Fig. 4. Understanding AI.

With regards to Understanding, the survey results leaned towards humans and AI working together rather than AI replacing humans: The survey results leaned more towards humans and AI working together. 42% of respondents believe *Computers will produce intelligence to augment humans' decisions*, and 21% believe *Computers will be used to execute tasks where humans struggle*. 34% leaned towards *people being replaced by AI*, where 23% think *Computers that think, act and decide on its own without any human interaction*.

In the next section AI and human interaction, the top chosen options included the three selections where machines are used to augment human decision-making activities:

- AI generates insights that human uses as an input to their decisions – 28.73%.
- Human creates scenarios and AI evaluates with additional insights on decisions – 28%.
- AI makes the recommendation for a human to select from – 22.18%.
- Few respondents chose the options where AI performs most or all of the activities:
- AI makes the decisions but human implements (i.e. approves/rejects) - 12.73%.
- AI decides and implements with no human review – 8.36%.

The responses indicate that most participants are of the opinion that AI requires human supervision and should not be left to make decisions by itself. The respondents indicated that AI could be used as a co-worker to contribute to activities that humans should perform. Additionally, one of the top three responses were that machines could be used to create scenario planning. Were humans create scenarios for machines to analyse. This illustrates the notion that machines are not yet trusted to perform general intelligence tasks and should be used for narrow automation where compute and number crunching outperforms humans.

4.3 Preparing for AI

This section discusses the results of respondents with regards to the preparation for AI impact on their businesses, and whether they believe they should be preparing for any impact that AI might have in the future. Over 85% of participants believe *AI will change how they deliver business value*, and this change is expected across all industries as depicted in Fig. 5.

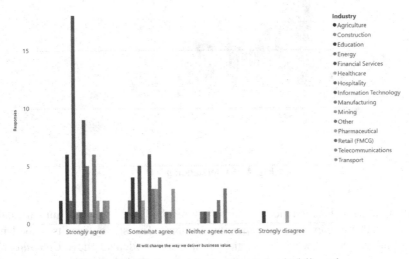

Fig. 5. AI will change how business value is delivered.

On the follow-up question, over 80% of participants also believed that AI will be crucial for the future and how they will compete in their markets. AI is seen as a crucial element in most industries, where 12/14 industries have most responses in 'strongly agree' or 'somewhat agree'. There were two other industries where the respondents were split between agreement and disagreement; in education, the view is that 54.54% agree that AI will be crucial and 45.46% disagreed or were unsure. In mining, 60% disagreed or were unsure, and 40% agreed that AI would be crucial to stay competitive.

On the third Likert statement, most respondents believe that it will take time to deploy AI in their businesses (over 73%). 11.43% were unsure and 15.23% respondents disagreed and believed that it will be quick to deploy AI in their organisations. Figure 6 depicts the distribution of responses across industries, where most participants believe it will take time to deploy and adopt AI. The interesting view is that even though very few participants have formal training, there are a few with experience in deploying AI in their organisations; these participants also agreed to the statement that it will take time to deploy and adopt AI.

According to the survey data there seems to be no impact when adding company size into the analysis, i.e. assuming smaller companies believe it might be simpler to adopt and deploy AI. All respondents' companies believe that it will take time to deploy and adopt AI, irrespective of company size. The only exception is IT consulting houses companies with 11–50 employees that are actively providing AI services.

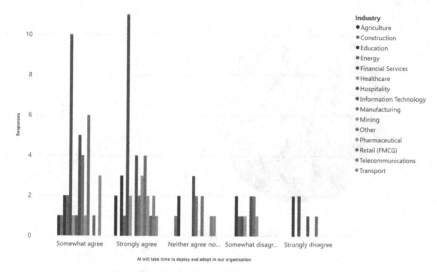

Fig. 6. AI will take time to deploy and adopt.

The last statement in this section checked whether participants think they are familiar with how AI will change their products and services. Over 65% of respondents believe that they are familiar with how AI will change their products and services. Close to 35% respondents are either uncertain or do not think they know what impact AI will have. Manufacturing, Education and Other are some of the industries that are not familiar with the impact that AI might have on their products and services.

In summary, this question illustrates that most industries are under the impression that AI will be critical to how they compete in the markets and will also change the way they operate in the markets in the future. Even though AI is perceived to play a crucial role, it is also seen as a complex technology to adopt and deploy. The majority of respondents believe they are comfortable with how AI will change their products and services, but they also believe it will take time to get ready for the changes.

The analysis indicated that over 85% of the respondents believe that AI will change the way business is done. However, most of the participants have not adopted AI, are unsure about any deployment, do not plan to deploy AI, or have only started with POCs/Pilots (58.49%). Figure 7 provides an overview of AI adoption from the participants: 41.51% of responses have deployed AI into some of their solutions, and only 4.72% have extensive AI deployment experience.

The next part of the survey explored whether businesses are actively preparing for AI deployments, skills development, or identifying the knowledge and experience needed for AI deployment/development. Most respondents are either not sure or disagree with the statement that product owners are familiar with AI. Close to 58% of respondents are either unsure or disagree that their product owners are familiar with AI and how it can advance their products. Furthermore, when analysing across industries, most industries are split between agreement, unsure, and disagreement. Only the telecommunications, pharmaceutical and construction agree that their product owners are not familiar with AI and how it can advance their products.

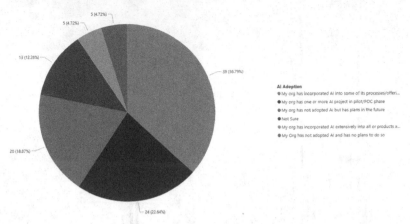

Fig. 7. AI experience.

The subsequent survey sections explore whether companies have identified the skills required to adopt AI technology and if they have started any preparation towards upskilling existing staff to learn more about AI. Even though 55% believe they have identified the skills needed, only 42% have upskilled staff to learn more about AI, 58% have not upskilled staff or are unsure, specifically, 21.5% of respondents were unsure about AI skills across their company.

The last question in this section has been structured to determine how companies are obtaining AI skills. Close to 30% of participants are reskilling current staff, and that close to 62% are going out to the market to obtain the skills require by either hiring new staff, purchasing AI solutions, or reaching out to vendors to build the required AI technology. These observations correlate to companies' skills preparation, where the skills have been identified, but employees have not been upskilled to deliver according to requirements.

4.4 Expected Impact of AI

The first question in this section included a Likert style question with eight statements. The eight statements were structured to determine the impact the respondents expect AI will have on their industry and markets. The statements included views on competitors, clients, suppliers, business processes, products, and competitive advantages (refer to the appendix). Most participants agreed with the statements that AI will have an impact on competitors, clients, suppliers, business processes, products, and competitive advantages, with 95% indicating that the effective use of AI will give a competitive advantage. The lowest score again alluded to the fact that AI adoption is lagging with only 44% of respondents agreeing with the statement that their suppliers have started to offer AI products. Figure 8 illustrates how the participants responded to the various statements.

Question	Agreement	Not Sure	Disagreement
Our Competitors will use AI.	80.38	10.28	9.34
Our clients will request AI-driven products.	66.36	14.95	18.69
Effectively using AI will give us a competitive advantage.	95.33	0.93	3.74
AI is required to reduce current pressures to reduce costs.	80.37	10.28	9.35
Newmarket entrants are using AI to offer improved products/services.	60.74	24.3	14.96
Our suppliers have started offering AI products and services.	43.92	33.64	22.44
Embracing AI will allow our organisation to improve our products	90.66	5.61	3.73
Embracing AI will allow our organisation to explore new markets.	86.91	5.61	7.48

Fig. 8. AI impact.

4.5 Alignment of Strategy

In the last section the survey explored the alignment of business strategy with AI. The aim was to determine if participants actively think about how they incorporate AI teams, priorities, and AI development guidelines into business strategy. In response to the first question most respondents believe their company is focusing on 'automating mundane tasks' (23.53%). This supports the data of the last question in the previous section where most participants also ranked automation as a top impact area for AI. The following areas of focus included 'improving customer experience' (21.11%) and then a tie between 'adding new features to customer-facing products' and 'improving decision-making for operational teams' with 18.69% each from the respondent selections.

In the results of the follow-up question, respondents indicate they will only deploy AI solutions once they understand the financial impact and will deploy with early more minor releases to get feedback from customers. However, 86.54% of participants believe that AI solutions will evolve and improve as we learn to understand the impact technology has on operations as depicted in Fig. 9.

The next question was concerned with team setup and development standards. The participants' responses favoured 'unsure' or 'disagreement'. The organisations are not sure exactly how AI teams have been set up or whether coding standards were created that developers need to follow when deploying AI technology.

The last question in this section was structured to get a view of if/how organisations have set up their AI adoption teams (Fig. 10). As depicted in Fig. 10, 31.51% of respondents were unsure if/how the company has set up AI teams, 26.03% have teams set up with AI roles and mandates. Only 21.23% have coding standards and frameworks to guide the deployment of AI solutions, and only 12.33% have created training curriculums to upskill their workforce. Noteworthy is that these responses do not align with responses in the previous section where close to 30% of respondents believe they can upskill existing staff. The last noticeable view is that less than 10% of participants have a dedicated AI leader.

Question	Agreement	Not Sure	Disagreement
We will only deploy AI models once we fully understand the financial impact	67.31	16.35	16.34
We will release early versions to get feedback from end-users	64.42	22.12	13.46
AI solutions will evolve from the original design and development as we better understand the impact	86.54	11.54	1.92
The organisation sets up cross-functional teams to work on AI requirements	50.96	29.81	19.23
The org has coding standards and pipelines for AI deployments.	39.42	25.96	34.62
AI teams are made up of members from multiple functional areas.	56.73	27.88	15.39

Fig. 9. AI deployment considerations.

Question	Agreement	Not Sure	Disagreement
Teams are clear on how to develop and deploy AI solutions	32.38	30.48	37.14
Business have backlogs of requests for more AI products/services	27.62	35.24	37.14
Back office divisions are aware of AI requests that have been prioritised	21.91	38.1	39.99
All areas across the organisation are aware of the org AI strategy	26.66	28.57	44.77

Fig. 10. AI adoption teams.

5   Discussion

AI is a technology that will arguably have a significant impact on business in the future [4, 5]. South Africa will also be affected, and it is therefore necessary to establish if South African businesses are preparing for the impact of AI on business to remain competitive, and this was the goal of the survey reported upon in this paper.

The survey results from 112 respondents across South African industries indicated that most respondents believe that AI is crucial to ensure that they remain competitive in the future. Furthermore, respondents believe that they will need AI skills and knowledge to participate in their respective markets and the survey results also indicate that business across South Africa believe that they are familiar with the opportunities and threats that AI could bring to their industry.

However, from the findings it is evident that even though respondents believe that it is a crucial technology to consider, there is no uniform definition of AI and no clear understanding of exactly what the technology should do and how to adopt or implement it. Significantly, the results from the survey also indicated how few companies are actively preparing and implementing AI currently. Further research is required to understand the reasons for this lack of adoption, as well as create possible mechanisms to assist with

AI deployment and to find proven approaches to integrate AI into existing products and technology. Additional focus could be placed on measuring the impact and new roles created to assist companies with a clear understanding of training required to upskill staff. Lastly, it will also be important to create business value measurements resulting from AI. These measurements should be used to truly monitor and measure the return on investments that AI brings to markets and indicate where it is most efficient.

6 Conclusion

The main research question was to determine whether South African companies are strategically preparing for the impact of AI on their business. The survey collected information on how companies define AI, whether they are actively preparing for AI impact (including the job content changes and related skills required) as well as information on the integration of AI into business strategy. In summary the results indicate that respondents agree that AI will be crucial in order to obtain a competitive advantage in the respective markets in the future, to build unique product/services and to enhance customer experiences. The respondents indicated that they are familiar and aware of the threats and opportunities AI could bring to their industries and markets. These views all align with the research presented in the literature review on how significantly AI could impact future markets. However, respondents indicated that initiatives to use or upskill and embed AI technology into businesses are still lacking. The main focus area for research now needs to shift to mechanisms to support business with AI training and upskilling, as well as adoption frameworks and methodologies that can help companies deploy AI solutions. The limitations of the study include that it is exploratory and only provides a view of a relatively small number of South African companies. In addition, more insight might be obtained focusing on specific industries as not all industries require and therefore adopt AI at the same pace. In addition, not all senior leaders are concerned with AI.

Appendix: Survey Questions

Focus area	Questions	Question type
Demographics	Please describe your experience with AI: a. I have experience using AI tools as an end-user of AI b. I have experience deploying AI solutions c. I have experience managing how end-users interact with AI d. I have formal training to develop AI solutions e. I have no AI experience f. Other (Text field)	Multiple Choice

(continued)

(*continued*)

Focus area	Questions	Question type
	Which best describes your role? a. AI Specialist/Data Scientist/ML Engineer b. Data/Software Engineer c. Director/Executive (non-IT) d. IT Executive/CIO/CTO e. IT Manager f. Manager (Non-IT) g. Other	Single Choice
	Please select your industry a. Agriculture b. Construction c. Education d. Energy e. Financial Services f. Healthcare g. Hospitality h. Information Technology i. Manufacturing j. Mining k. Pharmaceutical l. Retail (FMCG) m. Telecommunications n. Transport o. Other	Single Choice
	How many employees does your company have? a. 0–10 b. 11–50 c. 51–200 d. 201–500 e. 501–1000 f. 1001–5000 g. 5001–10000 h. >10000	Single Choice
	Company age? a. 0–2 b. 2–5 c. 5–10 d. 11–20 e. >20	Single choice

(*continued*)

(*continued*)

Focus area	Questions	Question type
	Where is your company clientele located? a. Local (City) b. Provincial c. National (SA) d. International (Africa) e. International (Global)	Multiple Choice
	In which Province(s) is your company located? a. Eastern Cape b. Free State c. Gauteng d. KwaZulu-Natal e. Limpopo f. Mpumalanga g. North West h. Northern Cape i. Western Cape	Multiple Choice
Understanding AI	What is your understanding of Artificial Intelligence (AI)? • Computers thinking on their own without human interaction • Computers producing intelligence to augment human decisions • Computers replacing people in the workforce • Computers executing tasks that humans struggle with • Other	Single Choice

(*continued*)

(*continued*)

Focus area	Questions	Question type
AI and human interaction	How do you describe the interaction between humans and AI? • AI makes the decisions, but human implements (i.e. approves/rejects) • AI decides and implements with no human review • AI generates insights that human uses as an input to their decisions • AI makes the recommendation for a human to select from • Human creates scenarios, and AI evaluates with additional insights on decisions	Multiple Answers
AI Preparation	To what extent has your organisation adopted AI? • My org has not adopted AI and has no plans to do so • My org has not adopted AI but has plans in the future • My org has one or more AI project in pilot/POC phase • My org has incorporated AI into some of its processes/offerings • My org has incorporated AI extensively into all or products and services	Single Choice

(*continued*)

(continued)

Focus area	Questions	Question type
	Has your organisation thought about what is needed for AI and the possible impact? Please indicate the degree to which you agree with the following statements about your organisation: • AI will change the way we deliver business value • We know how AI will change our product/services offerings • AI will take time to deploy and adopt in our organisation • AI is crucial to compete in our industry in the future • Has your organisation identified any AI skills required for future processes/product development? Please indicate the extent to which you agree with each of the following statements: • We have identified the skills and knowledge our org requires to adopt AI • We have upskilled staff in our organisation to learn more about AI • Our Product Owners are familiar with AI and how it can advance their products	Likert: Strongly disagree, Somewhat Disagree, Neither Agree/disagree, Somewhat agree, Strongly agree, Not sure
	How is your organisation obtaining the skills required to build and deploy AI? • Reskill current staff • Hire new staff from the market • Contract to vendors • Purchase third party solutions • Not sure	Multiple Answers

(continued)

(*continued*)

Focus area	Questions	Question type
AI Impact	Please provide a view of how much you agree to the following statements. Indicate how impactful AI could be to your industry and organisation • Our Competitors will use AI • Our clients will request AI-driven products • Effectively using AI will give us a competitive advantage • AI is required to reduce current pressures to reduce costs • Newmarket entrants are using AI to offer improved products/services • Our suppliers have started offering AI products and services • Embracing AI will allow our organisation to improve our products • Embracing AI will allow our organisation to explore new markets	Likert: Strongly disagree, Somewhat Disagree, Neither Agree/disagree, Somewhat agree, Strongly agree, Not sure
AI in Business and Strategy	Which of the following areas is your organisation focusing on? • Automating mundane tasks (Data capturing) • Improving decision-making for operation time schedules, stock control, marketing ROI • Adding new features to customer-facing products • Accelerating staff decision making with AI • Improving the customer experience (Personalisation) • Other	Multiple Answers

(*continued*)

(*continued*)

Focus area	Questions	Question type
	Regarding your organisation's approach to developing AI solutions, to what extent do you agree with the following statements? • We will only deploy AI models once we fully understand the financial impact • We will release early versions to get feedback from end-users • AI solutions will evolve from the original design and development as we better understand the impact • The organisation sets up cross-functional teams to work on AI requirements • The organization has coding standards and pipelines for AI deployments • AI teams are made up of members from multiple functional areas	Likert: Strongly disagree, Somewhat Disagree, Neither Agree/disagree, Somewhat agree, Strongly agree, Not sure
	Regarding your organisation's operating model for developing AI solutions, to what extent do you agree with the following statements? • Teams are clear on how to develop and deploy AI solutions • Business has backlogs of requests for more AI products/services • Back-office divisions are aware of AI requests that have been prioritised • All areas across the organisation are aware of the org AI strategy	Likert: Strongly disagree, Somewhat Disagree, Neither Agree/disagree, Somewhat agree, Strongly agree, Not sure

(*continued*)

(continued)

Focus area	Questions	Question type
	In your organisation which of the following describes it best: • In your organisation which of the following describes it best: • We have an appointed AI Executive • We have created training curriculums for AI • We have teams and departments with AI roles and mandate • We have standards and deployment frameworks or AI solutions • Not sure	Multiple Choice

References

1. Alsheibani, S., et al.: Artificial Intelligence Adoption: AI-readiness at Firm-Level 9 (2018)
2. Bob Hayes: Investigating Data Scientists, their Skills and Team Makeup. http://businesso verbroadway.com/2015/09/23/investigating-data-scientists-their-skills-and-team-makeup/. Accessed 24 Jan 2019
3. Bughin, J., et al.: Artificial Intelligence: The Next Digital Frontier. Mckinsey Global Institute (2017)
4. Carter, D.: How real is the impact of artificial intelligence? The business information survey 2018. Bus. Inf. Rev. **35**(3), 99–115 (2018). https://doi.org/10.1177/0266382118790150
5. Cartwright, N.: 3 ways to prepare your bank's workforce for AI. https://medium.com/fintech-weekly-magazine/3-ways-to-prepare-your-banks-workforce-for-ai-bee35ee7ff23. Accessed 29 Aug 2022
6. Chrisinger, D.: The solution lies in education: artificial intelligence & the skills gap. Horiz. **27**(1), 1–4 (2019). https://doi.org/10.1108/OTH-03-2019-096
7. Davenport, T.H., Ronanki, R.: Artificial intelligence for the real world. Harvard Bus. Rev. **96**(1), 108–116 (2018)
8. Denning, P.J., Denning, D.E.: Dilemmas of artificial intelligence. Commun. ACM **63**(3), 22–24 (2020). https://doi.org/10.1145/3379920
9. Di Vaio, A., et al.: Artificial intelligence and business models in the sustainable development goals perspective: a systematic literature review. J. Bus. Res. **121**, 283–314 (2020). https://doi.org/10.1016/j.jbusres.2020.08.019
10. Duan, Y., et al.: Artificial intelligence for decision making in the era of Big Data – evolution, challenges and research agenda. Int. J. Inf. Manag. **48**, 63–71 (2019). https://doi.org/10.1016/j.ijinfomgt.2019.01.021
11. Enholm, I.M., et al.: Artificial intelligence and business value: a literature review. Inf. Syst. Front. (2021). https://doi.org/10.1007/s10796-021-10186-w
12. Guggenheimer, S.: Emerging AI Patterns. https://docs.microsoft.com/en-us/archive/blogs/ste vengu/emerging-ai-patterns. Accessed 29 Aug 2022

13. Hunt, W., et al.: Measuring the impact of AI on jobs at the organization level: lessons from a survey of UK business leaders. Res. Policy **51**(2), 104425 (2022). https://doi.org/10.1016/j.respol.2021.104425
14. Jöhnk, J., Weißert, M., Wyrtki, K.: Ready or not, AI comes—An interview study of organizational AI readiness factors. Bus. Inf. Syst. Eng. **63**(1), 5–20 (2021). https://doi.org/10.1007/s12599-020-00676-7
15. King, J.L., Grudin, J.: Will computers put us out of work? Computer **49**(5), 82–85 (2016). https://doi.org/10.1109/MC.2016.126
16. Kolasa-Sokołowska, K.: Artificial intelligence and risk preparedness in the aviation industry. In: Regulating Artificial Intelligence in Industry. Routledge (2021)
17. Landgrebe, J., Smith, B.: Making AI meaningful again. Synthese **198**(3), 2061–2081 (2019). https://doi.org/10.1007/s11229-019-02192-y
18. Loureiro, S.M.C., et al.: Artificial intelligence in business: state of the art and future research agenda. J. Bus. Res. **129**, 911–926 (2021). https://doi.org/10.1016/j.jbusres.2020.11.001
19. McWaters, R.J., Galaski, R.: The new physics of financial services: understanding how artificial intelligence is transforming the financial ecosystem. In: World Economic Forum (2018)
20. Mikalef, P., et al.: Developing an artificial intelligence capability: a theoretical framework for business value. In: Presented at the International Conference on Business Information Systems, June (2019)
21. Nilsson, N.J.: Artificial Intelligence Prepares for 2001 (1983). https://ai.stanford.edu/~nilsson/OnlinePubs-Nils/General%20Essays/AIMag04-04-002.pdf
22. Nilsson, N.J.: The Quest for Artificial Intelligence: A History of Ideas and Achievements. Cambridge University Press, Cambridge (2009). https://doi.org/10.1017/CBO9780511819346
23. Oates, B.J.: Researching Information Systems and Computing. SAGE, Thousand Oaks (2006)
24. Provost, F., Fawcett, T.: Data Science for Business: What You Need to Know about Data Mining and Data-Analytic Thinking. O'Reilly Media Inc., Sebastopol (2013)
25. Sestino, A., De Mauro, A.: Leveraging artificial intelligence in business: implications, applications and methods. Technol. Anal. Strat. Manag. **34**(1), 16–29 (2022). https://doi.org/10.1080/09537325.2021.1883583
26. Wilson, H.J., et al.: The jobs that artificial intelligence will create. MIT Sloan Manag. Rev. **58**, 14 (2017)
27. Young, T.J.: Questionnaires and surveys. In: Hua, Z. (ed.) Research Methods in Intercultural Communication: A Practical Guide, pp. 165–180. Wiley (2016)

Responsible and Ethical AI

Answerability, Accountability, and the Demands of Responsibility

Fabio Tollon[1,2,3](✉) (iD)

[1] Department of Philosophy/GRK 2073 "Integrating Ethics and Epistemology of Scientific Research", Bielefeld University, Bielefeld, Germany
Fabiotollon@gmail.com
[2] Department of Philosophy, Unit for the Ethics of Technology, Center for Applied Ethics, Stellenbosch University, Stellenbosch, South Africa
[3] Center for Artificial Intelligence Research (CAIR), Pretoria, South Africa

Abstract. Knowing who or what should be held morally responsible when something goes wrong (or right) is an important part of human social relations. Some authors have suggested that certain AI-based systems pose a threat to our responsibility practices, which might lead to a 'responsibility gap'. Such 'gaps' occur when we have no fitting candidate who can be held responsible for some event that was caused by an AI-system. If such gaps do in fact exist, then it might make sense to have an outright ban on such systems. Conversely, if these gaps do not exist, then perhaps there is nothing to worry about. In this paper I do not aim to resolve this debate. Rather, I wish to make a modest contribution to this literature. I will argue that in two specific senses of responsibility, namely, answerability and accountability, there might be no such responsibility gaps. While AI-systems might make it harder to *know* who we should hold responsible, they do not make such ascriptions impossible. Responsibility gaps then, on my view, are simply epistemic, and thus do not call for *special* attention any more than our usual practices of holding one another morally responsible.

Keywords: Responsibility gaps · Accountability · Answerability · AI

1 Responsibility and AI

Recent work in moral philosophy has been concerned with issues of responsibility as they relate to the development, use, and impact of artificially intelligent systems. Oxford University Press even published their first-ever Handbook of Ethics of Artificial Intelligence (2020), which was devoted to tackling (then) current ethical problems raised by AI-driven systems, with the hope of mitigating future harms by advancing appropriate mechanisms of governance for these systems. The book is wide-ranging (featuring over 40 unique chapters), insightful, and deeply disturbing. From gender bias in hiring, racial bias in creditworthiness and facial recognition software, and sexual bias in identifying a person's sexual orientation, we are awash with cases of AI systematically heightening rather than reducing structural inequality [1–3]. These cases remind us that things often

© The Author(s), under exclusive license to Springer Nature Switzerland AG 2022
A. Pillay et al. (Eds.): SACAIR 2022, CCIS 1734, pp. 371–383, 2022.
https://doi.org/10.1007/978-3-031-22321-1_25

go wrong in our usage and implementation of AI-based systems. It is therefore important that we are able to hold specific parties responsible for what has gone wrong.

In this paper I will address the question of responsibility as this relates to AI-based systems. First, I outline, in general terms, what is at stake in debates on responsibility and AI, and why we might worry about a "gap" in responsibility. Second, I outline various senses of responsibility, and clarify that I will only be interested in responsibility as answerability and responsibility as accountability in this paper. Third, I show how in both of these cases we do not need to worry about a metaphysical "gap" in responsibility, but there might well be an epistemic gap. We might not know who to hold responsible, but there are many other cases where this is also true, and which do not involve technology. In other words, this problem is not unique to the implementation and use of AI-based systems.

This paper's contribution, therefore, is one of building a bridge between theoretical debates in philosophy and the practical question of responsibility gaps. I do not offer a unique solution to the problem, but instead try to articulate the counters of the recent debate in this field and link it to more traditional debates in moral philosophy.

2   Responsibility Gaps and AI

Under normal circumstances, the manufacturer or operator is typically held responsible for the consequences that follow from the operation of their machine [4]. This is because it is supposed that manufacturers should be able to reasonably foresee the possible consequences that their machines could bring about. They should also be able to control these machines, so that if something were to go wrong, they would be able to intervene. This is true in two senses, as manufacturers are both responsible for the *intended* and the *unintended*, yet foreseeable, consequences of their creations (bearing in mind that ought implies can).

However, machines that learn, due to their behaviour being *in principle* unpredictable from the perspective of their creators, might pose a unique challenge to our responsibility ascriptions. They make it the case that manufacturers are no longer in control, cannot predict, and do not have knowledge of, the future consequences of their machines. In such cases, it would be unfair to hold manufacturers morally responsible for their machines' behaviour if this behaviour cannot (in principle) be predicted or reasonably foreseen [4]. Robert Sparrow has argued that this means we should proscribe certain technologies on the basis that they lead to such gaps in responsibility. This would lead to a ban on any autonomous or "learning" machines [4].

For example, consider a self-driving car that kills a child as she is crossing the road. Suppose further that the car's sensors failed to detect the child, and if a human had been in control, they would have been expected to identify the child and avoid hitting her. If the car had been operated by a human moral agent, we would have little difficulty in identifying the driver as the source of the moral harm, and if negligence can be proven, *blaming* the driver by holding them to account. In the case of the self-driving car, however, there is no driver to blame and to blame those individuals responsible for manufacturing or designing the car might seem to stretch our responsibility concepts beyond recognition (if there was indeed no fault with the software). This might be because the algorithm learnt some new behaviour based on information acquired during its operation (and thus was not under the *control* of those who created and designed it).

Here we are confronted with three possible "sources" of responsibility (at least in principle): the car, the human(s) (this includes collections of humans, such as a company), or nobody. In the first case, by virtue of its features *qua* car, the car does not seem to be a fitting target for our responsibility responses. Additionally, friends and family of the victims might not take kindly to such a determination. In the second case, it might seem unfair to hold the manufacturers, designers, programmers, engineers, etc. responsible, as they could not have foreseen or predicted the outcome. They were not in *control* of the system at the time of the harm, nor did they *intend* for anybody to be harmed. Finally, then, we might say that due to the difficulties just raised, *nobody* is in fact responsible. Having nobody be responsible is surely undesirable. Something went wrong and a person died. It is natural for us to expect, in the absence of excuse or justification, that the agent responsible should be sanctioned or punished for their behaviour. Additionally, having nobody be responsible would be *unfair* from the perspective of the victims. They would have no closure and would be left in the dark as to who or what is the legitimate target of their anger. This brings us to the two horns of the responsibility-gap: to avoid "nobody" being held responsible we apparently either need to hold manufacturers, designers, programmers, engineers, etc. unfairly responsible or proscribe "learning machines" to avoid this outcome.

However, "responsibility" is not a monolithic concept, and so when we investigate whether a human agent might be held responsible, we need to be clear about *which sense* of responsibility we are interested in. Responsibility has many "faces". In what follows I will clear the conceptual ground by outlining a few ways in which the concept of responsibility might be used. This will help clarify exactly what is meant by our various responsibility practices. We will then be in a position to see where gaps may emerge due to AI, and how we might avoid unfair ascriptions of responsibility.

3 Types of Responsibility

What does it mean to be responsible? It seems that the concept of responsibility has a remarkably wide range of applications. We regularly say things like "she is a responsible person" (virtue), "he is responsible for the accident" (accountability) or "she is responsible for the safety of the passengers" (moral obligation). Consider these remarks from H.L.A. Hart:

> "As captain of the ship, X was responsible for the safety of his passengers and crew. But on his last voyage he got drunk every night and was responsible for the loss of the ship with all aboard. It was rumoured that he was insane, but the doctors considered that he was responsible for his actions. Throughout the voyage he behaved quite irresponsibly, and various incidents in his career showed that he was not a responsible person. He always maintained that the exceptional winter storms were responsible for the loss of the ship, but in the legal proceedings brought against him he was found criminally responsible for his negligent conduct, and in separate civil proceedings he was held legally responsible for the loss of life

and property. He is still alive and he is morally responsible for the deaths of many women and children" [5].

Hart himself argues that there are four different senses of responsibility at play in this passage: role-responsibility (X's duty to keep, as captain, to keep the ship safe) causal-responsibility (it is alleged that what occurred could be attributed to being caused by X), liability-responsibility (X, due to his actions or the consequences of his actions, must compensate the harmed parties), and capacity-responsibility (X was not insane, and thus a full moral agent with the right capacities) [5]. These different senses of responsibility make explicit its pluralism, but also show that each sense of the concept has its own distinct meaning and realm of application [5]. While Hart's analysis gives us a glimpse of what responsibility pluralism looks like, there are more than the four cases he alludes to.

In attenuating the problem of responsibility-gaps we must have a handle on the kinds of responsibility that are at stake. By getting clear on these types of responsibility and appreciating that some senses are descriptive rather than normative, we will be in a better position to train our sights on the target of responsibility-gaps. Ibo van de Poel has a very informative sketch of the different types of responsibility that might exist [8]. I will draw on his account in what follows but will also extend and revise it. Specifically, the first six cases I take directly from van de Poel, and the last three are borrowed from David Shoemaker.

In the first case, we have responsibility as *cause*, where we say things like "the moon is responsible for the tides". Second, there is responsibility as *task*, such as the bus driver who is responsible for driving the bus. Third, responsibility as *authority*, where someone is made responsible for a task or project. Fourth, responsibility as *capacity*, which is the ability to act in a responsible way (by, for example, reflecting on reasons and weighing potential consequences). Fifth, there is responsibility as *liability*, as in "they are liable to pay damages". Sixth, responsibility as (moral) *obligation*, which is essentially an obligation to ensure some desirable outcome (the ship captain in Hart's example above) [8]. Seventh, responsibility as *attributability*, which is concerned with the character of the agent in question. Eighth, responsibility as *answerability*, which is grounded in the agents' evaluative reasons for acting in this or that way. Lastly, we have responsibility as *accountability*, which is when an agent can be appropriately blamed for an action or event [6].

The first four of these cases are descriptive in nature. That is, they either are or are not the case, based on whether some conditions are met either by the agent or the environment [7]. They also tell us something about the relation between the agent and the states of affairs in question, but do not act as a guide as to what we *ought to do*. What we ought to do is the realm of normative ethics, which is how we should understand the last five conceptions. The last five senses of responsibility are normative, in that they guide us and what we ought to do. Each of these senses of responsibility imply a normative evaluation (or prescription), in that they are not so much concerned with what is the case (in a descriptive sense), but with ensuring desirable outcomes or evaluating the quality of the agent(s) in question.

However, and as will become clear throughout, the descriptive senses have a role to play in our normative evaluations. These descriptive senses often *inform* the normative senses of responsibility. For example, engineers have certain responsibilities due to the nature of the tasks they perform. There are engineering codes of conduct, and these play a role in determining the degree to which various engineers may be responsible in the normative sense(s). Thus, the descriptive senses of responsibility fashion us with a toolkit that we can make use of in order to be more precise in our understanding of the normative senses of responsibility.

How does this relate to a discussion on responsibility-gaps driven by technology? Due to the concept's pluralism, there is no single responsibility-gap. Rather, there may be many gaps, each with their own unique conditions of emergence and amelioration. My focus will be on the normative senses of responsibility, and whether in those cases a responsibility-gap emerges. This is because these senses of responsibility ask of us what we *ought to do* in cases where advanced technology seems to encroach upon our moral concepts and practices. Descriptively such gaps do not arise, as a key feature of responsibility-gaps is that they leave us with an indeterminacy in what we ought to do when they occur. In this paper I will specifically concern myself with responsibility as answerability and responsibility as accountability.

I will assume that the most pressing kinds of normative responsibility gaps are those that arise in the case of answerability and accountability. This is because, in each case, we expect something of the agent who we think ought to be held responsible: either they must satisfy and answerability demand, or they must be held to account.

4 Responsibility as Answerability

As a specific manifestation of our responsibility-practices, answerability is about assessing the quality of an agent's judgement by asking them to provide *reasons* for their decisions. Importantly, these reasons should not just be explanations but should be made in terms of *justification* [9]. For example, if I were to punch you and you asked me why acted as I did, it would be absurd for me to reference various neural firings that led to muscular contractions in my arm and the clenching of my fist. While this is perhaps an explanation of the action, such descriptions do not track what we mean when we demand that an agent answer for their action. When we make such answerability demands, we want to know what considerations the agent thought lent credence to their eventual decision. In the example above, you would demand that I provide you with reasons *for* why I thought it was a good idea to punch you. The ability to provide the right kind of reasons, in the form of justification for your actions, is thus necessary for answerability [9].

A responsibility-gap would emerge in the answerability sense when we are faced with a complex network of actors and decisions, and where we might not be able to discern for which particular reason a decision was made. For example, consider a doctor who uses a medical AI system in order to diagnose her patients. Such systems are often based on deep-learning algorithms, trained on massive datasets. The doctor, however, would not be privy to this training, and might be blind to the ways in which knowledge is represented and uncovered by the algorithmic system [10]. The capacity of the end user (in this case, the doctor) is therefore compromised by the design of the system and the way that it generates its outputs. The "gap" in answerability is therefore said to emerge because AI systems "make individual persons less able to understand, explain, and reflect upon their own and other agents' behaviour" [10].

4.1 The Wrong Kind of Reason

We might think, intuitively, that AI-based systems can satisfy answerability demands if they are made "explainable". That is, if they are made in such a way that they can provide useful outputs to human users or developers, then this might be construed as a kind of "reason".

This causal accounting, however, does not provide us with any *justification*, and thus would not satisfy an answerability demand, as this explanation is made for the wrong *kind* of reason. Even though insights into the inner causal workings of the AI might be informative for a user (in that she could use this information to justify her decision), the causal accounting, by itself, is not sufficient. While there is a "reason" that apples fall from trees (gravity), this is not the same as the reason I have for wearing a mask during a pandemic. In the case of the former, the reason explains physical interactions, whereas in the latter the reason comes with an associated pro-attitude. That is, my reasons for wearing a mask during a pandemic implies a normative endorsement of the decision and some situational awareness.

As Shoemaker puts it.

"When I assess your responsibility by demanding to know why you had or acted on attitude X, I am requesting that you answer for your seeming judgment of the value attached to X(-ing), not that you explain how various psychological elements causally contributed to your being motivated to have or act on X" [11].

The "reasons" a machine may have for making a decision are of the descriptive, not the normative kind. These reasons are, strictly speaking, merely technical explanations ("causal contributions" in the quote above). More specifically though, it is *humans* who are tasked with *interpreting* these explanations, *and then* providing reasons. This is a hint to why there is no gap in answerability: machines can only provide us with *technical* explanations (or perhaps these are not explanations at all, but I bracket this question for now), which are not at all the domain of our answerability practices.

To make the point clearer, it might be useful to draw on the different senses of responsibility outlined earlier. Recall that there were normative and descriptive sense of responsibility, and that responsibility gaps are concerned with *normative* senses. So, while the AI in question might make a causal contribution by means of its algorithmic process (and thus be "responsible" in this sense), there is no normative dimension to this responsibility as AI systems cannot give the right kind of reasons. The fact that AI systems are causal systems in this sense, however, does not in principle exclude them from discovering or having reasons. The issue, especially in the case of machine learning systems, is that they have the wrong kind of *information* that might qualify as a reason. Additionally, because we are interested in *moral* responsibility, our answerability demands must have a moral dimension, which, once again, technical explanations seem to lack. However, considering that AI systems can provide us with interesting information in our attempts to "track and trace" the sources of potential harm, it is worthwhile to consider the *collaborative* nature of many AI systems [8].

4.2 Collaborative AI

Instead of thinking of these systems as fully "outside" of ourselves, we should instead see that our interaction with them creates a kind of "hybrid" system, where human interactants are tasked with understanding and explaining the system's behaviour in conjunction with the system itself. This echoes the thought of Sven Nyholm, who argues that instead of seeing the agency of certain machines as being of an independent type, we should instead conceive of their agency in terms of human-robot collaborations [9]. Nyholm claims that we need to ask specific *questions* about the agency of the machine in question, and that our answers to these questions will determine how we should go about allocating responsibility [12]. More specifically, he claims that the agency exhibited by currently existing AI systems is of a *collaborative* kind, where the other parties in this collaborative structure are humans in various, hierarchically organized roles. Essentially, as long as people remain in control (in some relevant sense) and capable of updating the machines or robots in question, there is no worry of responsibility-gaps emerging [9, 10].

This "hybrid" approach, however, is not without its own problems. Specifically, it seems to have difficulty dealing with the so-called "problem of many hands" [14]. For example, we can easily imagine scenarios where the simplistic model of human-robot collaborations comes under strain. Imagine

> "an automated car could be executing the human driver's particular travelling goals (e.g., going to the grocery store), while the car-company determines the means by which that end is achieved (e.g., determining the route). In this case two sets of human–robot collaborations are involved, rather than an obvious form of shared collaboration. The "driver-car" collaboration and the "programmer-car" collaboration have their own goals and are not quite on the same team or part of one line of command" [10].

Here we see that different collaborative systems may have conflicting best interests, even if they have shared goals. This complicates any determination of who is to be held responsible in the distributed human-robot collaborative network. The difference between collective and individual agency becomes paramount, as it can be difficult to parse how individual actors specifically contributed to the outcome under question.[1]

4.3 The "Problem" of Many Hands

This problem (of many hands) is said to be especially pernicious in the context of technology which is dependent on software. This is because software is not developed in isolation by individual programmers. Rather, computer systems are developed by large teams, that are typically part of massive corporations. Individual programmers, designers, managers, engineers, etc. may not be entirely sure of their exact causal contribution to the final product, and thus apportioning responsibility is complicated, and satisfying answerability demands may prove difficult [14]. But herein lies the solution: again, our

---

[1] This complicates our responsibility ascriptions, as it is often unclear what the relationship is between joint responsibility and moral responsibility [17, 18, 19].

ability to hold individuals responsible may be complicated, but it would not be inappropriate to proceed with doing so, once all the known facts are in. Moreover, this problem is not *unique* to AI.

For example, consider the case of Therac-25, a radiation treatment machine that overdosed patients in at least six known instances.

"In the two-year period from 1985 to 1987, overdoses administered by the Therac-25 caused severe radiation burns, which in turn, caused death in three cases and irreversible injuries (one minor, two very serious) in the other three. Built by Atomic Energy of Canada Limited (AECL), Therac-25 was the further development in a line of medical linear accelerators which destroy cancerous tumours by irradiating them with accelerated electrons and X-ray photons. Computer controls were far more prominent in the Therac-25 both because the machine had been designed from the ground up with computer controls in mind and also because the safety of the system as a whole was largely left to software. Whereas earlier models included hardware safety mechanisms and interlocks, designers of the Therac-25 did not duplicate software safety mechanisms with hardware equivalents. After many months of study and trial-and-error testing, the origin of the malfunction was traced not to a single source, but to numerous faults, which included at least two significant software coding errors ("bugs") and a faulty microswitch. The impact of these faults was exacerbated by the absence of hardware interlocks, obscure error messages, inadequate testing and quality assurance, exaggerated claims about the reliability of the system in AECL's safety analysis, and in at least two cases, negligence on the parts of the hospitals where treatment was administered" [11].

Here we are confronted with a situation in which identifying an *individual* whose fault caused the harmful outcomes seems impossible. However, while this might make it difficult to *blame* certain individuals, it does not get in the way of us demanding *answers* from them. We can ask individual engineers, designers, technicians, etc. *why* they made this or that decision, why they avoided certain warnings, why they exaggerated claims about the safety of the system without evidence, etc. This process might be long and complicated, and involves untangling a messy causal knot, but that does not make it impossible. By demanding answers we can better apportion responsibility and get around a potential responsibility-gap by determining who was at fault, for which reasons they were at fault, and to what degree they had control or knowledge of their fault. The *means* by which we do this is by demanding answers.

A further issue might occur when, after demanding answers from individual agents, we find that individuals themselves had no idea what exactly their contribution might have been. This might be because of poor management decisions (keeping various agents in the dark for security reasons) or due to engineers having very specific technical skills and being unable to anticipate how their small role might be implicated in a larger problem. Significantly, however, we can see how careful reflection on the nature of gaps in answerability have revealed that it is not really anything new: the problem of many hands was around before the emergence of sophisticated AI, and so there might not be anything special about AI-based systems (in the sense that they somehow make answerability demands impossible, resulting in a gap).

As mentioned in the introduction, we can now see more clearly how the main issue here is epistemic: we are presented with an array of agents who each make small causal contribution to the outcome, but it is unclear which contributions "matter" more than others. Moreover, it seems *unfair*, due to these small causal contributions, to individually blame any of those involved. However, here we seem to have drifted into the territory of *accountability*, as we are interested in how to go about apportioning *blame*. While the problem of many hands started as a matter of answerability, once such answerability demands are met, it seems that in some cases we would want to hold the relevant parties to account. It is responsibility as accountability that I turn to next.

5 Responsibility as Accountability

Accountability is one of the most discussed kind of responsibility-gap in the literature [12–14]. This is perhaps due to its important link to *blame* and *praise*, as it seems that when we hold an agent to account, we deem them *blameworthy* or *praiseworthy*, and thus *culpable* or *on the hook* for what they have done [10]. Significantly, such practices are also thought to justify *sanctioning* behaviour [12]. In the absence of such a justification, demanding reparations or punishing supposed wrong doers would be unfair. How might such a gap emerge? Imagine an autonomous vehicle that runs over an innocent pedestrian. "Drivers" (or perhaps more accurately, passengers) in these cars might not be thought of as blameworthy for this event. This is because they would not satisfy our ordinary criteria of control, knowledge, or intention which are usually thought to render such a judgement appropriate. Similarly, it might seem unfair to attribute responsibility to the creators, designers, and manufacturers of the car. This could be because "agents operating in such a socio-technical system (designers, programmers, drivers, regulators, bystanders etc.) may more easily find themselves acting wrongly, for instance, causing an avoidable road crash, while at the same time having a legitimate excuse" [10]. We may find ourselves in a situation where we have *nobody* to legitimately blame, even though it would seem the situation demands the attribution of responsibility.[2] But what exactly are we doing when we *blame* an agent? What is the purpose of this blame? On one theory of punishment, we might say we are looking for *retribution*. Therefore, before discussing the features of AI that might generate a gap, I will first consider what features of our psychology (and their interaction with AI or robotic systems) that might create a gap in accountability. This has been termed the "retribution gap", but on my account would just be a specific manifestation of the more general accountability gap [18].

5.1 Retribution Gaps

What is retribution? We can take a hint from the word itself. The prefix 're', in this context means 'back' and 'tribuere' means 'to assign' or 'allot'. Thus, Retribution literally means 'payback', in the sense of dishing something out (an action or a thing) in response to, usually, a past harm. What is this 'payback'? In common usage, the payback is usually in the form of *personal revenge* taken by victims against those who they perceive to have

---

[2] Here I assume that the algorithm the car is fitted with is itself not problematic.

harmed them. When we are harmed or wronged our first response is often to seek out someone who is culpable, someone that we can *blame* and therefore *punish*. Increasing levels of automation in society, however, mean that AI and robotic systems may become responsible (in the causal sense, at least) for moral harms. These systems, at present, do not meet the conditions for retributive blame, and thus we see how a retribution gap may be opened [18].

Take a simple example, outside the domain of AI, which might also be able to illustrate this problem. Imagine that the driver of a vehicle for a delivery company hits a young girl, causing her severe physical injury. Further, we can anticipate that the delivery company would have certain rules in place for such an event: they might try to settle out of court, and thus offer up a significant amount of money to the aggrieved parents. It would be natural to think that the parents might not be satisfied with this outcome. Why? Because resolving the question of who will *pay* for the harm is different to determining who *deserves to be blamed* [10]. Should the situation warrant it, the parents might want the *individual* driver to be blamed for the incident (and to perhaps suffer a form of punishment that is not merely financial). More specifically, we might find that the parents have a natural disposition to seek *retribution* as a form of punishment. Deterrence or sanction might not be enough to satisfy their intuitions. This kind of issue is brought into even sharper relief when we consider it in light of AI systems.

For example, in 2018 Elaine Herzberg became the first pedestrian to be involved in a fatal accident with an Uber self-driving car in 2018. In the aftermath of the incident, Uber suspended all testing of its self-driving vehicles. The fallout from the accident, however, is quite instructive. After Hertzberg's death, there were numerous attacks on self-driving cars. In Chandler, Phoenix, for example, Waymo vans (a driverless car company owned by Google) had rocks thrown at them, their tires slashed, and, in one case, had a man wave a firearm at the backup driver [15]. In the latter case, the man specifically referred to Herzberg's death as one of the reasons he "despises" self-driving cars [15]. While it was clear that the autonomous car was *causally* responsible for the harm, the car itself cannot be held responsible in the full sense (due to it lacking certain capacities). However, there is still a worry that such an apportioning of responsibility would not *satisfy* many peoples' *retributive* intuitions. The question then becomes how common these intuitions are, and how they ought to matter in our moral theorizing.

According to John Danaher, retribution gaps emerge when the all-too-human desire to blame a guilty party cannot find an appropriate target. He claims that the emergence of such gaps is a threat to the rule of law and can lead to moral scapegoating. If the rule of law fails to accommodate common intuitions about fair and proportionate punishment, there is a worry that people might lose trust in the legal system [18]. Additionally, if appropriate targets of blame cannot be found, people might end up apportioning their retributive desires in improper ways. As seen in the Hertzberg case above, there is evidence that cases of "robotic" harm may lead people to *blame the vehicles* because it seems that blaming the manufacturers or programmers is inappropriate. Blaming the vehicles, however, is incoherent, and so such actions would in all likelihood not satisfy retributivist intuitions. The worry, therefore, is that this might lead to the unfair blaming of manufacturers or designers, who perhaps really were out of the loop and had done nothing wrong.

Danaher, however, seems to shift from the descriptive to the normative without sufficient justification. When he claims that retributivism is the thesis "that agents should be punished, in proportion to their level of wrongdoing, because they *deserve* to be punished", he is clearly locating the retribution-gap in the field of normative ethics [18]. However, the evidence he cites, from the human propensity to attribute agency, our proclivity to punish, and that when people punish, they justify said punishment in retributive terms, are all descriptive [18]. While such evidence is indeed convincing as an account of *why* we might have retributive tendencies, it does not explain what we *ought to do* with these intuitions. In other words, if there is to be a real retribution-gap (in the normative sense) then we require a normative argument for why agents *deserve* to be punished, which is exactly what Danaher *assumes* but does argue for convincingly. In order for there to be no *appropriate* targets for our innate retributivism, a normative case needs to be made for what we should do in light of this. The "gap" therefore,

> "arises not between retributive intuitions and unsuccessful attempts to find appropriate targets for moral blame (as in Danaher's account), but rather between retributive intuitions and how we act on them" [20].

Drawing on research done on implicit biases and their relation to moral character, Kraaijeveld argues that we can *take* responsibility for our retributive intuitions [20]. What he means here is that we can exercise a kind of *ecological control* over our unintentional and automatic responses (such as our retributive intuitions) to certain stimuli. It seems plausible that individuals would be capable of revising their intuitions should evidence to the contrary become available. Should a case arise where a robot has committed some moral harm, and our retributive intuitions cannot find an appropriate target for blame, "then one ought not to let them guide one's behaviour" [20]. In this way, we can see that the retribution-gap can be overcome, if not in the descriptive sense, but in the normative sense, where it matters most.

6 Conclusion

Although the multi-faceted nature of AI development complicates our ability to assign blame and make answerability demands, the important question is not just whether it is more *difficult* to apportion responsibility, but whether it is *philosophically problematic*. Mere difficulty in assigning responsibility does not yet amount to a responsibility-gap. For example, in complex legal proceedings it is often difficult to ascertain who the guilty party is, and some trials take longer and are more complicated than others. However, this is not a good motivation (on its own) to call for a revision of the way trails are conducted. Nor is it a good enough reason to claim that trials, because they are sometimes difficult, are an *inappropriate* mechanism with which to determine guilt or innocence. That is to say, while it might be difficult or take us some time to discern who or what was morally responsible for some outcome, this is only an *epistemic* problem, and one which we encounter all the time in our normal responsibility ascriptions. It can take time for all the relevant facts to become clear, and for an ascription of responsibility to be properly justified.

Of course, one could also object that this analysis has an important blind spot: the two senses of responsibility that I have focused on are *backwards looking* in the sense that they are concerned about what has already happened. What about forward looking responsibility? While I do not have the space in this paper to deal with this problem, I have addressed it in another paper (see [16]), where I argue that there might not (necessarily) be a gap in forward-looking responsibility, either.

References

1. Gebru, T.: Race and gender. In: Dubber, M., Pasquale, F., Das, S. (eds.) Oxford Handbook of the Ethics of AI, pp. 253–270. Oxford University Press, New York (2020)
2. Buolamwini, J., Gebru, T.: Gender shades: intersectional accuracy disparities in commercial gender classification. Proc. Mach. Learn. Res. **81**, 1–15 (2018). https://doi.org/10.2147/OTT. S126905
3. Nyholm, S., Gordan, J.-S.: Ethics of artificial intelligence. In: Fieser, J., Dowden, B. (eds.) Internet Encyclopedia of Philosophy (2021)
4. Sparrow, R.: Killer robots. J. Appl. Philos. **24**(1), 62–78 (2007). https://doi.org/10.1111/j. 1468-5930.2007.00346.x
5. Vincent, N.A.: A structured taxonomy of responsibility concepts. In: Vincent, N.A., van de Poel, I., van den Hoven, J. (eds.) Moral Responsibility, pp. 15–35. Springer, Dordrecht (2011). https://doi.org/10.1007/978-94-007-1878-4_2
6. Shoemaker, D.: Responsibility from the Margin. Oxford University Press, Oxford (2015)
7. van de Poel, I.: The relation between forward-looking and backward-looking responsibility. In: Vincent, N., van de Poel, I., van den Hoven, J. (eds.) Moral Responsibility. LOET, vol. 27, pp. 37–52. Springer, London (2011). https://doi.org/10.1007/978-94-007-1878-4_3
8. Santoni de Sio, F., Mecacci, G.: Four responsibility gaps with artificial intelligence: why they matter and how to address them. Philos. Technol. **34**, 1057–1084 (2021). https://doi.org/10. 1007/s13347-021-00450-x
9. Nyholm, S.: Attributing agency to automated systems: reflections on human–robot collaborations and responsibility-loci. Sci. Eng. Ethics **24**, 1–19 (2017) https://doi.org/10.1007/s11 948-017-9943-x
10. de Jong, R.: The retribution-gap and responsibility-loci related to robots and automated technologies: a reply to Nyholm. Sci. Eng. Ethics **26**(2), 727–735 (2020). https://doi.org/10.1007/ s11948-019-00120-4
11. Nissenbaum, H.: Accountability in a computerized society. Sci. Eng. Ethics **2**, 25–42 (1996)
12. Köhler, S., Roughley, N., Hanno, S.: Technologically blurred accountability? In: Ulbert, C., Finkenbusch, P., Sondermann, E., Tobias, D. (eds.) Moral Agency and the Politics of Responsibility, pp. 1–19. Routledge, London (2017)
13. Tigard, D.W.: There is no techno-responsibility gap. Philos. Technol. **34**, 589–607 (2020). https://doi.org/10.1007/s13347-020-00414-7
14. Matthias, A.: The responsibility gap: ascribing responsibility for the actions of learning automata. Ethics Inf. Technol. **6**(3), 175–183 (2004). https://doi.org/10.1007/s10676-004-3422-1
15. Romero, S.: Wielding Rocks and Knives, Arizonans Attack Self-Driving Cars. The New York Times, Chandler, Arizona, 31 December 2018 (2018)
16. Tollon, F.: Is AI a problem for forward looking moral responsibility? The problem followed by a solution. In: Jembere, E., Gerber, A.J., Viriri, S., Pillay, A. (eds.) SACAIR 2021. CCIS, vol. 1551, pp. 307–318. Springer, Cham (2022). https://doi.org/10.1007/978-3-030-95070-5_20

17. Pettit, P.: Responsibility incorporated. Ethics **117**(2), 171–201 (2007). https://doi.org/10.1086/510695

18. List, C., Pettit, P.: Group Agency: The Possibility, Design, and Status of Corporate Agents. Oxford University Press, New York (2011)

19. van de Poel, I., Fahlquist, J.N., Doorn, N., Zwart, S., Royakkers, L.: The problem of many hands: climate change as an example. Sci. Eng. Ethics **18**(1), 49–67 (2012). https://doi.org/10.1007/s11948-011-9276-0

20. Danaher, J.: Robots, law and the retribution gap. Ethics Inf. Technol. **18**(4), 299–309 (2016). https://doi.org/10.1007/s10676-016-9403-3

Does Counterfactual Reasoning Hold the Key to Artificial General Intelligence?

Ethan Vorster[1,2]([✉]) [ID]

[1] Department of Philosophy, University of Johannesburg, Johannesburg, South Africa
evorster333@gmail.com
[2] Centre for Artificial Intelligence Research (CAIR), Pretoria, South Africa

Abstract. In this paper, I argue that counterfactual reasoning is a necessary condition for the realization of artificial general intelligence (AGI). This position is similarly held by computer scientist Judea Pearl. However, Pearl's notions are often vague, misleading, and result in conflations. Thus, this paper serves two purposes. One, as a critique of Pearl's position. Two, to introduce a novel argument, namely, the Counterfactual Room Argument, which aims to present a clearer and more rigorous interpretation of the role of counterfactual reasoning in AGI.

Keywords: Artificial general intelligence · Counterfactuals · Judea Pearl · Causation · Understanding

1 Introduction

We appear to be in an artificial intelligence (AI) spring, an era of renewed interest, funding and research which is fuelling the push towards creating artificial general intelligence (AGI)[1]. Since the 1950s AI research has gone through cycles of overconfident predictions and promises of breakthroughs (springs) and disappointments and ceilings which are fundamentally insurmountable given the limitations of our hardware and software (winters) [13]. This time around though there appears to be new methods such as CovNets (convoluted neural networks); Generative Adversarial Networks; Deep Neural Nets which hint at finally achieving the goal of AGI. AI researchers have drawn from numerous fields such as developmental psychology, neuroscience, philosophy of mind etc. to try and construct synthetic intelligence. However, the nuance and intricacy of general intelligence is proving to be as elusive as ever. We hear predictions that AGI is around the corner whilst hearing sceptics claim that next to no progress has been made in the direction of genuine AGI. However close AI researchers appear to get to taking the next step towards building AGI a new hurdle reveals itself which hinders the development of reverse engineering intelligence.

---

[1] I define AGI as artificial intelligence that is generally intelligent across multiple domains. This means that one AI system can behave intelligently in a novel domain space without prior information/data about such a domain and learn or infer how best to approach tasks/challenges/problems set out in this novel domain space.

A. Pillay et al. (Eds.): SACAIR 2022, CCIS 1734, pp. 384–399, 2022.
https://doi.org/10.1007/978-3-031-22321-1_26

A central goal of AI research, whether implicit or explicit, has been to put a 'mirror' up to ourselves as human beings and reverse engineer the profound experience which we call consciousness. Our awareness, our qualia, our mind, has puzzled us for millennia, we now finally have a chance to gain some deeper understanding of it. This can happen through AI. If we can build (reengineer) consciousness artificially, then we might be able to answer questions which have plagued us for so long about the nature of our own state of being. This view is not often expressed in the field of AI. All too often it seems that we are plunging full steam ahead towards creating AGI without sufficiently fleshing out the reasons why this is a valuable and worthwhile idea. One possible reason for this trend is our desire to understand our state of consciousness. Building consciousness from the ground up will provide invaluable lessons and insights into the mechanics of our intelligence. From this perspective it seems inevitable that we as a species would head down this rabbit hole. The topic of this paper is in line with this tradition of uncovering fundamental tenets of AGI vis-à-vis our own consciousness.

This paper investigates the role that counterfactual reasoning can play in the development of AGI[2]. The central question which I seek to answer is: can research into the nature of counterfactual reasoning be a key steppingstone towards the development of AGI? And relatedly: is the ability to reason counterfactually one of the missing ingredients in the development of true AGI? I argue that it is. For reasons that will become clear in Sect. 3, I call my argument demonstrating this the Counterfactual Room Argument (CFRA), and it takes the following form:

- **Premise 1)** Consciousness is a necessary condition for imagination.
- **Premise 2)** Imagination is a necessary condition for the ability to reason counterfactually.
- **Conclusion 1)** Therefore, consciousness is a necessary condition for the ability to reason counterfactually.
- **Premise 3)** The ability to reason counterfactually is a necessary condition or an agent to demonstrate understanding.
- **Premise 4)** Understanding is a necessary condition for an agent to pass a requisite version of the Turing Test [7] for AGI.
- **Final conclusion)** Therefore, consciousness is a necessary condition for an agent to pass a requisite version of the Turing Test for AGI.

There is obviously a lot to unpack in this argument in terms of the content of the premises. This is the aim of this paper. Prima facie, counterfactual reasoning appears to be the ingredient we have been looking for in our search to create AGI. In their 2018 book *The Book of Why: The New Science of Cause and Effect* Judea Pearl and co-author Dana Mackenzie make a similar argument to that above. They claim that counterfactual reasoning is a crucial aspect of human intelligence because counterfactual reasoning demonstrates an understanding of something even more fundamental: causation. I agree with their line of reasoning. However, as we shall see, they formulate a rather vague and loose argument around the role of counterfactuals in AGI as well as human intelligence.

---

[2] AGI and strong AI refer to the same concept. I use them interchangeably throughout the paper. I will provide a definition of AGI and strong AI in the paper.

In response to Pearl and Mackenzie, the first section of this paper is dedicated to a critique of their argument in the hopes of gleaning any useful insights that they put forth. The second section of the paper is dedicated to developing the CFRA set out above. In order to do so I introduce two famous thought experiments from AI research, namely, the Turing Test (TT) and The Chinese Room [3]. I use the Chinese Room (CR), formulated by John Searle, to explore notions of understanding. Searle doesn't speak explicitly of counterfactual reasoning in his thought experiment. However, I use the mechanics of the thought experiment to differentiate between merely providing the correct answer to a question which requires counterfactual reasoning and actual understanding resulting from counterfactual reasoning. This distinction is ripe for application in the TT as counterfactual questions are a common form of questioning designed to separate man from machine. This is because, as I will elaborate on in the paper, counterfactuals are a surprisingly dense concept. To answer a counterfactual question convincingly requires an understanding of causation as well as imagination. The latter being a fundamentally conscious experience, as I will demonstrate later. Hence, any agent/being that can pass the TT must be conscious. However, I will tweak the conditions of the original TT in order to make it a more rigorous test of intelligence. The original conditions, although still presently difficult to overcome by AI researchers, can be surmounted by a *chance* correct answer to a counterfactually loaded question. We thus need to reframe the TT so that it does not rely on merely convincing a panel of human judges but rather requires a more thorough demonstration of intelligence.

2   A Critique of Judea Pearl and Dana Mackenzie

Judea Pearl claims to be a key figure in the "Causal Revolution" [14]. Along with several scientists, philosophers, engineers, historians [6] and economists Pearl has begun a trend of taking the notion of cause and effect seriously in their analyses. Causal analysis was supposedly looked down upon by any academic who took their work seriously [14]. However, in recent decades cause and effect has become a robust lens through which to approach numerous studies, ranging from economics, agriculture, climate science and recently, and most importantly for the purposes of this paper, AI and machine learning.

Before delving into a critique of Pearl and Mackenzie I want to pose a question. It is in fact the same question posed by Alan Turing, the famous computer scientist, mathematician, and code-cracker who is hailed as one of, if not the, godfather of AI. What can a causal reasoner do that a non-causal reasoner cannot? It appears that this question has the same allure for Pearl and Mackenzie, who through their revised TT, intend to answer it. Their revised TT is dubbed "the mini-Turing test" [14]. The original TT, or as Turing himself called it, "The Imitation Game" [18], is a test devised to determine whether a robot can convincingly display human-like conversational characteristics. In the third section of this paper, we will investigate the TT in more detail. For now, what is important is that the robot essentially has to carry out natural human language which has proven to be a significant hurdle for computer scientists working in AI. One of the many reasons why natural language has been so tricky to program is causation.

2.1 Pearl and Mackenzie's Mini-Turing Test

Let us consider Pearl and Mackenzie's mini-Turing test. They were inspired by Turning's insight into a possible path to human-level intelligence in AI. Turing proposed that instead of trying to impart the intelligence of an adult human, why not start by trying to develop child-like intelligence and then teach the robot from there so that it can grow, cognitively, towards adult human intelligence [15]. Both Turing and Pearl and Mackenzie make some bold claims about the nature of a child's brain make up and the means by which they acquire knowledge. Turing assumes that a child's brain is *tabula rasa* and can be moulded quite easily whereas Pearl and Mackenzie claim that their brains are already imparted with rich "mechanisms and prestored templets" [14]. Delving into the psychology of a child's brain could perhaps be a fruitful exercise for the purposes of this topic but doing so rigorously is beyond the scope of this paper. Nonetheless, the main point that Pearl and Mackenzie are trying to make is that an AI needs to master causation step-by-step in order to build a solid understanding of the phenomenon. Their main aim in their revised TT is "slightly less ambitious" [14] than the original. They formulate it as follows:

> How can machines (and people) represent causal knowledge in a way that would enable them to access the necessary information swiftly, answer questions correctly, and do it with ease, as a three-year-old child can? [...] The idea is to take a simple story, encode it on a machine in some way, and then test to see if the machine can correctly answer causal questions that a human can answer. It is "mini" for two reasons. First, it is confined to causal reasoning, excluding other aspects of human intelligence such as vision and natural language. Second, we allow the contestant to encode the story in any convenient representation, unburdening the machine of the task of acquiring the story from its own personal experience. [...] Obviously, as we prepare to take the mini-Turing test, the question of representation needs to precede the question of acquisition. Without a representation, we wouldn't know how to store information for future use. Even if we could let our robot manipulate its environment at will, whatever information we learned this way would be forgotten, unless our robot were endowed with a template to encode the results of those manipulations. One major contribution of AI to the study of cognition has been the paradigm "Representation first, acquisition second." Often the quest for a good representation has led to insights into how the knowledge ought to be acquired, be it from data or a programmer [14].

They are probably right to try and achieve lower-order cognition before moving on to full adult-level intelligence. I agree that such a method would be part of the developmental trajectory of AGI. Pearl and Mackenzie claim that a test of this capacity would demonstrate genuine understanding on the part of the AI. This is because the AI would not be able to come to the correct answer to a question about the causal nature of the story by keeping prestored answers. Doing so would amount to cheating. The authors claim that "cheating is not easy; in fact, it is impossible" [14]. Keeping prestored answers to questions would require the AI to store a list with "more entries that the number of atoms in the universe" [14]. The AI would need an "efficient representation and answer-extraction algorithm" [14] if it is to satisfactorily pass the mini-Turing test. They claim that such a representation exists, and it takes the simple form of a dot and arrow diagram. We will return to this topic of 'cheating' further on when discussing Searle's CR.

2.2 Pearl and Mackenzie's Conflations

The first conflation I wish to address deals with the following quote from the authors: 'representation first, acquisition second'. It is unclear if Pearl and Mackenzie are saying that for an AI to understand causation it needs to have a sense of conscious representation or just be able to pull up a causal diagram with dots, arrows and labels. The former is very different from the latter. The authors use a series of dot and arrow diagrams to explain a scenario wherein a court order dictates whether or not a pair of riflemen shoot a person.

What we need to think about in this case is the hard problem of representation. How do we as humans come to a representational understanding of causation? Pearl and Mackenzie state that we as humans use similar diagrams to understand causal puzzles [14]. I am somewhat doubtful of that. Perhaps they do not mean that we have diagrams appearing in our minds eye, or perhaps they do. However, to assume that we utilize simple dot and arrow diagrams to understand causation is an assumption on their part which is both unsubstantiated and has important implications for their theory of consciousness and intelligence. Again, it may be trivial to debate whether or not Pearl and Mackenzie believe we see causal diagrams, but it is consequential when analysing their theory of representation. One cannot conflate two very different conceptions of 'representation'. To consciously represent something in your mind's eye is a deep and profoundly conscious experience. It cannot be conflated with digitally pulling up a simple diagram and reading off the results of the dots and arrows. Where the line may blur is how the diagram is formulated to begin with. If the AI formulates the diagram on its own, it is very different to having diagrams presented and manually altered by a human-in-the-loop system. On the one hand, if the diagram is formulated on its own then there is a much stronger leaning towards human-level consciousness or strong AI. On the other hand, if the diagram is manually altered through constant human supervision, then the system is very far from anything like a conscious being who is able to understand causation on its own. The authors go on to say:

> If we want our computer to understand causation, we have to teach it how to break the rules. We have to teach it the difference between merely observing an event and making it happen. "Whenever you make an event happen," we tell the computer, "remove all arrows that point to that event and continue the analysis by ordinary logic, as if the arrows had never been there." Thus, we erase all the arrows leading into the intervened variable (A). We also set that variable manually to its prescribed value (true). The rationale for this peculiar "surgery" is simple: making an event happen means that you emancipate it from all other influences and subject it to one and only one influence—that which enforces its happening [14].

Further on they also state:

> From the computing perspective, our scheme for passing the mini-Turing test is also remarkable in that we used the same routine in all three examples: translate

the story into a diagram, listen to the query, perform a surgery that corresponds to the given query (interventional or counterfactual; if the query is associational then no surgery is needed), and use the modified causal model to compute the answer. We did not have to train the machine in a multitude of new queries each time we changed the story. The approach is flexible enough to work whenever we can draw a causal diagram, whether it has to do with mammoths, firing squads, or vaccinations. This is exactly what we want for a causal inference engine: it is the kind of flexibility we enjoy as humans [14].

In these excerpts Pearl and Mackenzie are clearly talking about manual intervention. They call them 'surgeries' which alter the details of the causal diagram so that the AI can adapt to the situation, environment, puzzle or question. This method may be applicable for training a budding AI, but it most certainly is not evidence of strong AI. Granted, this is their mini-Turing test but I'm not sure what it is accomplishing. If you are manually inputting a causal diagram and manually altering the details of the diagram then what are you accomplishing that cannot already be accomplished? They are also unclear about how this test functions as a possible building block towards more rigorous tests which start to then formulate strong AI. What would this process look like? Pearl and Mackenzie are unspoken on this issue.

Another point of critique which needs to be made against Pearl and Mackenzie is their use of counterfactuals. This is the main point of critique which I want to hedge against them. Counterfactuals are a key aspect of their theory of AGI development. They take the top spot of their Ladder of Causation as the highest form of intelligent reasoning that an agent can possess. Pearl and Mackenzie devote an entire chapter of their 2018 *Book of Why* to the topic of counterfactuals. In an attempt to elaborate on counterfactuals Pearl and Mackenzie cite David Lewis and delve into his possible-worlds theory but fail to do it justice. The authors oversimplify Lewis's theory by summing it up in a quote: "why not take counterfactuals at face value: as statements about possible alternatives to the actual situation?" [8]. Now, this line from Lewis is a cornerstone upon which a lot of his counterfactual theory unfolds but to distil Lewis down to this simple quote is to miss the nuance which Lewis brings to the table. Pearl and Mackenzie write:

> [...] but the upshot of the long story is that counterfactuals have ceased to be mystical. We understand how humans manage them, and we are ready to equip robots with similar capabilities to the ones our ancestors acquired 40,000 years ago [14].

If the authors had made this claim after providing a robust and watertight analysis of counterfactuals, then it would hold. However, such an analysis was not accomplished. Pearl and Mackenzie try to use Lewis to demonstrate that counterfactuals are straightforward instead of using Lewis to demonstrate their complexity. They neglect the wisdom that can be garnered from Lewis. This can be seen as an innocent oversight or a gross misrepresentation. Another point of contention which is notable from the above quote is that Pearl and Mackenzie claim that we now understand counterfactuals and that we understand how humans compute them. However, the pages leading up to this statement do not support such a bold claim. In fact, after reading their notion of counterfactuals, a critical reader might be left more confused than before. They weave together theories from Hume and Lewis to explain counterfactuals but both writers are in the end mis-represented and oversimplified [14]. It's not necessarily that what Pearl and Mackenzie write or think about counterfactuals is spurious per se but rather that they do not explain them sufficiently to justify making the claim that is quoted above. I go into more detail on counterfactuals and Lewis's theory in the following section of the paper. There we will get a taste of the detail that Lewis provides in his counterfactual theory which Pearl and Mackenzie obfuscate.

Pearl and Mackenzie state numerous times throughout their book that counterfactuals are imaginative phenomena. They write "counterfactuals can only be imagined" [14]. They are deeply conscious experiences, which I agree with. They are constituted by conscious representations in one's mind's eye. However, the authors are vague in their analysis of counterfactuals. Besides claiming that we know what counterfactuals are [14], which we have yet to fully figure out, they then use another form of counterfactual analysis and then conflate the two without pointing out the difference. Let me explain. The other form of counterfactual that they use is a form of problem-solving and not imagining. Pearl and Mackenzie use counterfactuals in a statistical way to work out certain problems. They run through an example of figuring out someone's would-be salary had their past been different. Without getting into too much detail they write:

> Now let's demonstrate how to derive counterfactuals from a structural model. To estimate Alice's salary if she had a college education, we perform three steps:
>
> 1. (Abduction) Use the data about Alice and about the other employees to estimate Alice's idiosyncratic factors, U_S(Alice) and U_{EX}(Alice).
> 2. (Action) Use the *do*-operator to change the model to reflect the counterfactual assumption being made, in this case that she has a college degree: ED(Alice) = 1.
> 3. (Prediction) Calculate Alice's new salary using the modified model and the updated information about the exogenous variables U_S(Alice), U_{EX}(Alice), and ED(Alice). This newly calculated salary is equal to $S_{ED\,=\,1}$(Alice) [14].

This three-step process may indeed be a clever way of solving a problem, but it does not involve imagination. It is problem solving with a few variable tweaks here and there. This process could not be further from having the profound conscious experience of placing oneself in an altered temporal state and imagining a new causal environment with potentially bizarre rules and characteristics. Now, it is not necessarily a bad thing that Pearl and Mackenzie have raised these two forms of counterfactuals. However, in claiming that we now know what counterfactuals are and that their mystical edge has dissolved, it cannot be overlooked that they then conflate the two conceptions of counter-factuals and pass them off as the same thing. There is a fundamental difference between counterfactual *reasoning* and using 'counterfactuals' to solve a problem. Counterfactual *reasoning* demonstrates a mental process wherein a conscious agent is imaginatively placing themselves in a temporally or spatially altered context, while reapplying causal understanding to the same non-existent context. Using 'counterfactuals' to solve a prob-lem, like I mentioned above, amounts to tweaking input variables in an algorithm for example and, as a process, does not even fall on the same cognitive spectrum as counterfactual *reasoning*. If Pearl and Mackenzie had stated at some point that these two conceptions are distinct and that we need to find a way to reconcile them then their claim might appear more insightful. However, they conflate these two conceptions of counterfactual *reasoning* and using 'counterfactuals' as problem solving tools and seems to claim that if we have one then we have the other. Essentially, they have started off by making, what appears to be, quite an interesting claim, that upon closer inspection starts to fall apart. If they are claiming that imagination is necessary for counterfactual reasoning and that counterfactual reasoning is necessary for human-level AI then this is quite profound. However, if they are claiming that counterfactuals are tools for a special type of problem-solving then it is not that interesting in the context of AI or strong AI.

3   The Counterfactual Room Argument

In this section I will expand on the premises of the CFRA (see Sect. 1 Introduction). I have chosen to call this argument the CFRA in the spirit of the CR proposed by John Searle. The difference between the two being the content of the enclosed hypothetical 'room'. In Searle's case he uses Chinese letters to make his argument, in my version I am using counterfactual reasoning. Let us begin by looking at premise one.

3.1   Premise One: Consciousness is a Necessary Condition for Imagination

This premise is fundamentally couched in an assumption. There is no way, within the scope of this paper that I can *prove* that one needs to be conscious[3] in order to imagine[4]. I am taking it to be an intuitive assumption. This could perhaps lead to the unravelling of the argument, but I am sure most people would agree that one needs to be conscious in order to imagine. I am not claiming that consciousness is sufficient for imagination merely that it is a necessary prerequisite. I am also aware of conditions such as aphantasia which prohibit the formation of mental images [4]. This is a tricky topic and deserves its own dedicated discussion as it deals intimately with conceptions within psychology and philosophy of mind. For now, let us agree that the ability to produce mental images, aka, to have a 'mind's eye' is not sufficient for imagination but merely one aspect of it. I raise this topic in order to foreclose any critiques which state that I am claiming that people with conditions such as or similar to aphantasia are not conscious or do not have an imagination. That is most certainly not the case. People with aphantasia can still reason causally and counterfactually without necessarily having psychovisual images of a given scenario.

---

[3] Consciousness has been a difficult concept to define for centuries and remains a contentious topic for philosophers, however, I define consciousness as: a state of sentient self-awareness which allows one to experience phenomena such as qualia, thoughts, emotions, introspections etc. As important as the concept of consciousness is for this paper, I unfortunately cannot expand on its definition further due to the scope of this paper.

[4] I define 'imagination' as a cognitive faculty which enables the formation of images, concepts and ideas which are not present to an agent's sensory environment.

3.2 Premise 2: Imagination is a Necessary Condition for the Ability to Reason Counterfactually

This premise is in line with Pearl and Mackenzie's claim that "counterfactuals can only be imagined" [14]. I agree with them on this point. However, it needs some clarification. A counterfactual alone does not necessarily need imagination as a prerequisite. Take for instance the latest text to image transformer models such as DALL-E 2 and Stable Diffusion. These models are able to generate entirely novel images from text prompts with sunning accuracy. Whether or not these models are engaging in 'creativity' remains a question, but they are, in a sense, formulating counterfactual images. Prompts such as "Stary night but drawn in the style of Picasso" or "Humans 200 years from now" are scenarios that don't exist and yet these models draw from billions of data points to create these images[5].

However, counterfactual *reasoning* does require a sense of imagination. I mentioned this distinction between using counterfactuals as a problem-solving tool vs reasoning counterfactually above. The former being a mere method of problem solving by tweaking input variables, the latter being a deeply conscious process. I wish to steer clear of the conflation that Pearl and Mackenzie make. I claim that counterfactual *reasoning* can only be seen as a deeply conscious phenomenon. Here again we can make reference to Lewis's possible world semantics in order to demonstrate this point. Lewis defines a counterfactual as follows:

> A counterfactual "If it were that A, then it would be that C" is (non-vacuously) true if and only if some (accessible) world where both A and C are true is more similar to our actual world, overall, than is any world where A is true but C is false [9].

Lewis himself does not detail the metaphysical or ontological nature of these 'worlds'. He simply assumes that they exist. He received some criticism for this point but that is not important for the purposes of this paper. I ground my theory of counterfactuals in Lewis's conception of other worlds. I utilise possible world semantics as a useful framework for the understanding of counterfactuals, but I will not necessarily commit to their reality as Lewis does with his Modal Realism [10]. It is through these other worlds which we can come to reason counterfactually. Other worlds, whether imaginative or real[6], range from very similar to our actual world, out to an infinite number of more and more dissimilar worlds. For example, we could ask a counterfactual query which functions through a simple yes or no disjunction. In this 'world' things would likely be not all that different from our current world. However, we could ask a counterfactual

---

[5] One might point out, and rightly so, that these are not necessarily counterfactual images but rather non-factual images. However, my aim in bringing up these transformer models was to expand on the state-of-the-art ML and the ways in which it is moving towards human-like processes such as creativity and imagination.

[6] The existence of these other worlds might have some actual validity within quantum physics. One popular theory within quantum physics is known at the 'many worlds' theory of quantum mechanics which posits branched realities wherein events unfold differently from one 'main reality' [2].

query in which we would have to radically reposition ourselves, mentally, in order to compute it. For example, we could ask, what if Hitler had never been born, or what if I had been born in North Korea? These questions require us to imagine a series of possible worlds, ranging from the most to the least similar, wherein countless implications play out. Importantly, these implications play out in our mind in a causal manner. In other words, we temporally and/or spatially place ourselves in these different worlds and apply our understanding of causation to these scenarios. Such a feat is a deeply conscious experience and, I agree with Pearl and Mackenzie's third rung here, requires imagination. As humans we seem to do this day in and day out with ease, but how so? We have yet to figure this out. Perhaps Lewis's worlds really do exist, and we enter into them with ease; in which case counterfactuals would be anything but demystified (refer to footnote 6).

I claim that this mental process is a form of *reasoning* because we can come to answer questions about such situations which we have limited information about. We can only assume how these different worlds play out, but our common sense allows us to make inferences about how events might unfold. Not only can we imaginatively place ourselves in these mental situations, but we can imagine what might happen next or what we might think if *x* were to happen. One way in which these scenarios play out in our minds is through the theory of understanding as simulation which states that we simulate mental models to help us understand counterfactual possibilities. Or as Mitchell states:

> I can easily imagine reading a story about, say, a car accident involving a woman crossing a street while talking on her phone, and understanding the story via my mental simulation of the situation. I might put myself in the woman's role and imagine (via simulation of my mental models) what it feels like to hold a phone, to push a stroller, to hold a dog's lead, to cross a street, to be distracted, and so forth [12].

3.3 Premise 3: The Ability to Reason Counterfactually is a Necessary Condition for an Agent to Demonstrate Understanding

As mentioned above, reasoning counterfactually allows us to place ourselves in different mental environments and perform a range of causal inferences. Such a task is crucial when we are talking about understanding. In the context of AI this is of particular relevance. We can think of understanding as "Representation Manipulability" [5]. Firt explains this idea of understanding as follows:

> [...] one understands when one possesses a representation of that which is understood that is sufficiently robust to be manipulable for inferential and practical purposes. In other words, understanding occurs when we have a robust mental representation of the thing to be understood. This robustness is expressed by the ability of the understander to manipulate and tweak this representation to examine inferences and take actions [...] understanding the relationships between relevant parts of a subject matter amounts to manipulating the system by changing parts of it and observing the impact on the overall system. [...] It allows the agent to make correct inferences about a world in which the relevant differences obtain [5].

The notion of "Representation Manipulability" [5] falls squarely in line with the idea of imagination (representation) being a necessary condition for reasoning counterfactually. Firt is clearly talking about counterfactual reasoning in this case. What an agent needs in order to gain a sense of understanding is the counterfactual theory of causation. If A had not happened would B have still taken place? This not only applies to retrospective counterfactuals but also future-oriented ones wherein we need to "imagine different possible futures" [12]. Being able to grasp a scenario in this way demonstrates clear causal understanding between elements of a story or a scene, for example. Take for instance the scene below in Fig. 1.

Fig. 1. A common every day urban situation.

In order to *grasp* what is going on in a scene such as Fig. 1, a mental model of some sort is required. As humans, we develop mental models which allow us to intuitively understand situations like this. For instance, that the people and cars probably aren't suddenly going to float off or that when a car is momentarily blocked from view by a group of people the car will still be there when they have passed or that if one of the women had to let go of their bags that the bag would fall to the ground and not hover. These models consist of various forms of intuition such as, what psychologists have termed, intuitive physics, intuitive biology and intuitive psychology [12]. Without getting into too much detail, these forms of intuition allow us to build cause and effect models of our world which play a central role in bestowing us with two of the most fundamental cognitive features of human intelligence, namely, general intelligence and common sense. We can define general intelligence as "autonomous unsupervised learning of any problem-domain, which includes the ability to transfer gained knowledge, i.e., to share knowledge between domains" [5]. This is the type of intelligence we are aiming to craft in AGI. As Firt explains it:

Instead of systems that should be taught how the world (or the problem-domain) looks like, AGI systems learn independently, *understand* the problem-domain, extract the essential features from it and are able to reuse gained knowledge and re-apply it by tweaking it to fit the new domain—they learn from experience [5].

3.4 Premise 4: Understanding is a Necessary Condition for an Agent to Pass a Requisite Version of the Turing Test for AGI

Now we can delve into the mechanics of the CFRA, the CR as well as the TT. Briefly, the CR is a thought experiment wherein someone (John Searle himself) is placed in a room with three stacks of paper and instructions which grant him the ability to symbol-match Chinese letters or as he calls them squiggles and squoggles [16]. Given a series of questions from someone on the outside of the room Searle can match letters and words together to the extent that a native Chinese speaker on the outside of the room would not be able to distinguish whether Searle spoke (understood) Chinese or not. The correct answers are simply being produced by the person on the inside [16]. The CR functions as a critique of *understanding* in AI literature. Searle is a reductionist, as opposed to a dualist, and believes that there is nothing beyond the physical realm. He therefore believes that if we simply learn how the brain works, we can in principle build it from the ground up artificially to create consciousness, as consciousness is nothing but the result of physical interactions. However, what the CR critiques is the simulation of intelligence and understanding. A simulation of intelligence or understanding, for Searle, does not equate to genuine intelligence or understanding. Or as he famously sums it up in the dictum "syntax is not identical with nor sufficient by itself for semantics" [17]. Searle is a monolingual English speaker and does not speak or understand any Chinese. However, through the mechanics of the enclosed room and the instructions for symbol-matching he can still produce answers to questions which render a Chinese speaker non-the-wiser whether he genuinely understands Chinese or not. Let us look at the example of a joke being told in the CR to drive the point home. If a joke was told to Searle in Chinese while in the room, he might respond appropriately through his symbol-matching, but he wouldn't *get* the joke in the same way he would *get* the joke if it were told in his native tongue. He wouldn't *feel* the humour or grasp the contextual meaning behind the words, in short, he wouldn't understand. Bishop [1] points out that "there is a clear "ontological distinction" between [the] two situations". Understanding is a deeply conscious experience and is different to a simulation of understanding which is all Searle claims AI is achieving at the moment. Now, there are lots of replies to Searle's position, some which he pre-empts himself. However, delving into these replies is unfortunately beyond the scope of this paper. What is important for now is the distinction between merely producing the correct responses to a set of questions, either through instructions or chance (syntax), and understanding words, concepts, contexts and language on a fundamental level (semantics). I apply this distinction in the CFRA.

The CFRA is a rather simple thought experiment. Take a proposed AGI and sit it down in front of a group of interrogators. In the CR the person inside the room was human but it was not clear, to someone on the outside, whether they understood Chinese or not. In the CFRA the 'room' is the opaque mind of the AGI and what is unclear is whether the intelligence of the AI is strong or weak. In other words, if the AI has a mind (is conscious) and can understand (strong AI) or if the AI is still merely functioning as an inert tool (weak AI). The possible AGI is then asked a counterfactual query. What answers might we expect? It is not too hard to imagine that an AGI might provide some satisfactory answers or answers that make sense. But this is where the crucial distinction made in the CR comes into play. How would we be able to determine if the

AGI is in fact demonstrating understanding in order to answer such queries or if they are simply 'cheating'? Pearl and Mackenzie claim that it would be impossible to cheat as the number of prestored answers would be far too large, potentially infinite. However, as the questioners, we cannot ask infinite questions or wait around for infinite answers. We can only ask a set number of questions. Yes, it might be a very large finite set of questions but not so large that today's digital storage systems wouldn't cope. This discussion might run us into technicalities surrounding the capabilities of memory cards and hardware, which is a very important discussion but one that deserves a dedicated investigation. For now, let us say that it is, in theory, possible for an AGI to 'cheat' in such a test. How then would you test whether an AGI is demonstrating genuine causal and counterfactual reasoning and engaging in imagination? This is where the TT comes in.

Before we go into the details of the TT, I must elaborate on the caveat regarding the nature of AGI which also functions as a segway into the TT. It can perhaps be understood as a fallacy to think of AGI as functioning in the same way as human general intelligence. We may create AGI one day which displays tenets of general intelligence, but it may carry out this task in a radically different way. Mitchell [13] calls this fallacy "The lure of wishful mnemonics". It has been a common trend in the history of AI research to over anthropomorphise elements of AGI with the use of words such as 'learn', 'read', 'understand' or 'goal' [13]. An AGI might one day display general intelligence but in a way that we as humans simply will not grasp. This does however run into a central problem in the philosophy of mind, namely, the problem of other minds [11]. Even as human beings we cannot know that the people around us think, feel, experience or understand things in the same way as one another. We merely infer that everyone around us inhabits the same state of mind (disregarding those with severe cognitive impairments) via the behaviour that we see from them. We might say to ourselves 'I act this way when x happens and so do you so we probably think in the same way and experienced x in the same or a similar way'. I mention this issue because the problem of other minds plays a key role in determining whether or not we can tell if an AGI is genuinely reasoning causally or counterfactually. We may be able to get to the bottom of this problem through the TT.

Let me provide a brief overview of the TT for the uninitiated. Turing initially called this thought experiment "The Imitation Game" [18]. In the original setup, a man and a woman were placed in two confined rooms and an interrogator on the outside would have to ask them a series of questions via a typewriter. The goal of the game was to ask questions in such a way so as to determine which contestant was a man and which was a woman. In the adapted setup, one of the contestants was replaced with a robot. The goal of this replacement was twofold. One, to see whether the interrogator would still guess incorrectly which contestant was which. Two, and most importantly, it was a fruitful way to reframe the question "can machines think?" [18]. The test was said to be passed by a machine if within five minutes of questioning the machine could fool interrogators at least 70% of the time. Turing predicted, in 1950, that such an achievement would take place in roughly 50 years (2000) [18]. It has still not happened. However, in 2014 at the Royal Society in London, a chatbot named Eugene Gootsman from Russia and Ukraine 'passed' the TT[7]. I say 'passed' because it was able to meet Turing's requirements. Although Eugene fooled the interrogators, he/it did so by keeping a "large set of sentence templates that can be filled in based on a set of programmed rules that are applied to the input text it gets from its conversation parameter" [12]. In other words, to Pearl and Mackenzie's dismay, Eugene cheated. Eugene most definitely is not conscious in a human sense. He/it is merely able to fool a group of interrogators. This problem has been a recurring critique of the TT, namely that all a robot has to do is act human enough to fool other humans. Not that this is an easy hurdle to clear, but it does not prove much more than human imitation. Here is where a more refined and stringent form of the TT comes into play. Perhaps a TT wherein there is no time limit and where there are any number of interrogators. Ideally, the interrogators would be composed of a table of experts such as philosophers, psychologists, engineers, computer scientists, language specialists etc. not merely a random group of people. Such a table of experts would devise questions which would hopefully pose enough of a challenge to the AI such that it would only be able to answer these questions were it demonstrating genuine understanding and therefore consciousness. Questions composed of complex counterfactuals, if answered convincingly would show a sense of imagination and causal reasoning which has yet to be demonstrated by an AI. Consider a meta question posed to an AI such as: 'if you were a human interrogator in a TT what questions would you ask to determine causal reasoning?'. Such a question would require a lot from an AI model. Questions like this might be the only way to really determine whether an AI is genuinely reasoning causally or counterfactually or has a sense of consciousness. As I stated above, we cannot really *know* because of the problem of other minds and their opacity. We can perhaps only come to conclude that an AI is conscious in this way.

I propose a more stringent form of the TT whereas Pearl and Mackenzie propose a more relaxed form. However, if the original TT has already been passed, as seen above with Eugene Gootsman, then I cannot see what is going to be accomplished by setting the bar even lower, as Pearl and Mackenzie have done, if we are hoping to devise a test for genuine consciousness on the part of an AI. If we take the accomplishments of GPT-3, a large language model with 175 billion parameters, as an example, we can already see how far we have come in natural language processing, yet we still have not achieved AGI in this domain. Although there are proponents of simply scaling up these models, there is still no consensus as to the methods needed to achieve true AGI.

3.5 Final Conclusion: Therefore, Consciousness is a Necessary Condition for an Agent to Pass a Requisite Version of the Turing Test for AGI

If we take the contents of the premises of this argument to be true, then we must conclude that in order to pass our revised and fortified TT an agent must be conscious. Such a

---

[7] Consider the technical difference between a chat bot such as Eugene and todays large language models such as GPT-3, LaMDA, Gato etc. These models do not use lookup tables with predefined answers. They are deep neural nets, sometimes with hundreds of billions of parameters and are far more complex. This tells us two things. One- the original TT is clearly and inadequate test for current and future AI models. Two- we still have a long way to go in the development of AGI given that even these large language models are not yet generally intelligent.

claim has several implications. For instance, if an agent passes the TT, then we might have to consider it a moral agent or even grant it personhood status given its state of consciousness. This is a topic for another paper.

Before wrapping up I want to tease out the relationship between the CR, the CFRA, and our revised TT. The CR functions as a hypothetical thought experiment into the nature of understanding. The CFRA similarly functions as a hypothetical thought experiment into the nature of understanding. Where Searle uses a man, a room and Chinese symbols to illustrate his point, I am using a supposedly generally intelligent AI, its opaque cognition and counterfactual queries respectively. The TT, however, can and has been put into practice and is a useful method of investigating the state of an AI's intelligence. Our revised version of the TT can also be put into practice. For now, it functions as a thought experiment, but it can be practically demonstrated. The CFRA has its fundamental limits as a thought experiment. It is at these limits where our revised TT takes over in order to transfer the experiment from the theoretical into practical realm in order to uncover, through questions that require understanding, a sense of consciousness.

4   Conclusion

Pearl and Mackenzie put forward many notable points; from their Ladder of Causation [14] to their mini-Turing test [14]. However, upon closer inspection, their contributions appear to lose their thrust. I have demonstrated this by elaborating on the conflations they make in their theories, most notably in their vague and diluted use of counterfactuals. However, one of the main themes of their book remains intriguing: the role of counterfactual reasoning in the development of AGI. In an attempt to sustain this line of inquiry I formulated the CFRA. In this argument I made the case that the TT, as it stands, is not sufficient for demonstrating genuine consciousness on the part of an AI given that it has been passed by a non-conscious chatbot (Eugene Gootsman). As such, it ought to be restructured so as to demand a much more rigorous demonstration of conscious intelligence. If an agent can pass this revised TT, then they must be conscious. This is because they would have to demonstrate understanding. If they can demonstrate understanding then they must be able to reason counterfactually, and as I have argued, reasoning counterfactually requires imagination. Further, although I assume that consciousness is a necessary condition of imagination, I take this assumption to be uncontentious.

What the findings of this paper demonstrate is that we are still some ways away from creating AGI. However, a promising line of research into the role of counterfactuals in AI is opening up in multiple disciplines, such as law, physics, philosophy, and medicine. These research tracks are encouraging as they are reframing the often overlooked, misunderstood and unappreciated world of counterfactuals.

References

1. Bishop, J.M.: Artificial intelligence is stupid and causal reasoning will not fix it. Front. Psychol. **11**, 513474 (2021). https://doi.org/10.3389/fpsyg.2020.513474
2. Boughn, S.: Making sense of the many worlds interpretation. arXiv arXiv:1801.0858 (2018)

3. Crane, T.: The Mechanical Mind: A Philosophical Introduction to Minds Machines and Mental Representations. Routledge, New York (2003)
4. Dawes, A.J., Keogh, R., Andrillon, T., Pearson, J.: A cognitive profile of multi-sensory imagery, memory and dreaming in aphantasia. Sci. Rep. **10**, 10022 (2020). https://doi.org/10.1038/s41598-020-65705-7
5. Firt, E.: The missing G. AI Soc. **35**(4), 995–1007 (2020). https://doi.org/10.1007/s00146-020-00942-y
6. Harari, Y.N.: Sapiens: A Brief History of Humankind. Vintage, London (2015)
7. Hodges, A.: Turing: A Natural Philosopher. Weidenfeld & Nicolson, London (2021)
8. Lewis, D.K.: Counterfactuals. Blackwell, Oxford (1973)
9. Lewis, D.K.: Counterfactual dependence and time's arrow. Noûs **13**, 4 (1979)
10. Lewis, D.K.: On the Plurality of Worlds. Blackwell, Oxford (1986)
11. Merlo, G.: The metaphysical problem of other minds. Pac. Philos. Q. **102**, 633–664 (2021). https://doi.org/10.1111/papq.12380
12. Mitchell, M.: Artificial Intelligence: A Guide for Thinking Humans. Pelican Books, London (2019)
13. Mitchell, M.: Why AI is harder than we think. arXiv arXiv:2104.12871 (2021)
14. Pearl, J., Mackenzie, D.: The Book of Why: The New Science of Cause and Effect. Basic Books, New York (2018)
15. Rafetseder, E., Schwitalla, M., Perner, J.: Counterfactual reasoning: from childhood to adulthood. J. Exp. Child Psychol. **114**, 389–404 (2013). https://doi.org/10.1016/j.jecp.2012.10.010
16. Searle, J.: Minds, brains and programs. In: Hofstadter, D.R., Dennett, D.C. (eds.) The Mind's I: Fantasies and Reflections on the Self and Soul. Penguin Books, New York (1980)
17. Searle, J.: Yin and Yang strike out. In: Rosenthal, D.M. (ed.) The Nature of Mind. Oxford University Press, New York (1991)
18. Turing, A.M.: Computing machinery and intelligence. In: Hofstadter, D.R., Dennett, D.C. (eds.) The Mind's I: Fantasies and Reflections on the Self and Soul. Penguin Books, New York (1950)

Author Index

Printed in the United States
by Baker & Taylor Publisher Services